NATO ASI Series

Advanced Science Institutes Series

A series presenting the results of activities sponsored by the NATO Science Committee, which aims at the dissemination of advanced scientific and technological knowledge, with a view to strengthening links between scientific communities.

The Series is published by an international board of publishers in conjunction with the NATO Scientific Affairs Division

A Life Sciences B Physics	Plenum Publishing Corporation London and New York
C Mathematical and Physical Sciences D Behavioural and Social Sciences E Applied Sciences	Kluwer Academic Publishers Dordrecht, Boston and London
F Computer and Systems Sciences G Ecological Sciences H Cell Biology I Global Environmental Change	Springer-Verlag Berlin Heidelberg New York London Paris Tokyo Hong Kong Barcelona Budapest

PARTNERSHIP SUB-SERIES

1. Disarmament Technologies	Kluwer Academic Publishers
2. Environment	Springer-Verlag
3. High Technology	Kluwer Academic Publishers
4. Science and Technology Policy	Kluwer Academic Publishers
5. Computer Networking	Kluwer Academic Publishers

The Partnership Sub-Series incorporates activities undertaken in collaboration with NATO's Cooperation Partners, the countries of the CIS and Central and Eastern Europe, in Priority Areas of concern to those countries.

NATO-PCO DATABASE

The electronic index to the NATO ASI Series provides full bibliographical references (with keywords and/or abstracts) to about 50000 contributions from international scientists published in all sections of the NATO ASI Series. Access to the NATO-PCO DATABASE compiled by the NATO Publication Coordination Office is possible in two ways:

- via online FILE 128 (NATO-PCO DATABASE) hosted by ESRIN, Via Galileo Galilei, I-00044 Frascati, Italy.

- via CD-ROM "NATO Science & Technology Disk" with user-friendly retrieval software in English, French and German (© WTV GmbH and DATAWARE Technologies Inc. 1992).

The CD-ROM can be ordered through any member of the Board of Publishers or through NATO-PCO, Overijse, Belgium.

Series I: Global Environmental Change, Vol. 38

Springer
Berlin
Heidelberg
New York
Barcelona
Budapest
Hong Kong
London
Milan
Paris
Santa Clara
Singapore
Tokyo

Evaluation of Soil Organic Matter Models

Using Existing Long-Term Datasets

Edited by

David S. Powlson
Pete Smith
Jo U. Smith
IACR-Rothamsted
Harpenden, Hertfordshire AL5 2JQ
UK

Springer
Published in cooperation with NATO Scientific Affairs Division

Proceedings of the NATO Advanced Research Workshop "Evaluation of Soil Organic Matter Models Using Existing Long-Term Datasets", held at IACR-Rothamsted, Harpenden, UK, May 21–26, 1995

Library of Congress Cataloging-in-Publication Data

Evaluation of soil organic matter models using existing long-term datasets / edited by David S. Powlson, Peter Smith, Jo U. Smith.
p. cm. -- (NATO ASI series. Series I, Global environmental change ; vol. 38)
Includes bibliographical references and index.

1. Humus--Mathematical models--Evaluation--Congresses. I. Powlson, D. S. (David S.) II. Smith, Pete, 1965- . III. Smith, Jo U., 1964- . IV. NATO Advanced Research Workshop "Evaluation of Soil Organic Matter Models Using Existing Long-term Datasets" (1995 : IACR-Rothamsted) V. Series.
S592.8.E93 1996
631.4'17--dc20 95-45659
CIP

ISBN-13: 978-3-642-64692-8 e-ISBN-13: 978-3-642-61094-3
DOI: 10.1007/978-3-642-61094-3

Softcover reprint of the hardcover 1st edition 1996

Typesetting: Camera ready by authors/editors
Printed on acid-free paper
SPIN: 10475257 31/3137 - 5 4 3 2 1 0

PREFACE

This volume is a product of the NATO Advanced Research Workshop entitled "Evaluation of Soil Organic Matter Models Using Existing Long-Term Datasets" held at IACR-Rothamsted, Harpenden, U.K. on 21-26 May 1995.

Soil organic matter (SOM) represents a major pool of carbon within the biosphere, estimated at about 1400 x 10^{15}g globally, roughly twice that in atmospheric CO_2. Changes in climate are likely to influence the rates of accumulation and decomposition of carbon in SOM both directly through changes in temperature and moisture, and indiectly through changes in plant growth and rhizodeposition. Other changes, especially in land use and management, will have even greater effects whether caused directly by climate change or by other environmental factors, or by changes in agricultural, economic or social policies.

SOM may act as both a source and sink of carbon during global environmental change and research into predicting the effects of global environmental change on soil organic matter has been identified as a high priority within the Global Change and Terrestrial Ecosystems (GCTE) programme of the International Geosphere-Biosphere Programme (IGBP).

SOM models embody our best understanding of soil carbon dynamics and are needed to predict how global environmental change will influence soil carbon stocks. These models are also required for evaluating the likely effectiveness of different mitigation options. The first important step towards systematically evaluating the suitability of SOM models for these purposes is to test their simulations against real data. Since changes in SOM occur slowly, long-term datasets are required. This Workshop brought together leading SOM model developers and leading long-term experimentalists from around the world to begin the process of testing SOM models using long-term datasets from diverse ecosystems, land-uses and climatic zones within the temperate region.

This Workshop has initiated a large body of work comprising SOM model evaluation and comparison which will continue during the remainder of 1995. It is intended that the results of this work will be published separately in the near future as a special issue of *Geoderma*.

The Workshop marked the launch of the GCTE Soil Organic Matter Network (SOMNET), a network of soil organic matter (SOM) model developers and experimentalists working on long-term expriments that include measurements of soil carbon.

We are grateful to NATO for sponsoring the Workshop and the publication of this volume.We would very much like to thank the Workshop Steering Committee; Prof. Martin Körschens, Prof. Nico Van Breemen, Dr Ted Elliott, Dr Peter Grace, Dr Keith Paustian, Dr Bent Christensen and Dr John Kimble for their invaluable help and Dr Margaret Glendining, Mr Paul Poulton, Mr Kevin Coleman and Mrs Nicky Bradbury for help during the Workshop.

David S. Powlson, Pete Smith & Jo U. Smith
Soil Science Department
IACR-Rothamsted,
Harpenden, Herts. AL5 2JQ, U.K.

WORKSHOP PARTICIPANTS

From left to right: Back Row: Peter Gregory, Ed Gregorich, Jean-Alex Molina, Uwe Franko, Nicky Bradbury, Jon Arah, Paul Poulton, John Gaunt, Bent Christensen, Keith Paustian, David Jenkinson, Ellis Benham, Dan Richter, Bill Parton, Mac Post, George Ayaga, Luc Dendooven, A.P. Gupta, Luke Sanger, Max Beran.

Centre row: John Catt, John Ingram, Eva-Maria Ritskowski, Elena Yakimenko, Annett Müller, Huub Klein Gunnewiek, Aroon Parshotam, Jo Smith, Peter Grace, Marc Frankinet, Paul Rasmussen, Jan Klír, Paul Hargreaves, David Powlson, Mick Wheelan, Oleg Chertov, Gregory Buyanovsky.

Front row: Martin Körschens, Cesar Izarraulde, Pete Smith, Nico Van Breemen, Lars S. Jensen, Graham Crocker, Margaret Glendining, Kevin Coleman, Bill McGill, Changsheng Li, Matias Vanotti, Marja Plentinger, Katalin Debreczeni.

CONTENTS

PLENARY PAPERS

Why evaluate soil organic matter models?

David S. Powlson
Soil Science Department
IACR-Rothamsted
Harpenden
Herts AL5 2JQ, UK

Soil is essentially a non-renewable resource on which natural ecosystems and all forms of agriculture depend. It acts as a buffer or interface between terrestrial systems, water and the atmosphere. The organic matter present in soil (SOM) has a profound influence on many soil properties so changes in SOM content, composition or dynamics can have far reaching impacts. Some changes occur slowly over periods of decades so long-term experiments are necessary to measure them. Models are the only practical means of making projections at places other than the sites of long-term experiments.

Agricultural impacts of organic matter in soil

The value of organic matter in enhancing the quality of soil, from an agricultural standpoint, has been recognised for many years. A soil that has been managed to maintain or increase its organic matter content is likely to have superior physical properties compared to an otherwise similar soil where organic matter has declined. Some examples include greater formation or stabilisation of aggregates, ease of cultivation, decreased risk of compaction or other physical damage and improved water holding characteristics (Haynes *et al.*, 1991). Other factors, such as risk of erosion, are associated with organic matter but more with its management and location (e.g. surface mulching with crop residues) rather than its content within the bulk soil. It is sometimes difficult to demonstrate the value of increased organic matter content through its direct effect on crop yields although there are examples of this (e.g. Johnston & Powlson, 1994). Perhaps more significant is the increased stability of soil

NATO ASI Series, Vol. I 38
Evaluation of Soil Organic Matter Models
Edited by David S. Powlson, Pete Smith and Jo U. Smith

structure which SOM produces and this can act as a buffer against other adverse growing conditions for crops. A slow decline in SOM content may have no perceivable impact on crop growth for many years, either because there really is no impact or because and underlying adverse effect is masked by improvements in management practices. However, with some combinations of soil type, landform and climate, a serious decline in soil structure may occur once SOM declines beyond a threshold value and precipitate a rapid and large decline in crop yields. As some forms of physical degradation are essentially non-reversible, this aspect of SOM is crucial to the sustainability of agricultural systems. It is probably impossible to define a universally applicable index of sustainability; see Powlson & Johnston (1994) for a discussion of this. However measurements of SOM, when considered in the context of soil type, land use and climate, represent an important component in any assessment of sustainability and provide some degree of integration of soil properties.

A further value of SOM in relation to agricultural productivity is its role as a reservoir of plant nutrients. In part this function can be replaced by the use of inorganic fertilizers, and these will often be essential if acceptable crop yields are to be achieved. There are, however, situations where the maximum yield of certain crops can only be attained if soil organic matter is maintained at a high value (Johnston & Powlson, 1994: Körschens & Müller, 1995). In situations where farmers cannot afford the costs of inorganic fertilizers in the necessary quantities it is important for the long-term stability of the agroecosystem that management practices are adopted that maintain soil organic matter as a source of nutrients, even if the short term benefits are not apparent.

Environmental implications of organic matter in soil

Estimates of the quantity of carbon held in SOM globally show it to be about twice the 750 Pg present in the atmosphere as CO_2. Eswaran *et al.* (1993) estimated the global stock of carbon in SOM to a depth of 1 m to be 1576 Pg and Post *et al.* (1982) estimated 1394 Pg. Even small a change in this large stock could cause a significant change in atmospheric CO_2 concentration. For an overview of issues relating to the global terrestrial carbon cycle see Smith *et al.* (1993). Jenkinson *et al.* (1991) used the Rothamsted Carbon Model, one of the models tested in the model evaluations conducted at this Workshop, to calculate the additional release of CO_2 from soil as result of several global warming scenarios. The aim was to see the extent to which a positive feedback might occur in which global warming would be reinforced as additional CO_2 is released from soil. The result indicated a significant, but not catastrophic, feedback. Assuming no changes in land use or rainfall, an average increase in temperature of 0.03°C per year was calculated to cause the release of an additional 61 Pg of C over 60 years. This is 19% of the evolution of C from fossil fuel burning over the same period if present production continues unchanged. The simplifying assumptions made in the above calculation would tend to give a conservative estimate of CO_2 release. In addition to the direct effects of climate change on the flux of CO_2 between soil and atmosphere considered by Jenkinson *et al.* (1991), other changes in land use and management may be even more significant, whether driven by climate change or by socioeconomic factors. Continued clearing of tropical forests will cause further release of carbon, as will the cultivation of high organic matter soils in northern latitudes if global warming permits some northward movement of cultivatable land.

There is considerable discussion on the extent to which the increasing concentration of atmospheric CO_2 might be mitigated through alterations to land management practices. For example the CENTURY model, another of those included in the model evaluations at this workshop, has been used to simulate the decline in SOM in the Great Plains of North America following cultivation (Cole *et al.*, 1989) and the increases in SOM that might be expected from various land management options in Sweden (Paustian *et al.*, 1992). A particular mitigation option being considered worldwide is the afforestation of significant areas of land currently used for agriculture. This is a real option in North America and Western Europe where there are currently surpluses of many agricultural products although this option is dependent on a continuation of intensive agricultural practices on the remaining agricultural land. In addition to carbon sequestered in tree biomass, the additional carbon held in soil as SOM increases is potentially significant and requires quantification. A related issue concerns the possible role of biofuels in decreasing net CO_2 evolution from anthropogenic sources. As many biofuel crops are perennials the decreased cultivation that would accompany their widespread introduction may lead to significant increases in soil carbon stocks.

Carbon from SOM and any plant or animal residues that enter soil provide substrate for the soil microbial population. Many of the microbially mediated processes in soil influence the formation or destruction of trace greenhouse gases, especially nitrous oxide and methane (Goulding *et al.*, 1995; Mosier *et al.*, 1991; Bouwman, 1990). Changes in land management practices can alter the quantity, composition and location of substrates in soil and, thus, indirectly influence the source/sink balance of these gases. An improved understanding of soil processes and the controlling factors is required if rational judgements are to be made on the relative merits of different mitigation options.

Slow changes in SOM and the use of models

Some effects of a change in soil management or land use are apparent very quickly: for example the suite of chemicals exuded into soil from plant roots will change immediately following the introduction of a new crop and could cause an immediate change in the species composition of the rhizosphere. However changes in the total quantity of carbon present in soil tend to occur slowly. In temperate regions it is common for changes in SOM content to be undetectable within 1 or 2 decades of even a major change in agricultural practice. Powlson *et al.* (1987) gave examples of this taken from two experiments in Denmark on light textured soils in which straw from spring barley was either burned or incorporated into soil after harvest. Even though the quantity of straw incorporated was substantial (approximately 4 t ha^{-1} of dry matter each year) there was no statistically significant difference in the total organic carbon content of soil between the two treatments 18 years after they started. This is partly because the changes occur against a relatively large background content of soil carbon and partly because spatial variability makes changes in any soil property difficult to measure.

Long-term field experiments are the ideal means of detecting and quantifying slow changes in SOM and other soil properties but it is not feasible to conduct them under the full range of climates and situations of interest. Furthermore, information on the agricultural or environmental consequences of management practices is currently required by land users and policy makers within a short time period. Whilst the setting up new long-term experiments is an extremely important investment for the future, it is not an option for providing answers to *current* questions. Hence the reason for this workshop on "Evaluation of soil organic matter models using existing long-term datasets". Three statements can be made:

1. It is necessary to use *existing* long-term experiments to address current research and policy issues related to SOM (and other issues) because it is these that are available.
2. SOM *models* are an essential tool in making projections of likely trends in SOM and related properties and processes in soil because many of the changes are slow. It is generally impossible to conduct experiments that will provide results within a reasonable timescale. Similarly it is impossible to conduct experiments at sites covering all possible combinations of soil type, climate, land use, management option, climate change scenario etc. Thus models will often be the only possible way of extrapolating from current knowledge in both time and space.
3. *Evaluation* of models using *existing long-term data* is essential as it will usually be impossible to check whether or not a particular model projection is correct. The best that can be done is to test any model against as much existing data as possible, covering as wide a range of conditions as is likely to be encountered when it is used to make serious projections.

The need for model evaluation

Some models are specifically intended as research tools to help an investigator to organise knowledge, identify gaps or test hypotheses. In such cases the modeller can argue that the numerical results of a model run are not of prime importance. However SOM models, and many others such as those concerned with climate change, atmospheric composition or crop growth, are increasingly seen by policy makers as a means of making serious projections of future conditions. Despite scientists protesting that "more work is needed" before a particular modelling task is undertaken, the policy maker tends to take the view that *some* degree of prediction is better than none at all so there is a risk that inappropriate models will be used. The results of model runs will be used in formulating policies concerned with land use, agricultural practice or environmental protection that will influence the lives and livelihood of many millions of people. For example, SOM models are already being used to consider possible mitigation options to sequester carbon in soil or vegetation as a means of decreasing the rate at which atmospheric CO_2 concentration rises. The options under consideration include the afforestation of land currently used for agriculture. SOM models are also being used to compare the impacts of different land management practices in tropical areas where there is an urgent need to increase primary production in sustainable ways. In some cases there may be conflicts between the aim of conserving natural resources and that of increasing production of food, fibre or energy. This is a particular focus of the GCTE programme on "Alternatives to slash and burn" (GCTE, 1992). To use a model in this context that has not been well tested is certainly unscientific. In view of the possible impacts on the lives of people who have no control over the formulation or use of the model it could also be argued that it is immoral.

By testing SOM models using datasets from a wide range of situations it is possible to better define the limits of a particular model and identify those situations where it is not likely to function well. This is of considerable importance as models developed for one purpose are increasingly being used in new ways. For example, the Rothamsted Carbon Model (Jenkinson *et al.*, 1987; Jenkinson, 1990), originally designed to simulate SOM changes in arable soil, has been used to estimate global net primary production (Jenkinson *et al.*, 1992) and the impact of climate change on global soil carbon stocks and CO_2 release (Jenkinson *et al.*, 1991). Ojima *et al.* (1993) used the CENTURY model (Parton, 1995), developed to simulate plant:soil dynamics in North American grasslands, to examine possible changes in carbon storage in 7 ecoregions throughout the world using climate change scenarios generated by two General Circulation Models. There is also a need to consider the impacts of changes in land use on the stocks of carbon in soil and vegetation so it is necessary to know whether a model originally designed to simulate carbon dynamics in one land use system (e.g. arable or managed grassland) will give correct simulations in another (e.g. forest or natural grassland). This can only be done within a reasonable timescale by using data from existing long-term sites.

Problems with long-term soil and plant data

As existing long-term experiments were not generally designed to answer today's questions, it is not surprising that they often do not contain the treatments that would be most appropriate. Consequently ingenuity and intelligence are required in using and interpreting the data in new ways. It may be necessary to use one treatment as proxy for another and to use supplementary information from other experiments to assess the reliability of this.

Alternatively it may be necessary to use data from one climatic zone to draw conclusions on SOM changes in another if a suitable experiment does not exist in the zone of interest. In this case it will be necessary to first test the responses of the model compartments or parameters to climatic factors, probably using data from elsewhere. In addition there are many practical issues that can seriously affect the quality of data from long-term experiments including soil movement between plots, erosion, changes in cropping or sampling practices with time and changes in analytical methods. These are considered by Glendining & Poulton (1995).

Most of the conceptual pools in all current SOM models do not correspond directly with experimentally measurable fractions of SOM. Achieving a greater degree of matching between model pools and measurable fractions would greatly increase the opportunities for veriryimg model simulations as well as being a significant scientific advance in the fundamental understanding of SOM dynamics. This topic is discussed in detail by Christensen (1995). Long-term experiments are ideally suited for providing experimental material (soil and plant samples) and data for such research.

Conclusions

The dynamics of SOM are central to wide range of urgent environmental and agricultural issues. Many of the issues are not amenable to direct experimentation because of the long time periods involved, because they involve consideration climate change scenarios that cannot be reproduced on a large scale or because broad conclusions are required based on data from a limited number of sites. Modelling SOM changes is the only practical approach in many instances. A large number of models are now published, of which some are well validated but not necessarily in the widest possible range of environments. If SOM models are to be used in formulating land use policies and comparing scenarios it is essential that they are first evaluated to test their robustness and their applicability in terms of soil types, climates and land use systems. To fail to do this would not only be unscientific but also morally wrong in view of the possible impacts on people in the areas concerned. As SOM changes are often very slow, and the full impact of a change in land management practice may well take decades to become apparent, a long-term perspective is required. Existing long-term experiments, or well monitored sites under a specific form of vegetation, are a valuable source of data for evaluating SOM models; indeed they are usually the only source of data. The GCTE-SOMNET initiative is an attempt to bring together long-term data and models, firstly to facilitate model evaluation and development and, secondly, to act as a resource and stimulant for new research in this area that is a vital component of global efforts to understand environmental change.

References

Bouwman, A.F. (ed) (1990) *Soils and the Greenhouse Effect*, Wiley, Chichester, UK, 575 pp.

Christensen, B.T. (1995) This volume.

Cole, C.V., Stewart, J.W.B., Ojima, D.S., Parton, W.J. & Schimel, D.S. (1989) Modelling land use effects of soil organic matter dynamics in the North American Great Plains. In: *Ecology of Arable Land. Perspectives and Challenges* (eds M. Clarholm & L. Bergström), Kluwer Acad. Publ., Dordrecht, The Netherlands: 89-98.

Eswaran, H., Van Den Berg, E. & Reich, P. (1993) Organic carbon in soils of the world. *Soil Science Society of America Journal* **57**: 192-194.

GCTE (1992) Global Change and Terrestrial Ecosystems *The Operational Plan* (eds W.L. Steffen, B.H. Walker, J.S.I. Ingram & G.W. Koch), Global Change Report No 21, International Council of Scientific Unions, Stockholm, 95 pp.

Glendining, M. & Poulton, P.R. (1995) - this volume.

Goulding, K.W.T., Hütsch, B.W., Webster, C.P., Willison, T.W. & Powlson, D.S. (1995) The effect of agriculture on methane oxidation in soil. *Philosophical Transactions of the Royal Society London A* **351**: 313-325.

Haynes, R.J., Swift, R.S. & Stephen, R.C. (1991) Influence of mixed cropping rotations (pasture-arable) on organic matter content, water stable aggregation and clod porosity in a group of soils. *Soil & Tillage Research* **19**: 77-87.

Jenkinson, D.S. (1990) The turnover or organic carbon and nitrogen in soil. *Philosophical Transactions of the Royal Society London B* **329**: 361-368.

Jenkinson, D.S., Adams, D.E. & Wild, A. (1991) Model estimates of CO_2 emissions from soil in response to global warming. *Nature* **351**: 304-306.

Jenkinson, D.S., Harkness, D.S., Vance, E.D., Adams, D.E. & Harrison, A.F. (1992) Calculating net primary production and annual input of organic matter to soil from the amount and radiocarbon content of soil organic matter. *Soil Biology and Biochemistry* **24**, 295-308.

Jenkinson, D.S., Hart, P.B.S., Rayner, J.H. & Parry, L.C. (1987) Modelling the turnover of organic matter in long-term experiments at Rothamsted. *Intecol Bulletin* **15**: 1-8.

Johnston, A.E. & Powlson, D.S. (1994) The setting-up, conduct and applicability of long-term, continuing field experiments in agricultural research. In: *Soil Resilience and Sustainable Land Use* (eds D.J. Greenland & I. Szabolcs), CAB International, Wallingford, UK: 395-421.

Körschens, M. & Müller, A. (1995) - this volume.

Mosier, A.R., Schimel, D.S., Valentine, D., Bronson, K. & Parton, W. (1991) Methane and nitrous oxide fluxes in native, fertilized and cultivated grasslands. *Nature* **350**: 330-332.

Ojima, D.S., Parton, W.J., Schimel, D.S., Scurlock, J.M.O. & Kittel, T.G.F. (1993) Modeling the effects of climatic and CO_2 changes on grassland storage of soil C. In:

Water, Air, and Soil Pollution (eds J. Wisniewski & R.N. Sampson), Kluwer Academic Publishers, The Netherlands, pp 643-657.

Parton, W.J. (1995) - this volume.

Paustian, K., Parton, W.J. & Persson, J. (1992) Modeling soil organic matter in organic-amended and nitrogen-fertilized long-term plots. *Soil Science Society of America Journal* **56**: 476-488.

Post, W.M., Emmanuel, W.R., Zinke, P.J. & Stangenberger, A.G. (1982) Soil carbon pools and world life zones. *Nature* **298**: 156-159.

Powlson, D.S., Brookes, P.C. & Christensen, B.T. (1987) Measurement of soil microbial biomass provides an early indication of changes in total soil organic matter due to straw incorporation. *Soil Biology and Biochemistry* **19**: 159-164.

Powlson, D.S. & Johnston, A.E. (1994) Long-term field experiments: their importance in understanding sustainable land use. In: *Soil Resilience and Sustainable Land Use* (eds D.J. Greenland & I. Szabolcs), CAB International, Wallingford, UK: 367-393.

Smith, T.M., Cramer, W.P., Dixon, R.K., Leemans, R., Neilson, R.P. & Solomon, A.M. (1993) The global terrestrial carbon cycle. In: *Water, Air, and Soil Pollution* (eds J. Wisniewski & R.N. Sampson), Kluwer Academic Publishers, The Netherlands, pp 19-37.

The Contribution of the Soil Organic Matter Network (SOMNET) to GCTE

P.J. Gregory[1] *and J.S.I. Ingram*

GCTE Focus 3 Office,

Department of Plant Sciences,

Oxford University,

South Parks Road,

Oxford OX1 3RB, U.K.

Summary

The Global Change and Terrestrial Ecosystems (GCTE) Core Project of IGBP includes a Task aimed at determining the effects of global change on soil organic matter (T3.3.1). This paper describes the rationale, structure and objectives of GCTE, and in particular those of Task 3.3.1. The paper also introduces the GCTE soil organic matter research network (SOMNET).

Introduction

It is now well appreciated that the quantity of carbon stored in soils is substantial (estimated at 1580 Gt; IPCC, 1995) and larger than the combined totals of the atmosphere and vegetation (1360 Gt). The amounts in the atmosphere and terrestrial ecosystems are small, however, when compared with those in the oceans but the current anthropogenic emissions of carbon have disturbed the equilibria between the various sources and sinks and several exchanges in the carbon cycle have become unbalanced (IPCC, 1995). The burning of fossil

[1] *Department of Soil Science, The University of Reading, PO Box 233, Whiteknights, Reading, RG6 6DW, U.K.*

NATO ASI Series, Vol. I 38
Evaluation of Soil Organic Matter Models
Edited by David S. Powlson, Pete Smith and Jo U. Smith

fuel and the manufacture of cement, together with the removal of perennial vegetation and its replacement by agriculture have all transferred carbon to the atmosphere. Although the amounts transferred to the atmosphere through changes in land use are small (122 ± 40 GtC between 1850 and 1990; Houghton, 1994) compared to the changes wrought by industrialisation (230 GtC between 1751 and 1991; Andres *et al.*, 1994), significant exchange between the land and atmosphere has occurred as a consequence of intensified land use by mankind.

The importance of this transfer of carbon from soils to the atmosphere lies not only in the global carbon cycle and the consequences of global warming, but also in the potential of soils to produce food, fibre, fuel and building materials for people (Greenland, 1994). Of concern, particularly, are the changes in land use predicted for tropical regions where the human population is increasing most rapidly. During the nineteenth century (until the 1940s), carbon emissions from terrestrial ecosystems were dominated by the expansion of agriculture in the middle and high latitudes (Houghton & Skole, 1990). However, since the 1950s, emissions from the tropics have been increasing as a consequence of changes in land use such that emissions during the 1980s averaged 1.6 ± 1.0 GtC/yr (Dixon *et al.*, 1994). Such losses from soils that are already of low inherent fertility and in which biological activity has a major role in sustaining fertility (Woomer & Swift, 1994) are clearly of concern in relation to future productivity and the well-being of the indigenous population.

There have been several major publications on soils and global change in recent years (eg. Bouwman, 1990; Anderson, 1992; Rounsevell & Loveland, 1994). To a large extent these have aimed to highlight the importance of soils and the processes occurring in them for environmental change; rates of emission and absorption of radiatively active gasses have been documented and possible consequences for productive capacity have been explored. Tinker & Ingram (1994) make an important point about the global impact of changes in soils which are largely a consequence of pervasive human activity on a global scale. In some sciences (eg. atmospheric and oceanographic) the word "global" implies effects that, while they may originate at a local or regional scale, are transmitted around the world, such as the changes in nitrous oxide concentrations in the atmosphere. However, in soil, because transport processes are usually limited in extent, for soil scientists "global" usually means some process or effect which occurs so frequently in several different ecosystems that it can be considered to have global significance. For example, the mineralisation of organic matter consequent from the change from forest or pasture to arable cropping. Both definitions of "global" bring with them their own scientific problems. For soil scientists, used to obtaining results from location-specific experiments and to dealing with the complexities of the wide variety of agricultural and terrestrial ecosystems, a major problem is how to "scale up" the results so that they have wide application.

This workshop is a step towards the synthesis of existing knowledge on the transformations of soil organic matter through the use of models and long-term datasets. This paper seeks to place this workshop in the wider context of global change research and to show how its outputs will link with other elements of that research.

Global change research and GCTE

The International Geosphere-Biosphere Programme (IGBP) has focussed its work on climate change on three major driving variables: atmospheric composition, climate and land use. The world's terrestrial ecosystems are being subjected to changing environmental conditions unprecedented both in their rate and in their geographical extent. The ability of human societies to ameliorate, adapt to, and benefit from these rapid changes requires fundamental knowledge of the responses of terrestrial ecosystems to the forces of global change. In response to this challenge, IGBP established a Core Project on Global Change and Terrestrial Ecosystems (GCTE).

The objectives of GCTE are:

1. To predict the effects of changes in climate, atmospheric composition, and land use on terrestrial ecosystems, including (i) agriculture, forestry and soils, and (ii) ecological complexity.

2. To determine how these effects lead to feedbacks to the atmosphere and the physical climate system.

GCTE's operational definition of "global change" encompasses far more than just predicted climate change alone (IGBP, 1992). It includes changes in atmospheric composition, such as the concentration of CO_2 and other greenhouse gases, which have increased and will continue to increase and which have direct impacts on vegetation with or without climate change. It also includes change in land use, as driven by demographic, economic, technological, and social pressures. Over the next few decades this human dimension of global change will have a more profound influence on the fate of terrestrial ecosystems than will changes in climate and atmospheric composition; this aspect will be particularly important in tropical regions.

GCTE research is based on a set of focused scientific questions with global repercussions. Although GCTE has no money to commission research of its own, it has structured its programme to ensure cohesiveness and to facilitate the flow of people, ideas, expertise and information among its component parts. The research is organised into a hierarchy of *Foci*, *Activities*, and *Tasks*, which become more specific as one goes down the hierarchy. The four Foci are:

1. Ecosystem Physiology

2. Change in Ecosystem Structure

3. Global Change Impact on Agriculture, Forestry and Soils

4. Global Change and Ecological Complexity

The first two Foci attempt to provide a fundamental understanding of the impacts of global change on ecosystem function, composition and structure, and their feedback effects. The third and fourth Foci are designed to examine the more specific impacts of global change on systems of great importance to humans, i.e. for the production of food and fibre, and for the maintenance of the Earth's diversity and ecosystem complexity.

Global Change Impact on Agriculture, Forestry and Soils (GCTE Focus 3)

Many agroecosystems are threatened by damage to soil and water resources through sub-optimal technology. Major land-use changes, which are part of global change, may greatly increase this stress, driven by increasing demands for agricultural and forest products from a growing population. Climatic and atmospheric changes will have further impacts on these stressed and rapidly changing systems in ways we cannot yet predict with any accuracy. The ability to capitalise on the beneficial effects of global change, while avoiding or reducing adverse effects, will require a strong predictive capability.

While the lack of knowledge over the precise climate in the future makes exact predictions impossible, it is important to be able to predict the consequences of defined scenarios and to identify the most sensitive components of managed systems. To this end, GCTE Focus 3 has initiated interdisciplinary research to improve our general predictive ability for selected agronomic species, grasslands and rangelands; for multi-species cropping systems; for managed forests; for major pests, diseases and weeds; and for soils. As with other GCTE Foci, the main directions for research are in monitoring, experimentation and modelling; this Focus differs from other Foci by including a management component and emphasising the importance of harvestable products of economic importance.

Four major types of production system have been identified: monocrop agriculture, pasture and grazing systems, multi-species cropping (including agroforestry and rotational systems), and forestry. These have been grouped into three Activities:

1. Key agricultural systems - including a representative selection of monoculture food crops, together with improved pastures and rangelands and emphasising harvestable products.

2. Multi-species agroecosystems - the norm in much of the world. Modelling such complex agricultural systems is in its infancy.

3. Managed forests - including intensively-managed plantations to more natural forests which are utilised by humans.

Two further Activities crosscut these three major areas: one addresses the effect of global change on pests, diseases and weeds and the other addresses the effect of global change on soils (Figure 1).

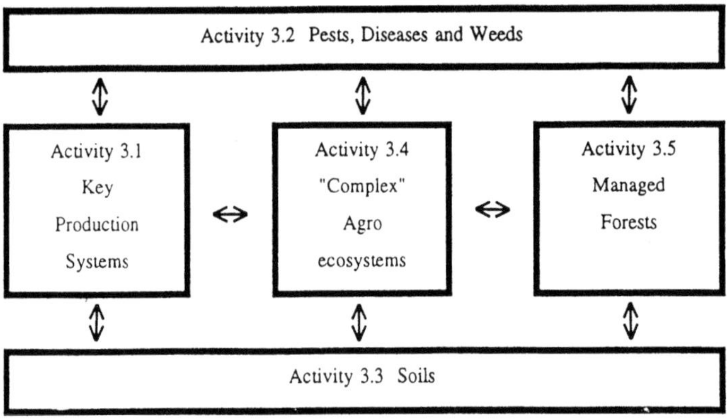

Figure 1 *The Activities that comprise GCTE Focus 3*

Initially, much effort was placed into establishing networks for experimenters and modellers in forestry and key crop production systems. This approach has been very effective at bringing together scientists from a wide range of countries to focus their attention on common methods of experimentation, validation of models so that the effects of the drivers of global change on harvestable products can be quantified, and collation of datasets so that regional and global outputs from models can be attempted. The wheat network has already published some comparisons between various models (Goudriaan *et al.*, 1993) and detailed datasets are being collated. Formation of this network has also promoted the more detailed investigation of particular physiological processes such as the influence of temperature and daylength on the appearance of leaves (Jamieson *et al.*, 1995). Similar networking activities are underway for other crops and multi-species agroecosystems; pastures and rangelands have published their detailed plans for future work (Stafford Smith *et al.*, 1995). Managed forests have a network of experimenters initially collecting data in the northern boreal forests that will provide results on the effects of elevated CO_2 on forest productivity and carbon partitioning (IUFRO, 1995).

Effects of Global Change on Soils (GCTE Activity 3.3)

Climate change will have significant impacts on soils directly (e.g. changed potential erosion from changed rainfall patterns); or indirectly via land-use change (e.g. if arable cropping

Table 1 *The Structure and Objectives of the GCTE Soil Activity*

Task 3.3.1 Global Change Impact on Soil Organic Matter

Objectives

To develop an increased understanding of the relationship between environmental controls and the processes responsible for the formation and loss of SOM across the range of soils and climates of the world.

To apply this understanding to assess global change impacts on soil productivity and environmental degradation and to help in the development of strategies for the mitigation of greenhouse gas emissions.

Task 3.3.2 Soil Degradation Under Global Change

Objectives

To design and undertake experimental and monitoring programmes to provide a predictive understanding of the impacts of changes in climate and land-use on soil erosion.

To refine and adapt current erosion models for use in global change studies from plot to regional scales.

Task 3.3.3 Global Change and Soil Biology

Objectives

To quantify the role of soil populations and communities in organic matter decomposition, nutrient availability and trace gas emissions.

To define the extent to which soil populations and communities are affected by global change; and to estimate how such changes will affect their function with respect to decomposition, nutrient release and trace gas emission.

potential edges northwards into boreal forests). Table 1 shows the three principal Tasks of the Soils Activity and outlines their major objectives.

Much attention has already been focussed on the possible effects on the C, N, S and P fluxes and pools in soils (Tinker & Ineson, 1990: Anderson, 1992) and given the importance of soil organic matter (SOM) for both physical stability and fertility, this must remain a key area for research activity. Soil biological processes underpin many aspects of present and predicted soil changes, hence alterations in these processes (as a result of changes in climate, CO_2, and possibly other atmospheric constituents) is another key area for research. Particular emphasis is placed on soil fauna activities, because of their importance in relation to decomposition and soil structure. The degree of biological diversity required to sustain particular ecosystems and systems of agricultural production is largely unknown but of increasing urgency particularly as global changes occur (Kennedy & Smith, 1995).

Soil properties vary from the transitory and rapidly variable, such as nitrate content, to the virtually permanent, such as texture. The number of visible properties that will be altered directly by changes in temperature, rainfall or CO_2 concentration is fairly small, though a few soils are morphologically unstable, and may alter rather rapidly. Changes in climate will affect soils in a terrain, in the sense that their erosion potential will alter with rainfall, plant cover and cultivation; and changes in atmospheric CO_2 concentration may lead to changed soil organic matter quantity and type, via its impact on vegetation. The topics shown in Table 1 are those where global change is likely to have a direct or indirect impact on soils, and where this impact has important practical consequences.

Global Change Impact on Soil Organic Matter (GCTE Task 3.3.1)

Soil organic matter (SOM) is crucial to terrestrial ecosystem processes. It plays key roles in nutrient release, maintenance of soil structure, in greenhouse gas emissions, and is a major pool in the global carbon cycle. Global change will affect the soil's capacity to retain and recycle organic carbon as SOM, so it is essential to gain a sound predictive understanding of its likely impacts on soils, and in particular on SOM dynamics.

SOM levels are maintained by a delicate balance between inputs and decomposition processes. SOM is not however a simple substance, but constitutes a very diverse range of materials, both chemically and physically. To simplify the situation, SOM can conceptually be divided into discrete pools which can be identified on the basis of their approximate residence times (e.g. Parton *et al.*, 1988). Although this classification is simple, the dynamics of each pool is governed by a large and complex range of processes, which occur over widely different time periods. The overall goal of this Task is to gain a better understanding of the interactive effects of global change on these processes, and two objectives have been specified (Table 1).

One area of special interest for GCTE is how elevated CO_2 will alter the amount and kind of dead plant material entering the soil and the subsequent effect on decomposition and SOM formation. Also, the allocation of inputs into above- versus below-ground compartments may significantly alter the proportion of plant residues forming new SOM compared with that respired by microbes, especially as more or less of the primary production flows into root exudates and exfoliates. For example, elevated CO_2 generally appears to increase root biomass which may translate into greater rhizodeposition (Rogers *et al.*, 1994). The indirect influences of plants on soil water, soil temperature and erosion must also be taken into account as climates change or management alters the relationship between the regional macroclimate and the soil microclimate. An experimental programme is being designed to address these issues, and will form one of the GCTE Core Research Programmes addressed by this Task.

Changes in CO_2 and climate are essentially issues concerning future scenarios. One dimension of global change has however been ongoing for many years - change in management. Management is a key factor affecting SOM which, under most situations, causes faster rates of change than would changes in CO_2 or climate. However, since change in climate may result in altered management, interactions between management and climate may be the most significant factor which needs to be studied. Conservation tillage has been touted as a means of reducing losses, or even increasing SOM levels, yet experiments including this practice are just now becoming mature enough to yield reliable information - this highlights the fact that the temporal dimension is a major consideration in SOM experiments of this type; the paucity of datasets suitable for determining the effects of management on SOM is currently a serious impediment to progress. The GCTE SOM network (SOMNET) will play a major role in overcoming this, by identifying, categorising and evaluating long-term experiments world-wide of value to SOM research.

Because of the feedbacks among components of the ecosystem, including soils, and interactions among the exogenous driving variables, the complexity of the system must be accounted for so that we may make robust predictions of the impact of global change on SOM. Application of simulation models provides a way to account for the complexity and should be used as part of a systems approach to understanding SOM. Of particular interest is the range of SOM modelling approaches currently being developed and refined. As important as the models themselves are, datasets from long-term experiments are needed to validate (or refute) a given modelling approach for a given set of conditions. But how robust are these models for global change studies? We need to compare the models with data sets obtained from a wide range of ecosystems, with the objective of identifying for which set of conditions each model is best suited. SOMNET will provide the mechanism for undertaking this work.

Contribution of this Workshop to GCTE

This workshop will be a major step forward in assessing the different modelling approaches, and in determining which types of long-term experiment are most suitable for model development. It will also see the firm establishment of SOMNET, including the formal launch of the European component.

As already shown, SOMNET is only one part of the SOM task. However, it is the initial Core Research Project (CRP) and will assist in linking the results of experiments dealing with CO_2 and climate effects with the wider SOM community. With this in mind, part of this workshop will design experimental protocols to help validate those models identified through this workshop (and subsequent ones) as being most widely applicable for elevated CO_2 studies etc. A comparable approach has been adopted by Task 3.3.2 (Soil Degradation) where a worldwide network of scientists working on soil erosion has already been established and three CRPs with key questions have been identified. Task 3.3.2 CRPs are:

1. Linking erosion processes across temporal and spatial scales.

2. Identification and quantification of key thresholds for soil resource erosion.

3. Feedback relationships between erosion and global change.

We hope that an appropriate list of CRPs will emerge from this workshop which can then be usefully linked with other aspects of the GCTE programme. Each CRP would bring together appropriate existing (or planned) individual research projects, and should have clearly defined aims and criteria for success but they need not necessarily start simultaneously. An energetic convener for each CRP will need to be identified who will be instrumental in finalising the respective components of the Task's Implementation Plan.

SOMNET and other GCTE Activities

The influence of global change on SOM cannot be studied in isolation. Rather, it must be assessed within a framework that accounts for it as a component of the ecosystem. The approach outlined above requires interfaces with other components of GCTE.

Many models of SOM show marked responses to temperature and to rainfall so can be used to estimate the response to climate change if it is assumed that the quantity of organic matter returned to soils is unaltered (eg. Bradbury & Powlson, 1994). Quantity and quality of organic matter inputs are major determinants of model output yet these are the two factors that may change most under conditions of enhanced CO_2. Moreover the adaptation of human societies to climate change through the modification of their agricultural systems will also change these two inputs. Links with Focus 1 and the experiments on the response of crops to an atmosphere enriched by CO_2 will provide estimates of the quantity and quality of organic matter inputs that might be expected. Similarly, links to the emerging biogeochemistry Activity will enable the work to be extended into a wider range of ecosystems.

Vegetation, climate and soils interact so that there will be an increasingly important role for soil scientists to play in the emerging models of global vegetation (Woodward, 1992). Current models of global vegetation are unconstrained by soil limitations but SOM dynamics clearly have a part to play and there is, of course, an interaction in terms of the quantity and quality of organic materials returned to soils as vegetation changes. Finally, as demonstrated in Figure 2, there is a very close interaction between soil organisms, SOM and the production of both nutrients and gaseous products.

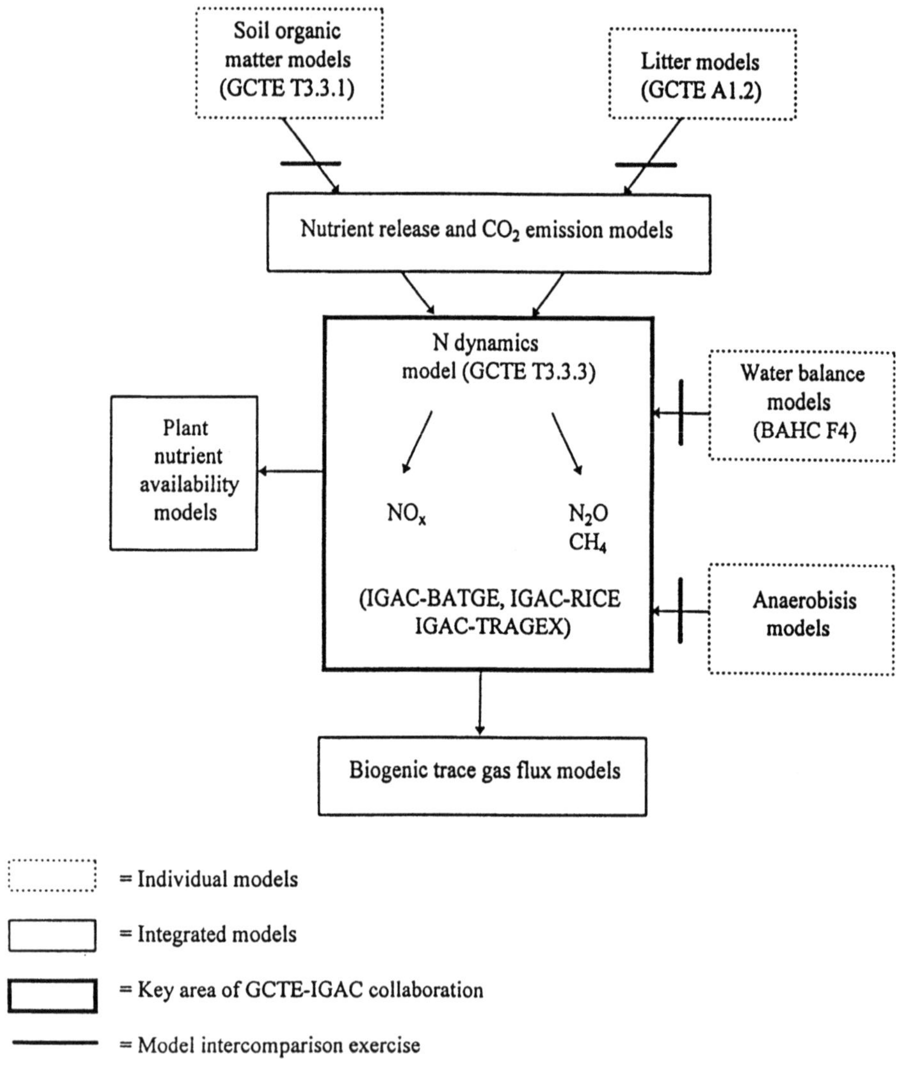

Figure 2 *Interactions between Activities of GCTE Focus 3 and other GCTE, IGAC and BAHC Activities*

Atmospheric chemistry is the responsibility of the International Global Atmospheric Chemistry Project (IGAC) but a joint IGAC/GCTE group will link the edaphic and atmospheric components together; SOMNET clearly has a major role to play in this endeavour.

Conclusions

The Soils Activity of GCTE is emerging as an important programme that cuts across many of the complex issues of predicting the response of ecosystems to global change. In an agricultural context, the changes in land use and intensification required to support an increasing population will require a more thorough understanding of the role of SOM in maintaining soil fertility especially on alfisols and oxisols where inputs are currently low. If, as seems increasingly likely, climate is changing with the temperate latitudes becoming wetter and tropical regions becoming drier, then we shall need to have predictive ability of the effects on SOM so that adapted or new systems of food production can be introduced. For natural and less intensively managed ecosystems, the interaction between vegetation and soils is important in determining the cycling of nutrients and the adaptation of vegetative systems that can occur. Soil organic matter has, then, a crucial role to play in understanding the phenomena resulting from global changes to the environment and SOMNET is a first step in bringing together SOM researchers on a global scale.

References

Anderson, J.M. (1992). Responses of soils to climate change. *Advances in Ecological Research* **22** 163-210.

Andres, R.J., Marland, G., Boden, T. & Bischoff, S. (1994). Carbon dioxide emissions from fossil fuel consumption and cement manufacture 1751 to 1991 and an estimate for their isotopic composition and latitudinal distribution. In *The Carbon Cycle* (Eds T.M.L. Wigley & D. Schimel). Cambridge University Press.

Bouwmam, A.F. (Ed.) (1990). *Soils and the Greenhouse Effect*. 575 pp. John Wiley, Chichester.

Bradbury, N.J. & Powlson, D.S. (1994). The potential impact of global environmental change on nitrogen dynamics in arable systems. In *Soil Responses to Climate Change* (Eds M.D.A. Rounsevell & P.J. Loveland). pp. 137-153. NATO ASI Series: Global Environmental Change, Vol 23. Springer-Verlag, Berlin.

Dixon, R.K., Brown, S.A., Houghton, R.A., Solomon, A.M., Trexler, M.C. & Wisniewski, J. (1994). Carbon pools and flux of global forest ecosystems. *Science* **263** 185-190.

Goudriaan, J., van de Geijn, S.C. & Ingram, J.S.I. (1993). GCTE Focus 3 Modelling and experimental data comparison workshop report. Lunteren, Netherlands.

Greenland, D.J. (1994). Land use and soil carbon in different agroecological zones. *Advances in Soil Science*.

Houghton, R.A. (1994). Emissions of carbon from land-use change. In *The Carbon Cycle* (Eds. T.M.L. Wigley & D. Schimel). Cambridge University Press.

Houghton, R.A. & Skole, D.L. (1990). Carbon. In *The Earth as Transformed by Human Action* (Eds B.L. Turner II, W.C. Clark, R.W. Kates, J.F. Richards, J.T. Mathews & W.B. Meyer). pp. 393-408. Cambridge University Press, Cambridge.

Intergovernmental Panel on Climate Change (1995). *Climate Change 1994: Radiative Forcing of Climate Change*. 339 pp. Cambridge University Press, Cambridge.

International Geosphere-Biosphere Programme (1992). *Global Change and Terrestrial Ecosystems: The Operational Plan*. Report No 21. ICSU Stockholm, Sweden.

International Union of Forestry Research Organizations (1995). *Effects of global change on Managed Forests*. GCTE Report No. 4. IUFRO, Vienna, Austria.

Jamieson, P.D., Brooking, I.R., Porter, J.R. & Wilson, D.R. (1995). Prediction of leaf appearance in wheat: a question of temperature. *Field Crops Research* **41** 35-44.

Kennedy A.C. & Smith, K.L. (1995). Soil microbial diversity and the sustainability of agricultural soils. *Plant and Soil* **170** 75-86.

Parton, W.J., Stewart, J.W.B. & Cole, C.V. (1988). Dynamics of C, N, P and S in grassland soils: a model. *Biogeochemistry* **5** 109-131.

Rogers, H.H., Runion, G.B. & Krupa, S.V. (1994). Plant responses to atmospheric CO_2 enrichment with emphasis on roots and the rhizosphere. *Environmental Pollution* **83** 155-189.

Rounsevell M.D.A. & Loveland P.J. (Eds) (1994). *Soil Responses to Climate Change*. pp. 3-11. NATO ASI Series: Global Environmental Change, Vol 23. Springer-Verlag, Berlin.

Stafford Smith M., Campbell, B., Steffen, W., Archer, S. & Ojima, D. (1995). *GCTE Task 3.1.3 Global change inpacts on pastures and rangelands. Implementation plan*. GCTE Report No. 3. GCTE, Canberra, Australia.

Tinker, P.B. & Ineson, P. (1990). Soil organic matter and soil biology in relation to climate change. In *Soils on a Warmer Earth* (Eds H.W. Scharpenseel, M. Schomaker & A. Ayoub). pp 71-88. Springer-Verlag, Berlin.

Tinker, P.B. & Ingram, J.S.I. (1994). Soils and global change - an overview. In *Soil Responses to Climate Change* (Eds M.D.A. Rounsevell & P.J. Loveland). pp. 3-11. NATO ASI Series: Global Environmental Change, Vol 23. Springer-Verlag, Berlin.

Woodward, F.I. (1992). Predicting plant responses to global environmental change. *New Phytologist* **122** 239-251.

Woomer P.L. & Swift M.J. (Eds) (1994). *The Biological Management of Tropical Soil Fertility*. John Wiley & Sons: Chichester.

For further information about GCTE Task 3.3.1 please contact:

Dr Ted Elliott or Dr Vern Cole (Task 3.3.1 Co-Leaders), Natural Resource Ecology Lab, Colorado State University, Fort Collins, CO 80523, USA, TEL +1 303 491 5645, FAX +1 303 491 1965, tede@hordeum.nrel.colostate.edu or *John Ingram (Focus 3 Officer) GCTE Focus 3 Office, Department of Plant Sciences, Oxford University, South Parks Rd, OXFORD OX1 3RB, UK, TEL +44 1865 275079, FAX +44 1865 275060, john.ingram@plant-sciences.ox.ac.uk*

Why Site Networks?

Edward T. Elliott and Keith Paustian
Natural Resource Ecology Laboratory
Colorado State University
Fort Collins, Colorado, 80523,
U.S.A.

Summary

Long term experiments reveal the dynamics of ecosystems over relatively long time periods and agricultural field experiments comprise probably the largest long-term spatial data base of ecosystem experiments available. Networks of these long-term sites have a greater value than the individual sites alone for the following reasons. Networks can increase our understanding of ecosystems by combining data from a range of sites so allowing investigators to increase numbers of observations and examine how well the data can be generalized. Furthermore, a comparison of differences between experimental designs and locations may lead to new insight about agroecosystems and fundamental knowledge about ecosystem distributions can be tested and used to address important practical problems. Another advantage of site networks is that responses can be summed for a region based upon the distribution of driving variables and they also allow us to test our understanding embodied in ecosystem models and determine the level of confidence in the ability of the model to reproduce ecosystem behavior. Networks also provide data to support defensible regional projections and additional testing power is gained when time series of model output is compared with field data obtained over a period of years. It should be noted that responses of ecosystem components include interactions and feedbacks inherent in whole system dynamics. Currently, extensive networks are organized, or being organized in the U.S., Australia, Western and Eastern Europe (see this volume) and, to a lesser extent, in Africa. Additional efforts should be mounted, especially in the tropics.

NATO ASI Series, Vol. I 38
Evaluation of Soil Organic Matter Models
Edited by David S. Powlson, Pete Smith and Jo U. Smith

Introduction

One of the great challenges for ecological researchers will be to collate, organize and maintain data obtained from long-term field research sites. These data are especially valuable because they can reveal the dynamics of ecosystems over relatively long time periods compared with more commonly studied phenomena, which are studied over the course of a growing season or a few years. These longer-term response variables may be key for understanding the functioning of many aspects of ecosystems; soil organic matter (SOM) turnover, and the relationship between primary production as it correlates to decadal variations in climate being prime examples. A difficulty often encountered with this long-term data is the ability to interpret the results because they lack experimental treatments. An exception to this problematic characteristic is data obtained from agricultural field experiments, which comprise probably the largest long-term spatial data base of ecosystem experiments available. Until recently, this data set has been essentially untapped by ecosystem research scientists for the purposes of understanding large-scale and long-term system dynamics (Elliott *et al.*, 1994a; Paul *et al.*, in press; Paustian *et al.*, 1995).

Over the course of many years and at a large number of agricultural research sites, an enormous amount of information has been collected. Much of it has been published, but some is still waiting analysis and reporting. While the need for establishing new agricultural field experiments is justified where new practices are being examined, or at locations not having had previous experiments, there is a tremendous potential for increasing our understanding of ecosystems with already existing data (Paustian *et al.*, this volume). Even for locations that have excellent records of publication, the data can be further used as it is combined with data from other sites. The value of data may increase simply because investigators have an opportunity to increase numbers of observations and examine how well the data can be generalized. In addition, comparison of differences between experimental designs and locations may lead to new insight about agroecosystems not easily obtained by examining information from a single site.

A difficulty lies in interpretation of such cross-site data. Often, the design and the exogenous factors controlling the behavior vary simultaneously. This is not an unusual situation but rather the norm, and is analogous to natural systems where vegetation and soils change as climate varies, for example. For the experimentalist, this creates an untenable situation: multiple factors covarying and creating confounding results. So, what can be learned from networks of research sites? An approach to answering this question is that developed by ecosystem scientists and is a form of systems analysis. As will be argued, such site networks provide extremely valuable information for certain kinds of questions. First, we will examine these classes of questions and then give a quick description of a systems approach to the use of such data. Then, we will link this approach with the use of simulation models, which are tools which allow us to capture the complexity of ecosystems. Finally, we will place the use of site networks within the perspective of making regional projections of how changes in management and climate might affect important ecosystem components such as SOM and system plant productivity.

Regional Analysis

There are at least two reasons why analysis at regional scales is important; one is more fundamental while the other is more applied but based upon the fundamental knowledge. In general, scientist make observations about the distribution of ecosystems and their properties and are curious about what causes them. They develop hypotheses about their observations and then design experiments, or derive other tests based upon further observations. They compare new observations against those predicted from the hypotheses and evaluate how well their concept explains the observed patterns derived from their experiments. The concepts are re-evaluated and updated until a general framework is developed which can be more accurately used to make predictions of the observed patterns of ecosystem distribution. In this general sense, information about patterns within the landscape and region are analyzed and categorized. Development of this fundamental information is far from just academic when applied to such questions as "Will this crop plant do well here, or, if not, where will it do well?" or "What level of yield might I expect over this landscape given only part of it is suitable for growing this crop?". Fundamental knowledge about ecosystem distributions can be tested with site networks and used to address important practical problems.

Integration of information to the regional level is becoming an important approach to developing information used for making management and policy decisions because knowledge is required at the level of "geopolitical" areas. For example, knowledge of soil C storage at the level of an entire country is needed by international policy makers who are trying to limit emissions of carbon dioxide to the atmosphere. Information derived from individual sites, especially that derived from experiments, is valuable for understanding the processes controlling the dynamics of important ecosystem components. However, because response of components change as exogenous drivers vary over the landscape and region, one must sum responses based upon the distribution of drivers to obtain an accurate assessment for a region. Also, within a region one may want to sum responses over particular categories such as climate, crop type, soil type, landscape position or age class. With the increasing use of geographic information systems (GIS) and accumulation of useful GIS databases, it is becoming more possible to work effectively at the regional level. A general conceptual framework exists for work at this level of spatiotemporal resolution (Barnwell *et al.*, 1992; Elliott & Cole, 1989).

Knowledge about what controls the distribution of soil and vegetation types has been an underlying basis of soil science and pedology. The same concepts which serve as the underpinning of pedology (Jenny, 1941; 1980) also serve as the template to explain the function and distribution of ecosystems, including those used for agricultural purposes (Elliott *et al.*, 1993; 1994b). This fundamental paradigm allows us to predict the broad occurrence of soils and vegetation, although more work remains to be done so that we may predict these occurrences at a particular spot. The ability to predict these distributions can be focussed on more specific and applied problems related to the current distribution of ecosystems (including those that are agriculturally based), the regional properties of these systems, and how they might be altered with changes in management and climate. The primary way that we test our predictions is to compare them with results obtained from the array of research sites distributed within the region of interest; a site network. Because of the complexities involved, interpretation of site network results requires the use of a general conceptual framework.

One framework for evaluating regional information is the driving variable-process-property (DVPP) paradigm that is one of the fundamental underpinnings of ecosystem science (Elliott *et al.*, 1994b). Within this framework, driving variables control rates of processes, which determine the flow patterns of nutrients and energy among components. These patterns determine the structure and function of ecosystems. When considering ecosystem development within a region, the sets of driving variables are climate, organisms, relief, and parent material which act over time to determine ecosystem development. The "state factor" equation developed by the Hans Jenny (1941), is similar to this description except that the concept of processes has been added.

All ecosystems contain the same set of processes, (e.g. mineralization, photosynthesis). Variations in driving variables alter the rates of these processes, thus generating different ecosystems. A change in even one of the driving variable may result in an altered ecosystem. These variations are often expressed as boundaries or ecotones. To sort out problems related to differentiating cause and effect, experiments may be conducted where ecosystems are manipulated so that single variables may be isolated.

The DVPP paradigm may be applied to ecosystems at other scales in time and space but what constitutes driving variables and response variables (ecosystem components) is determined by the level of resolution at which one is working. Driving variables are independent of each other, at a given scale, but may have complex interactive effects on response variables, while response variables are capable of changing significantly within the given time period, and their behavior is usually linked to each other through feedbacks within the system. Even the driving variables (state factors) defined by Jenny (1941) interact with each other if long enough time scales are considered.

For any given set of driving variables that exist in nature, a resultant ecosystem exists. The distribution of these ecosystems over the landscape and region are determined by variations in the driving variables. Therefore, the distribution of ecosystem types represented by sites within a network, and their responses to experimental manipulation, can be described by the driving variables extant at each site. Other factors come into play, such as site history and vagarious perturbations, but each of these may be assigned to one of the driving variables (Elliott *et al.* 1994).

Simulation Models

The importance of simulation models in interpretation and extension of geographically distributed information is derived from the existence of interactions and feedback in ecosystems which are difficult to account for by other means such as statistical and linear equation constructs. Also, the time dependant and dynamic nature of systems are hard to describe with explicit accounting of pools and flows, as is done with sets of differential equations and state variables. Simulation models are tools which are valuable for quantitatively capturing these system characteristics and can be used to investigate new system configurations, projections into the future or back calculating to a prior state. They are particularly valuable for making regional projections when combined as part of a regional system analysis (Barnwell *et al.*, 1992; Donigian *et al.*, 1994; Elliott & Cole, 1989; Peterson *et al.*, 1994).

Interactions

Interactions are fundamental properties of ecosystems that must be understood and accounted for if systems behavior is to be captured. In a general sense, interaction between two components creates a mutual influence. We often hear of soil-plant interactions or plant-animal interactions. Used in this way the term is of general use, but not useful for quantitative treatment. The term is also used in a statistical sense to describe how two independent variables influence a response variable differently when considered together than when the individual responses to the independent variables are taken individually. It is with this usage that we will consider interactions within our ecosystem conceptual framework. Adhering to our ecosystem definitions described for the DVPP paradigm above, the driving variables are the independent variables among which interactions occur and the ecosystem properties are the response variables for which the interactions are expressed. Therefore, factors such as temperature and precipitation may interact in ways that result in soil N mineralization rates that would not be predicted from knowledge of the independent variables acting alone. We capture our expression of these interactions in the equations we derive that explain the influence of the driving variables on individual processes. These equations include factors that describe the multiplicative or additive effect (more complex expressions are sometimes used) of two or more driving variables on the rate of nutrient or energy flow between two components. For example, as they are multiplied together to modify the rate constant of the C oxidation by microorganisms, temperature and water content of the soil may vary over a range of values that might be expressed as a bell-shaped curve varying from zero to one, with both optima and minima.

At the spatial and temporal level of resolution of large-scale ecosystem development, we may need to be concerned with the interactive effects of climate, organisms, relief and parentmaterial (Jenny's state factors) as they act over time. It is not surprising that our ability to predict regional responses as these variables change over a region is a challenging task, but one which is aided by the use of simulation models. Site networks set up within a region allow us to test our understanding embodied in these generalized models.

Feedbacks

Feedbacks are other important ecosystem characteristics that complicate simple interpretation of the influence of driving variables on ecosystem properties. They occur between components within the system, such that an influence of one component on another has a subsequent influence back on itself in a way which modifies, corrects or strengthens itself. Feedbacks can be propagated through a number of different components before the effect influences the originating behavior. For example, under elevated CO_2, the plant C/N ratio may widen and cause a lower quality litter. As this litter falls to the soil, microbes may immobilize mineral N as they decompose the low N litter and thereby reduce the soil N available to the plant, which subsequently reduces the C/N ratio of the plant material thus completing the cycle. This cycle may eventually limit the ability of the ecosystem to continue to respond to the elevation of CO_2. The effect of this negative feedback loop, which cascades throughout a number of different components, will eventually bring the system into balance under elevated CO_2. Feedbacks may also be positive, as often exemplified by the process of erosion.

Projections of Large-Scale and Long-Term Changes in Ecosystems

Projections of agroecosystem response to changes in management and climate over a region can be made by integrating a series of approaches (Elliott & Cole, 1989). Simulations embody our best understanding of how ecosystems function (Metherell *et al.*, 1995). They are built from information derived from studies of their structure, which determine what components of the system will be included, and studies of their function, which most often take the form of process studies and describe what controls the rate of transfer of nutrients or energy between components. Studies of the processes (e.g. decomposition, mineralization, crop production, transpiration) are used to determine the functional form and quantitative description of this transfer of material between components. While many of the studies used to describe these processes may be done out of context of the whole system, they are integrated through construction of the model in a way that subsequently includes the interactions and feedbacks described above. It is through consideration of the whole system that one can account for the complexities of interactions and feedbacks that make simulation models such powerful ecological tools.

Regional projections can be made if information exists on the distribution of the appropriate driving variables over the region. The distribution of driving variables can be effectively organized using GIS technology. The GIS database of driving variables may consist of several layers, one each for the different variables. When these databases are overlain, they produce a set of polygons, each of which represents a set of driving variables with which the model can be run. If one of these layers is the current management, one may make a projection for the region of what the values of a particular response variable might currently be, or if the model is run for a number of years, one may predict how current management may affect the ecosystem at some point in the future. In addition, one might alter the management data layer to what might be expected for the future and make such projections. Likewise, climate can be altered to make projections. Simultaneous alterations of both climate and management can be made for what might be the most realistic projections.

When information generated from the model simulations is projected back onto the polygons from which the sets of driving variables are derived, the response of the entire region can be analyzed in a number of ways, including summing up this complex response for the entire region. Thus, a shift in the ecotonal boundary for the wheat and corn growing regions of the central U.S. following a change in climate might be evaluated as it impacts relative yield obtained by the farmer. However, before such projections can be made, their response must be compared to measured values and the level of confidence in their ability to reproduce ecosystem behavior must be evaluated. It is in this role that site networks are essential.

The Value of Site Networks

As stated earlier, site networks are important for testing fundamental hypotheses. Moreover, in the context of making defensible regional projections, they are crucial. Model output must be tested within the region for which projections wish to be made. One of the chief difficulties in evaluating information for regions is determining what constitutes representative data for that region. Information obtained from a single site, even though carefully replicated at that site, is essentially pseudoreplicated (Hurlert, 1984) if that site is intended to represent an entire region. Yet, if one moves too far from the original site to obtain a replicate site, driving variables such as soil texture, temperature or precipitation may change significantly, resulting in lack of good replication. In some sense, it is not possible to have replicated ecosystems. It is arbitrary as to what class of driving variable values are similar enough to represent a region. Even if enough samples are taken within a region (i.e., a site network) to adequately represent it, we may not know if that region is representative of all potential regions developing under that specific set of driving variables. At the very least, we should be reluctant to extrapolate our predictions to regions developing under sets of driving variables that fall outside the range of variable values for the region against which the model was compared.

A difficult aspect of comparing model output with site data is knowing the history of the site with enough certainty that one can accurately model events leading up to the point of comparison. With natural ecosystems, the model may be run to steady-state to make the comparisons. Often with agricultural research sites, the land on which the experiment has been conducted has had a series management regimes leading up to the current experiment. These must be modeled if the best fit to the data are to be obtained. Across the sites in the networks, there may be as many pre-experiment histories to account for as there are sites. Despite this complexity, additional testing power is gained when time series of model output can be compared with field data obtained over a series of years.

A final, but key reason which makes site networks valuable tools is their ecosystem context. Models may be validated by comparing the response of components with data obtained from laboratory or greenhouse experiments. While these data may have been obtained under much more strictly controlled conditions than possible in the field, they are conducted outside the domain of the ecosystem and therefore lack the interactive and feedback controls inherent within intact ecosystems. It is important to distinguish data used for validation which is obtained from an intact ecosystem from that derived from laboratory experiments.

References

Barnwell, T.O., Jackson R.B., Elliott E.T., Burke I.C., Cole C.V., Paustian, K., Paul, E.A., Donigian, A.S., Patwardhan, A.S., Rowell, A., & Weinrich, K. (1992) An approach to assessment of management impacts on agricultural soil carbon. *Water, Air and Soil Pollution* **64**: 423-435.

Donigian, A.S., Barnwell T.O., Jackson, R.B., Patwardhan, A.S., Weinrich, K.B., Rowell, A.L., Chinnaswamy, R.V., & Cole C.V. (1994) Assessment of alternative management practices and policies affecting soil carbon in agroecossytems of the central United States. EPA/600/R-94/067, 193 pp.

Elliott, E. T. & Cole, C. V. (1989) A perspective on agroecosystem science. *Ecology* **70**:1597 1602.

Elliott, E. T., Cole, C. V., & Cambardella, C. A. (1993) Modification of ecosystem processes by management and the mediation of soil organic matter dynamics. In: *The dynamics of soil organic matter in relation to the sustainability of tropical agriculture.* (ed: R. Mercx & K. Mulongoy), John Wiley and Sons, Chichester: 257-268.

Elliott, E.T., Burke, I.C., Monz, C.A. , Frey, S.D., Paustian, K.H., Collins, H.P., Paul, E.A., Cole, C.V., Blevins, R.L., Frye, W.W., Lyon, D.J., Halvorson, A.D., Huggins, D.R., Turco, R.F., & Hickman, M.V. (1994a) Terrestrial carbon pools in grasslands and agricultural soils: Preliminary data from the Corn Belt and Great Plains regions. In: *Defining soil quality for a sustainable environment.* (ed: J.W. Doran, D.C. Coleman, D.F. Bezdicek & B.A. Stewart), SSSA Special Publication 35, Madison, WI: 179-191.

Elliott, E. T., Janzen, H. H., Campbell, C. A., Cole, C. V., & Myers, R. J. K. (1994b) Principles of ecosystem analysis and their application to integrated nutrient management and assessment of sustainability. In: *Sustainable land management for the 21st century.* (ed: R. C. Wood & J. Dumanski), University of Lethbridge, Lethbridge, Canada: 35-57.

Hurlbert, S.H. (1984) Pseudoreplication and the design of ecological field experiments. *Ecol. Mongr.* **54**:187-211.

Jenny, H. (1941) *Factors of soil formation.* McGraw Hill, New York.

Jenny, H. (1980) *The soil resource: Origin and behavior. Ecological Studies 37.* Springer Verlag, New York.

Metherell, A.K., Cambardella, C.A., Parton, W.J., Peterson, G.A., Harding, L.A., & Cole, C.V. (1995) Simulation of soil organic matter dynamics in dryland wheat-fallow cropping systems. In: *Soil management and greenhouse effect.* (ed: R. Lal, J. Kimble, E. Levine & B.A. Stewart), Adv. Soil Sci., CRC-Lewis Publishers, Boca Raton: 259-270.

Paul, E.A., Paustian, K., Elliott, E.T. & Cole, C.V. (eds) (1995) *Soil organic matter in temperate agroecosystems: Long-term experiments in north america.* CRC-Lewis Publishers. (in press).

Paustian, K., Elliott, E.T., Collins, H.P., Cole, C.V., & Paul, E.A. (1995) Use of a network of long-term experiments in North America for analysis of soil C dynamics and global change. *Aust. Jn. Exp. Ag.* (in press).

Peterson, G.A., Westfall, D.G., & Cole, C.V. (1993) Agroecosystem approach to soil and crop management research. *Soil Sci. Soc. Am. J.* **57**:1354-1360.

The North American Site Network

Keith Paustian [1], Edward T. Elliott [1], Eldor A. Paul [2], Harold P. Collins [2], C. Vernon Cole[1] and Serita D. Frey [1]

[1] Natural Resource Ecology Laboratory
Colorado State University
Fort Collins, Colorado, 80523
USA

Summary

A network of long-term experiments in North America was organized to develop a regionalized database for evaluating environmental and management controls on soil organic matter and to assess the potential for C sequestration in agroecosystems. The network is comprised of 39 field experiments at 33 locations, concentrated in the Great Plains region of the U.S. and Canada and the U.S. corn belt. Historical data on climate, productivity, soils and management have been compiled for the sites. An extensive cross-site sampling of soils and analysis of soil organic matter fractions has been carried out for most of the sites. Information from the long-term experiments is being used to support a wide variety of research, ranging from detailed process studies to large-scale model-based assessments of agroecosystem C balances and CO_2 mitigation strategies. Networks of existing long-term experiments can play an important role in ecosystem and global change research.

---

[2] *Department of Crop and Soil Science and W.K. Kellogg Biological Station, Michigan State University, East Lansing, Michigan, 48824, USA*

NATO ASI Series, Vol. I 38
Evaluation of Soil Organic Matter Models
Edited by David S. Powlson, Pete Smith and Jo U. Smith

Introduction

Agricultural science has a rich history of long-term field experimentation, as exemplified by the ongoing "classical" experiments here at Rothamsted, the site of this workshop. These and other long-term experiments have made great contributions to the agricultural (and other) sciences over the years. Current concerns about global environmental changes and their potential effects on terrestrial ecosystems, including agroecosystems, are yet another area in which long-term field experiments can provide vital information.

Soil organic matter (SOM) constitutes the largest terrestrial C pool (Post *et al.*, 1990) and thus changes in soil C storage, induced by climatic and/or land use changes could have significant impacts on atmospheric CO_2 concentrations (Jenkinson *et al.*, 1991, Schimel *et al.*, 1994, Kirschbaum, 1995). In addition, SOM is an important factor regulating the emission of greenhouse gases such as N_2O (Robertson, 1993) and CH_4 (Schimel *et al.*, 1993). However, changes in SOM occur relatively slowly, requiring time periods of several years to decades for significant changes to occur. Thus, long-term field experiments in which SOM has been measured repeatedly over time provide one of the best sources of information on the rate of change of SOM. Many of the factors which influence SOM dynamics, such as plant productivity, degree of soil disturbance, microclimate and N availability, vary as a function of management practices such as crop rotation, tillage, and fertilization. Thus, data from these kinds of experiments can help to elucidate fundamental controls on SOM behavior.

In 1991 we began a research project entitled "Agroecosystem Carbon Pools and Dynamics", which included the participation of many of the long-term agricultural experiments ongoing in North America. The objective of the project was "...to develop defensible data to evaluate carbon pools and dynamics for agroecosystems, to identify critical soil and climatic parameters that might affect those dynamics, and to validate models to assess the impact of alternative agricultural management practices on carbon storage and fluxes in agroecosystems."

This paper describes the organization and data resources of the site network and some of the current applications of the site information in global change research.

Site network description

A total of 39 field experiments at 33 separate locations are involved in the network (Figure 1). Geographically, the sites are concentrated in the Great Plains region of the U.S. and Canada, and the Corn Belt region of the midwestern U.S. Some additional sites in Oregon, Georgia and Pennsylvania are included from outside of these regions. The project was originally designed to focus on the Great Plains and Corn Belt regions, which is why a number of other long-term sites, especially in the southeastern U.S., were not included.

The sites span a wide range of climatic conditions (see Appendix 1), with the coldest and driest sites in Saskatchewan (Melfort, 0.8°C; Swift Current, 328 mm) and the warmest and wettest in Georgia (Griffin 16.6°, 1267 mm). The distributions of mean annual temperature and annual precipitation are shown in Figure 2.

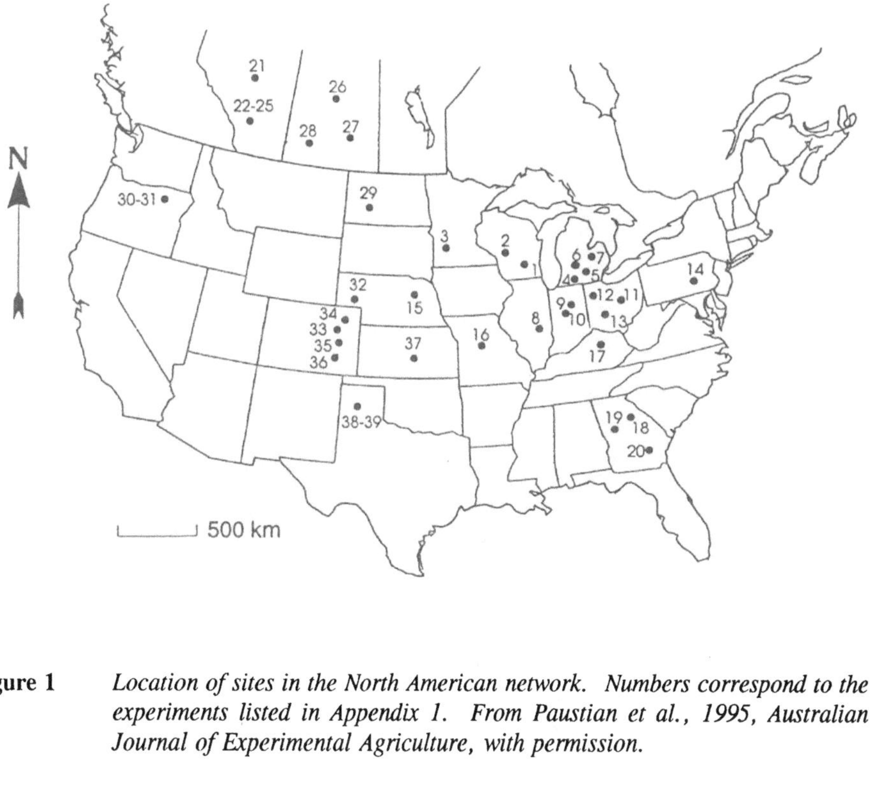

Figure 1 *Location of sites in the North American network. Numbers correspond to the experiments listed in Appendix 1. From Paustian et al., 1995, Australian Journal of Experimental Agriculture, with permission.*

The soil types (Soil Survey Staff, 1975) are dominated by grassland-derived Mollisols in the Great Plains and forest-derived Alfisols in the eastern sites, with more highly weathered soils (Ultisols) in the southeastern sites (Appendix 1). The most common soil textures are loam and silt loam, although nearly every soil textural class, from loamy sands to clay soils, is represented.

The sites can be roughly segregated according to the dominant crops grown. Wheat is the main crop in the drier sites, west of about 97°W longitude, while row crops such as maize and soybean are the main crops in the eastern, more humid areas. Collectively, the sites represent a rich array of different experimental treatments, including tillage, crop rotation, fertilization, organic amendments and pest control. The most common treatment variables are tillage, crop rotation and fertilization. Nineteen experiments have tillage comparisons, 26 have crop rotation comparisons, and 23 have different rates of fertilizer application. Nineteen of the sites also have nearby areas with native vegetation and untilled soils which can be used to evaluate historical effects of cultivation. In all, there are 439 different treatment combinations represented in the network. Thus there is an enormous wealth of information on how management practices influence soil organic matter, over a wide range of climate and soil conditions.

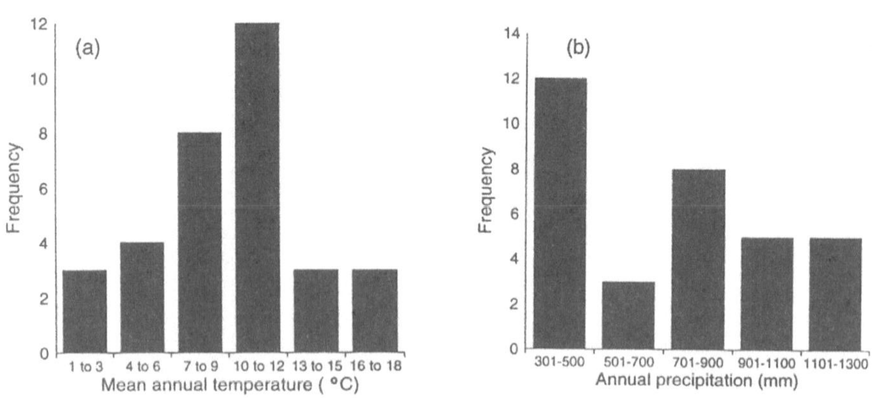

Figure 2. *Distributions of mean annual (a) temperature (°C), and (b) precipitation (mm) for the long-term experiment sites. From Paustian et al., 1995, Australian Journal of Experimental Agriculture, with permission.*

Site network information

Two sets of information from the field experiments have been compiled (Table 1). First, existing data have been compiled by individual scientists at the sites and organized in a common format. This compilation includes background information on site history, native vegetation, geomorphology, long-term climate averages and experimental design. Comprehensive data compilations include crop yields (and in several cases, crop residue amounts and nitrogen (N) concentrations in grain) and soil properties, including soil organic matter C and N, soil texture, bulk density and pH. Scientists at 30 sites (encompassing 35 different field experiments) presented synthesis papers based on this information at a workshop in 1992. Papers from the workshop will be published in a forthcoming book (Paul *et al.*, 1995).

To complement the existing site data, a cross-site sampling was carried out at 29 of the 33 site locations in the network. The purpose of this sampling was to have a consistent set of soil organic matter measurements, including several diagnostic organic matter fractions and samples taken to 1 m depth. By using a consistent sampling protocol and a single suite of analytical methods and equipment, sampling and measurement errors were minimized, providing the best possible conditions for intersite comparisons. Samples for the western sites were taken by a team from Colorado State University (CSU) and eastern sites were sampled by a team from Michigan State University (MSU). Individual laboratory assays of samples from all sites were then made at a single facility, either at MSU or CSU. In addition, subsamples from all collected samples were air-dried and archived. The sampling and analytical procedures used are summarized in Table 1 and a complete description of methods is provided by Elliott *et al.* (1994).

Overview of soil organic matter trends

The historical data compilations and the data from the cross-site sampling can be used to test and validate simulation models and to analyze the relative effects of climatic, edaphic and management factors using statistical models. While the data analysis is not yet complete, we will present some results for surface (0-20 cm) soil C measurements taken in the cross-site sampling to illustrate the variation across sites and to examine general patterns associated with climatic and management controls on SOM.

Table 1. *Data compiled from the long-term experiments.*

Historical databases

(1) climate (temperature, precipitation)
(2) soils (taxonomy, profile descriptions, soil C and N texture, bulk density, pH)
(3) production (dry weight of grain, forage, stover; some N data)
(4) management (planting and harvest dates, tillage practices, fertilizer and pesticide application)

Cross-site sampling

Field sampling

(1) Soil cores (6.4 cm dia.) taken to 1 m depth, 6 per plot (divided into 5 depth increments: 0-20, 20-25, 25-50, 50-100 cm, plus 0-2.5 cm samples in no-till only)
(2) Surface litter from 0.25 m^2 subplots (excluding current year's residue)

Lab analyses

All soil samples

(1) Total C and N (dry combustion in Carlo-Erba analyzer)
(2) Inorganic C
(3) Bulk density
(4) Textural analysis
(5) pH (10g soil:20 ml H_2O)

0-20 cm increment only

(1) $CHCl_3$-labile C and N (by direct extraction after 24 h fumigation; Brookes *et al.*, 1985)
(2) Particulate organic matter C and N (> 53 μm fraction with wet sieving; Cambardella and Elliott, 1992)
(3) Aerobic incubations (> 200 d) for C and N mineralization (Boyle and Paul, 1989)

For selected sites and samples

(1) $^{12}C/^{13}C$ in bulk soil and mineralized C in incubations
(2) ^{14}C dating (whole soil and SOM fractions)

Surface soil organic C ranged from about 1 kg C m^{-2} to 7 kg C m^{-2} for the network sites, with the exception of one location, Melfort, Saskatchewan, where soil C was in excess of 10 kg m^{-2} (Figure 3). It is well known that climate exerts a strong influence of SOM levels, and the data show the expected trend of increasing soil C going from hot dry conditions to cooler moister environments, as expressed by a simple quotient of mean annual precipitation and temperature. However, there is considerable variability which is not explained by these annual climatic variables. A variety of other factors, including differences in soil physical and chemical properties and differences related to previous land use (e.g., native vegetation, time under cultivation, soil erosion, cropping practices), are likely to be important sources of variability. More complex, multifactored statistical approaches can undoubtably help to account for some of this variability. However, we believe the use of simulation models, which are able to integrate the effects of numerous factors and their interactions, may be the best means for understanding and interpreting regional patterns of SOM distribution.

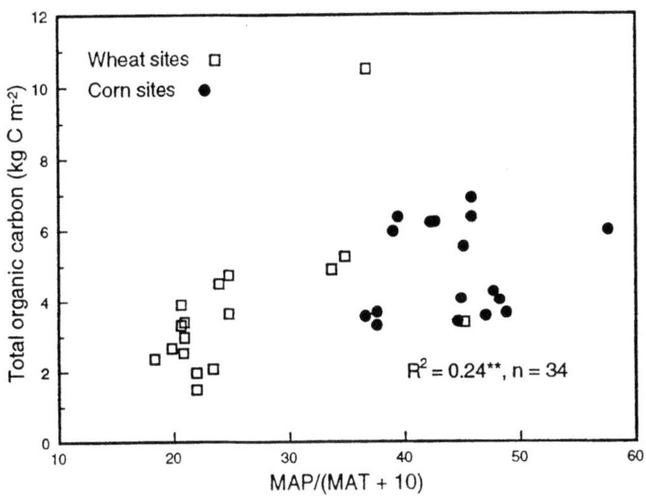

Figure 3. *Soil organic C levels (0-20 cm) as a function of mean annual temperature (MAT - °C) and mean annual precipitation (MAP - mm), for sites involved in the cross-site sampling. Soil C values are means for all treatments at the site (excluding native soil reference plots).*

For within-site comparisons, macro-scale effects of climate and edaphic conditions and land use history are eliminated, enabling an evaluation of management effects alone. However, by having data on similar treatment comparisons within a network of sites, the generality of a treatment effect and its variation as a function of environmental conditions or other management factors can be evaluated.

As an example, differences in soil C with no-till versus moldboard plow tillage are shown in Figure 4. The values represent the relative difference in soil C in the top 20 cm

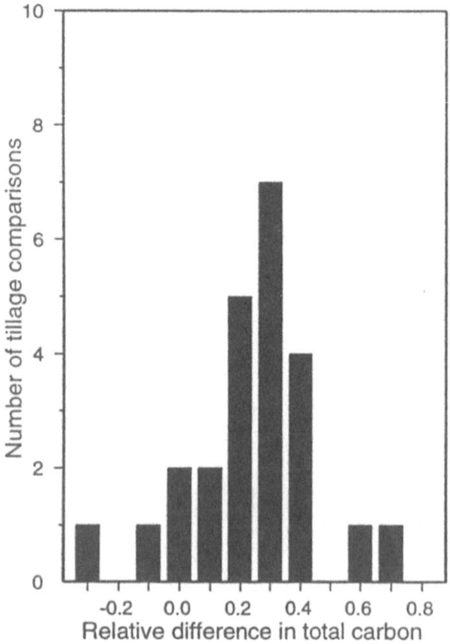

Figure 4. *Frequency distribution of relative differences between no-till(NT) and moldboard plow (MP) paired treatments. Relative differences were calculated as (NT-MP)/MP, for g C m^{-2} in the top 20 cm, adjusted to an equivalent soil mass (see explanation in text).*

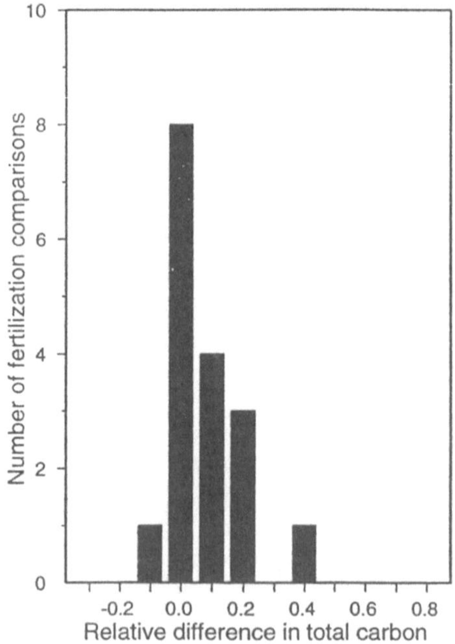

Figure 5. *Frequency distribution of differences between N fertilized (NT) and unfertilized (UF) paired treatments. Relative differences were calculated as (NF-UF)/UF, for g C m^{-2} adjusted to an equivalent soil mass (see explanation in text).*

(calculated as g C m^{-2} in no-till, minus C in moldboard tillage, divided by C in the moldboard treatment). The values were corrected to represent equal masses of soil, so that differences in bulk density for different tillage treatments would not bias the comparison. This was done by first calculating the mean C amount and soil mass for the top 20 cm. Then, for the treatment having the lower soil mass in the top 20 cm, part of the 20-25 cm increment (and its corresponding C content) was added to reach an equivalent soil mass for the comparison. There were a total of 24 paired comparisons having no-till *vs* moldboard treatments. Data from intermediate tillage treatments (e.g., stubble mulch, chisel plow) were not included.

A preliminary screening of the data, by plotting the frequency distribution of the pair-wise treatment differences, suggests that use of no-till usually increased total C levels relative to conventional tillage, for the entire depth of plowing (between 15 to 20 cm for these sites) and after accounting for the effect of tillage-induced changes in bulk density. However, it is clear that the response of systems to no-till *vs* moldboard plowing varied considerably from site to site. Further analysis and significance testing will be carried out to evaluate the underlying factors which control tillage responses.

A similar data screening was performed comparing N fertilization effects on soil C levels (Figure 5). Paired unfertilized and N fertilized treatments were selected and relative differences in soil C for the top 20 cm were calculated as described above for the tillage comparisons. Where more than one fertilization level occurred within a particular experiment, the fertilization level closest to the normal practice for the area was used. There were a total of 17 paired comparisons of unfertilized and N fertilized treatments.

The overall response of soil C to N fertilization was much less pronounced than was the case for tillage (Figure 5). Treatments at most sites showed little or no effect, with a few treatments exhibiting larger positive responses. Because fertilization effects on SOM include many indirect effects, such as influences on plant production and soil water relations, simulation models can be useful in interpreting such field results (e.g., Paustian *et al.*, 1992).

Applications of network information

The subject of this workshop involves the use of long-term experiment data to test and compare SOM models and this is certainly one of the major applications and justifications for the North American site network. Aside from the general benefits of having multiple data sets with which to test the generality and robustness of models, we believe that site networks are a key ingredient in large-scale analyses. There is an increasing need for improved regional and global-level assessments of ecosystem response to global change and the environmental impacts of land management.

The main approach used in such regional analyses is to scale-up "ecosystem-level" models using regionally-distributed driving variables, such as climate, soil properties and landuse practices (Elliott and Cole, 1989; also see Elliott *et al.* and Post *et al.*, this volume). Regional level model output is obtained by summing outputs for a number of spatial aggregations of driving variables (e.g., grid cells).

In the context of regional analyses, long-term sites represent "nodes" where the existence of high quality data can help ensure that the model successfully represents ecosystem responses for sets of climate, soil and management conditions (Elliott and Cole, 1989). The availability, distribution and representativeness of long-term experiments do set constraints for their use as validation points. However, they still represent one of the few kinds of "ground-truth" data that can be used, together with information from remote-sensing, to validate regional-level model analyses.

At present we and other collaborators in the site network are using the field sites and the information that has been compiled to support a variety of research, ranging from basic process studies of SOM turnover to regional assessments of the soil C balance in agricultural systems (Table 2). Efforts to promote the establishment of new site networks and closer ties between modelers and data holders, such as at this workshop, will undoubtably create new opportunities for research in soil, ecosystem and global change science.

Table 2. *Research projects utilizing North American site network information.*

* Assessment of C sequestration potential in agricultural soils.
* Assessment of agricultural ecosystem adaptation to climate and CO_2 change.
* Regional assessment of C sequestration on U.S. Conservation Reserve Program land.
* Regional assessment of impacts of conservation tillage on SOM.
* Regional soil quality assessment and monitoring.
* IPCC chapter on "Mitigation options in agriculture".
* Process studies of environmental and management impacts on soil structure and SOM dynamics.
* Process studies of tillage effects on microbial community structure and SOM stabilization.
* Carbon-dating and $^{12}C/^{13}C$ analyses of SOM turnover.
* Management and soil effects on microorganism biodiversity.

References

Boyle, M. & Paul, E.A. (1989) Carbon and nitrogen mineralization kinetics in soil previously amended with sewage sludge. *Soil Science Society of America Journal* **53**: 99-103.

Brookes, P.C., Landman, A., Pruden, G. & Jenkinson, D.S. (1985) Chloroform fumigation and the release of soil nitrogen: A rapid direct extraction method to measure microbial biomass nitrogen in soil. *Soil Biology and Biochemistry* **17**: 837-42.

Cambardella, C.A. & Elliott, E.T. (1992) Particulate soil organic matter changes across a grassland cultivation sequence. *Soil Science Society of America Journal* **56:** 777-783.

Elliott, E.T. & Cole, C.V. (1989) A perspective on agroecosystem science. *Ecology* **70:** 1597-1602.

Elliott, E.T., Burke, I.C., Monz, C.A., Frey, S.D., Paustian, K.H., Collins, H.P., Paul, E.A., Cole, C.V., Blevins, R.L., Frye, W.W., Lyon, D.J., Halvorson, A.D., Huggins, D.R., Turco, R.F. & Hickman, M.V. (1994) Terrestrial carbon pools in grasslands and agricultural soils: preliminary data from the Corn Belt and Great Plains regions. In: Doran, J.W., Coleman, D.C., Bezdicek, D.F., and Stewart, B.A. (eds.) *Defining Soil Quality for a Sustainable Environment.* SSSA Special Publication Number 35, SSSA, Madison, WI, pages 179-192.

Jenkinson, D.S., Adams, D.E. & Wild, A. (1991) Model estimates of CO_2 emissions from soil in response to global warming. *Nature* **351**: 304-6.

Kirschbaum, M.U.F. (1995) The temperature dependence of soil organic matter decomposition, and the effect of global warming on soil organic C storage. *Soil Biology and Biochemistry* **27:** 753-760.

Paul, E.A., Paustian, K., Elliott, E.T. & Cole, C.V. (eds) (1995) *Soil Organic Matter in Temperate Agroecosystems. Long-term Experiments in North America.* Lewis Publishers (in press).

Paustian, K., Elliott, E.T., Collins, H.P., Cole, C.V. & Paul, E.A. (1995) Use of a network of long-term experiments for analysis of soil carbon dynamics and global change: the North American model. *Australian Journal of Experimental Agriculture* (in press).

Paustian, K., Parton, W.J. & Persson, J. (1992) Modeling soil organic matter in organic-amended and nitrogen-fertilized long-term plots. *Soil Science Society of America Journal* **56**: 476-88.

Post, W.M., Peng, T-H., Emanuel, W.R., King, A.W., Dale, V.H. & DeAngelis, D.L. (1990) The global carbon cycle. *American Scientist* **78:** 320-326.

Robertson, G.P. (1993) Fluxes of nitrous oxide and other nitrogen trace gases from intensively managed landscapes: A global perspective. In: *Agricultural Ecosystem Effects on Trace Gases and Global Climate Change*. ASA Special Publication Number 55, pages 95-108.

Schimel, J.P., Holland, E.A. & Valentine, D. (1993) Controls on methane flux from terrestrial ecosystems. In: *Agricultural Ecosystem Effects on Trace Gases and Global Climate Change*. ASA Special Publication Number 55, pages 167-182.

Schimel, D.S., Braswell, B.H., Holland, E.A., McKeown, R., Ojima, D.S., Painter, T.H., Parton, W.J. & Townsend, A.R. (1994) Climatic, edaphic and biotic controls over storage and turnover of carbon in soils. *Global Biogeochemical Cycles* **8**: 279-93.

Soil Survey Staff (1975) *Soil Taxonomy. A Basic System of Soil Classification for Making and Interpreting Soil Surveys*. USDA Handbook No. 436, U.S. Government Printing Office, Washington, D.C.

Acknowledgments

Support for research described in this paper was provided by U.S. Environmental Protection Agency, U.S. Department of Energy and the Agriculture Research Service and Natural Resource Conservation Service of the U.S. Department of Agriculture. Collaborators in the long-term experiment site network: C.A. Campbell, R.P. Zentner (Ag Canada-Swift Current), G.P. Lafond, (Ag Canada-Indian Head), A.P. Moulin (Ag Canada-Melfort), H.H. Janzen, C.W. Lindwall (Ag Canada-Lethbridge), R.W. Blevins, W.W. Frye (Univ. of Kentucky), W.A. Dick (Ohio State Univ.), J.R. Brown, G.A. Buyanovsky, G.H. Wagner (Univ. of Missouri), R.G. Darmody, T.R. Peck (Univ. of Illinois), S. Peters (Rodale), J.L. Havlin (Kansas State Univ.), G. Lesoing, J. Doran, D.J. Lyon, (Univ. of Nebraska), D.R. Christenson, R.E. Lucas, F.J. Pierce, K.S. Pregitzer, G.P. Robertson, M.L. Vitosh (Michigan State Univ.), D. Huggins, (Univ. of Minnesota), P.F. Hendrix, (Univ. of Georgia), N.G. Juma, J.A. Robertson (Univ. of Alberta), G.A. Peterson, D.G. Westfall (Colorado State Univ.), M.B. Vanotti, L.G. Bundy (Univ. of Wisconsin), D. Griffith (Purdue Univ.), M. Hickman (USDA/ARS-W. Lafayette), R.R. Bruce (USDA/ARS-Watkinsville), D.V. MacCracken (USDA/ARS-Griffin), O.R. Jones, B.A. Stewart, P.W. Unger (USDA/ARS-Bushland), M.F. Vigil, A.H. Halvorson (USDA/ARS-Akron), A. L. Black, D.L. Tanaka (USDA/ARS-Mandan), P.E. Rasmussen and R. Smiley (USDA/ARS-Pendleton), are gratefully acknowledged for their contributions.

Appendix 1. Description of long-term field experiments in the North America site network. From Paustian *et al.* (1995), Australian Journal of Experimental Agriculture, with permission.

Site Location	Soil Type[1]	MAT[2] (°C)	Precip. (mm)	Experiment Started	Cropping Systems	Tillage	Treatments	No. of Treatments
1) Arlington, Wisconsin	Argiudoll	7.6	793	1958	Corn	Conventional	fertility	3
2) Lancaster, Wisconsin	Hapludalf	7.8	832	1967	Corn Alfalfa Corn-Soybean-Corn-Oat-Alfalfa Corn-Corn-Corn-Alfalfa-Alfalfa Corn-Corn-Oat-Alfalfa-Alfalfa Corn-Alfalfa	Conventional	rotation fertility	21
3) Lamberton, Minnesota	Haplustoll	6.2	632	1960	Corn	Conventional	fertility	18
4) Kellogg Biol. Stat, Michigan[3]	Hapludalf	9.2	920	1986	Corn-Soybean Never-tilled grassland	Conventional No-till	tillage fertility	7
5) E. Lansing, Michigan	Hapludalf	8.5	728	1963	Corn Grain Corn Silage	Conventional	fertility	10
6) E. Lansing, Michigan	Ochraqualf	8.5	728	1980	Corn Corn-Clover Corn-Rye	Conventional No-till	tillage rotation	4
7) Saginaw, Michigan	Haplaquept	8.5	788	1972	Corn-Sugar Beets Navy Beans-Sugar Beets Oats-Navy Beans-Sugar Beets Corn-Corn-Corn-Sugar Beets Corn-Corn-Navy Beans-Sugar Beets Oats-Alfalfa-Navy Beans-Sugar Beets	Conventional	rotation	6

Appendix 1. (Continued).

Site Location	Soil Type[1]	MAT[2] (°C)	Precip. (mm)	Experiment Started	Cropping Systems	Tillage	Treatments	No. of Treatments
8) Champaign-Urbana, Illinois	Argiudoll	11.0	940	1876	Corn Corn-Oats Corn-Oats-Hay	Conventional	rotation fertility	15
9) W. Lafayette, Indiana[3]	Haplaquol	10.8	952	1981	Corn Corn-Soybean Soybean Corn-Soybean-Wheat Soybean-Wheat	Conventional Conservation No-till	rotation tillage weed control	45
10) W. Lafayette, Indiana[3]	Haplaquol	10.8	952	1975	Corn-Soybean	Conventional Chisel Ridge-till No-till	tillage	4
11) Wooster, Ohio	Fragiudalf	9.5	733	1962	Corn Corn-Soybean Corn-Oats-Meadow	Conventional Minimum No-till	rotation tillage	9
12) Hoytville, Ohio	Ochraqualf	9.8	835	1963	Corn Corn-Soybean Corn-Oats-Meadow	Conventional Minimum No-till	tillage rotation	9
13) S. Charleston, Ohio	Ochraqualf	10.4	910	1963	Corn	Conventional Minimum No-till	tillage	3
14) Kutztown, Pennsylvania	Fragiudalf	10.5	1180	1981	Corn-Wheat-Hay-Corn-Soybean Corn-Soybean-Wheat Corn-Corn-Soybean-Corn-Soybean	Conventional	rotation fertility	3

Appendix 1. (Continued).

Site Location	Soil Type[1]	MAT[2] (°C)	Precip. (mm)	Experiment Started	Cropping Systems	Tillage	Treatments	No. of Treatments
15) Mead, Nebraska	Argiudoll	10.2	680	1975	Corn-Soybean-Corn-Oat/Clover Corn	Conventional	rotation fertility	4
16) Columbia, Missouri	Ochraqualf	12.4	814	1888	Wheat Alfalfa-Bromegrass Corn Oats Red Clover 3-,4-,5-, and 6-year rotations	Conventional	rotation fertility	39
17) Lexington, Kentucky	Paleudalfs	13.1	1127	1970	Corn	Conventional No-till	tillage fertility	8
18) Horseshoe Bend, Georgia	Kanhapludult	16.5	1246	1978	Sorghum/Soybean-Rye Sorghum/Soybean-Clover	Conventional No-till	tillage rotation	4
19) Watkinsville, Georgia	Kanhapludult	16.0	1252	1982	Clover-Sorghum Sorghum Soybean	No-till Conventional	rotation irrigation tillage	18
20) Griffin, Georgia[3]	Kanhapludult	16.6	1267	1976	Soybean-Wheat-Sorghum	Conventional Minimum No-till	tillage	6
21) Breton, Alberta	Cryoboralf	2.1	547	1930	Wheat-Oat-Barley-Forage Wheat-Fallow	Conventional	rotation fertility	9
22) Lethbridge, Alberta	Haploboroll	5.5	384	1911	Wheat Fallow-Wheat Fallow-Wheat-Wheat	Conventional Minimum	rotation fertility	12

Appendix 1. (Continued).

Site Location	Soil Type[1]	MAT[2] (°C)	Precip. (mm)	Experiment Started	Cropping Systems	Tillage	Treatments	No. of Treatments
23) Lethbridge, Alberta	Haploboroll	5.5	384	1951	Wheat Fallow-Wheat Fallow-Wheat-Wheat Lentil-Wheat Lentil-Wheat-Wheat Fallow-Wheat-Wheat-Hay-Hay-Hay	Conventional	rotation fertility	12
24) Lethbridge, Alberta	Haploboroll	5.5	384	1911	10-year Alfalfa-Cereal-Tuber Alfalfa-Alfalfa-Alfalfa-Wheat-Barley Corn-Wheat-Corn-Wheat-Barley Faba Bean-Wheat-Faba Bean-Wheat-Barley	Conventional	rotation fertility	15
25) Lethbridge, Alberta	Haploboroll	5.5	384	1955	Wheat-Fallow	Conventional Minimum No-till	tillage	5
26) Melfort, Sakatchewan	Udic Boroll	0.8	396	1957	Fallow-Wheat Fallow-Wheat-Wheat Clover-Wheat-Wheat Wheat Fallow-Wheat-Wheat-Hay-Hay-Wheat	Conventional	rotation fertility	8
27) Indian Head, Sakatchewan	Udic Boroll	2.5	435	1957	Wheat Fallow-Wheat Fallow-Wheat-Wheat Sweetclover-Wheat-Wheat Fallow-Wheat-Wheat-Hay-Hay-Hay	Conventional No-till (1990)	rotation fertility	9

Appendix 1. (Continued).

Site Location	Soil Type[1]	MAT[2] (°C)	Precip. (mm)	Experiment Started	Cropping Systems	Tillage	Treatments	No. of Treatments
28) Swift Current, Saskatchewan	Haploboroll	3.7	328	1966	Fallow-Wheat-Wheat Fallow-Flax-Wheat Fallow-Rye-Wheat Wheat Fallow-Wheat Wheat-Lentil	Conventional	rotation fertility	9
29) Mandan, North Dakota	Argiboroll	5.0	402	1984	Spring Wheat-Fallow Spring Wheat-Winter Wheat-Sunflower	Conventional Minimum No-till	rotation fertility tillage cultivar	36
30) Pendleton, Oregon	Haploxeroll	10.2	417	1931	Fallow-Wheat	Conventional	fertility residue burn	9
31) Pendleton, Oregon	Haploxeroll	10.2	417	1940	Fallow-Wheat	Conventional Reduced till	fertility tillage	18
32) Sidney, Nebraska	Haplustoll	8.2	381	1970	Fallow-Wheat Native Sod	Conventional Stubble-mulch No-till	tillage	4
33) Akron, Colorado	Paleustoll	9.3	424	1967	Fallow-Wheat	Conventional Reduced till No-till	tillage	3
34) Sterling, Colorado	Argiustoll	9.3	451	1985	Fallow-Wheat Wheat-Corn-Fallow Wheat-Corn-Millet-Fallow	No-till	rotation fertility	6

Appendix 1. (Continued).

Site Location	Soil Type[1]	MAT[2] (°C)	Precip. (mm)	Experiment Started	Cropping Systems	Tillage	Treatments	No. of Treatments
35) Stratton, Colorado	Argiustoll	10.7	410	1985	Fallow-Wheat Wheat-Corn-Fallow Wheat-Corn-Millet-Fallow	No-till	rotation fertility	6
36) Walsh, Colorado	Ustochrept	11.9	400	1985	Fallow-Wheat Wheat-Sorghum-Fallow Wheat-Sorghum-Sorghum-Fallow Sorghum	No-till	rotation fertility	8
37) Manhattan, Kansas	Haplustoll	12.8	835	1974	Sorghum Soybean Wheat Sorghum/Soybean Wheat/Soybean	Conventional Reduced till No-till	rotation tillage	15
38) Bushland, Texas	Paleustoll	12.7	473	1982	Wheat Sorghum Wheat-Fallow Wheat-Sorghum-Fallow	Stubble-mulch No-till	tillage rotation terrace	16
39) Bushland, Texas	Paleustoll	12.7	473	1981	Wheat-Sorghum-Fallow Wheat	Stubble-mulch No-till	tillage rotation	3

[1] U.S. Department of Agriculture classification system (Soil Survey Staff 1975).

[2] MAT is mean annual temperature.

[3] These four experiments were included in the cross-site sampling only.

The Australian Site Network

Peter R. Grace

Cooperative Research Centre for Soil and Land Management and University of Adelaide. Adelaide, South Australia

Summary

There are currently 32 trials listed as active participants in the Australian Network of Long-Term Agronomic Experiments. The primary objective of this network is to establish a central database of information for use by researchers actively involved in developing and testing soil and plant models for sustainable wheat and livestock production. The network sites are the primary source of data for two federally funded soil carbon storage and turnover studies to determine the long-term impact of agricultural management on soil C dynamics in agro-ecosystems. The three oldest cropping trials in the network (Dooen, Glen Innes and Urrbrae) range from 70 to 83 years in duration and are found on markedly different soil types (black clay loam, grey clay, and sandy-loam respectively). However, the physically protected pool, as measured by high energy ultra-violet photo-oxidation, has been consistently measured as 31% of the total soil organic carbon pool for samples taken during the course of the trials from a wide variety of crop and ley farming treatments. The majority of this protected organic matter fraction has been identified by ^{13}C NMR as charcoal, a largely inert form of carbon.

Introduction

In recent years, the usefulness of long-term agronomic experiments has been the topic of controversy in Australia and abroad. The considerable cost in maintenance, lack of change in management in line with present day strategies and a failure to adequately disseminate

NATO ASI Series, Vol. I 38
Evaluation of Soil Organic Matter Models
Edited by David S. Powlson, Pete Smith and Jo U. Smith

information has caused both governmental and industry-based funding bodies to question their continued benefit. Since the 1900's there have been at least fifty trials in Australia which have run in excess of 10 years, of which half were maintained at least 25 years or more. Fourteen of these "quarter-century" trials are still in existence and form of the core of the Australian Network of Long-Term Agronomic Experiments. Membership of the network is not restricted by trial duration, even though a minimum period of 10 years is usually envisaged. A number of cereal-ley farming trials initiated by the grains industry in 1994 and 1995 form the new wave of agronomic trials with long-term aspirations and are associated with the network.

Historically, more than 30% of Australia's total agricultural production comes from the temperate semi-arid zone on the cooler edges of the deserts extending from Geraldton, north of Perth, across southern Western and South Australia and into the Wimmera, Mallee and northern regions of Victoria and south-central New South Wales. The majority of trials in the site network can be found in this semi-arid zone which is dominated by winter rainfall with a typically Mediterranean climate (Figure 1). The soils range from deep, coarse-textured sands over loams to duplex soils with coarse-textured sands and sandy loams over clays and fine-textured red-brown earths, more commonly known as Alfisols (Soil Survey Staff, 1975). The widespread occurrence of inherent deficiencies in soil phosphorus and nitrogen have played a major part in the development of dryland cropping practices and the management imposed on the medium to long-term trials in Australia.

As outlined by Clarke & Russell (1977), long-term agronomic trials in Australia have historically been established to answer four key questions in relation to crop (predominantly wheat) and soil properties. These are the effect of fallows; fertilizers (principally superphosphate); pasture phases of various periods in the rotation and the effect of cereal crop rotations, including grain legumes. Donald's (1963) review of wheat yields for the period 1870-1960 (Figure 2) is testimony to the fact that these factors have played dominant roles in Australia's agronomic history.

Figure 1. *Geographic location of trial sites in the Australian Network of Long-term Agronomic Experiments.*

Network structure

There are currently 32 trials in the site network (Table 1) which is co-ordinated by the Cooperative Research Centre for Soil and Land Management. Fourteen of the trials are at least 25 years in duration, with six of these having been in existence at least 50 years. The network directory (Grace & Neill, 1994) also includes a supplementary listing of experiments which have been terminated since the mid-1950's but where comprehensive yield and soil data had been collected for at least 10 years. This supplementary listing includes trials at Kybybolite, Werribee, Narayen and Katherine. At a number of these supplementary sites, historical soil samples and in some cases original plots are still able to be identified.

The oldest trial in the network is the Permanent Topdressing (ley farming) trial at Rutherglen Research Institute, having been established in 1913. Cropping trials are still being maintained at Dooen (1917) and the Waite Agricultural Research Institute (1925). The cropping trials have maintained the same management strategies which were used in these early days (e.g. fallowing, crop residue removal, conventional cultivation and grass pastures).

Trials established post-1970 (e.g Wagga Wagga and Condobolin) have tended to rely more on tillage and stubble retention treatments with little or no fallowing. Plot size has also tended to be significantly smaller and the design more complex in the latter trials. There has also been a shift in emphasis from grass to improved (or legume dominant) pastures in the rotation sequence and the inclusion of alternative crops e.g. canola and brassicae to act as both a disease break and an opportunity to diversify. The use of grain legumes (particularly peas) as an alternative source of nitrogen has been a part of most of the longer-term trials with a shift to newer varieties (e.g. narbons) in the more extreme environments.

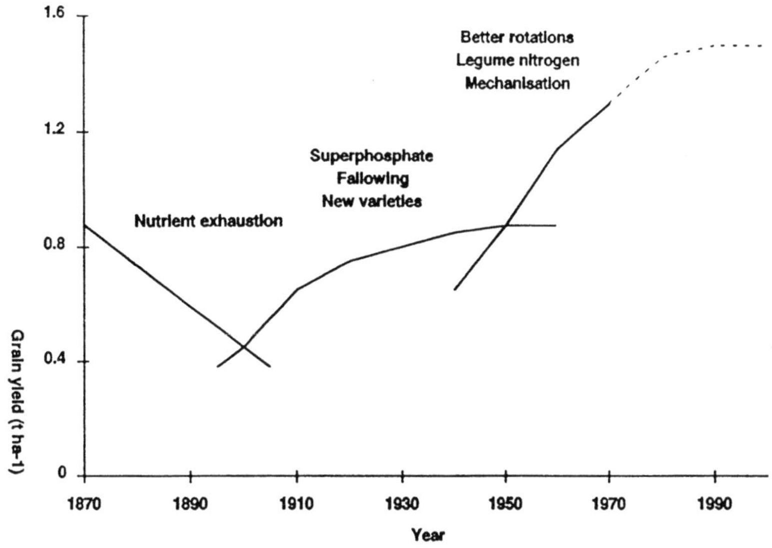

Figure 2. *The mean decennial yields of wheat (1870-1960) in Australia. Reprinted from Donald (1964).*

The network was initiated in October 1994 at what was the first Australian workshop on long-term agronomic experimentation (see Grace & Martin, 1994) and attended by over sixty researchers (national and international) and funders. At this time information was gathered on data availability and discussions on future direction, "preferred" requirements for minimum datasets, data usage and dissemination. The recommendations of this workshop were that long-term experiments were particularly critical in providing data on soil physical and chemical processes which are inherently slow in response to land management.

Table 1. *Agronomic trials actively participating in the Australia Site Network*

Trial and location		Established	Soil type[1]	Rainfall (cm)
Rutherglen (Victoria)	ley	1913	Natrixeralf	593
Dooen (Victoria)	crop-ley-fallow	1917	Torrert	425
Glen Innes (NSW)	crop	1921	Natrustalf	796
Urrbrae (South Aust)	crop-ley-fallow	1925	Rhodoxeralf	627
Walpeup MM1 (Victoria)	crop-fallow	1940	Calciorthid	330
Rutherglen (Victoria)	ley	1946	Natrixeralf	593
Armidale (NSW)	ley	1964	Haplustalf	750
Tamworth (NSW)	crop-ley	1966	Pellustert	425
Merredin (West Aust)	crop-ley	1966	Paleustalf	302
Newdegate (West Aust)	crop-ley	1967	Natrustalf	390
Chapman (West Aust)	crop-ley	1967	Palexeralf	412
Esperance (West Aust)	crop-ley	1968	Paleaquult	496
Salmon Gums (West Aust)	crop-ley	1968	Calciorthid	337
Warwick (Queensland)	crop-fallow	1968	Pellustert	717
Salmon Gums (West. Aust)	crop-ley	1973	Calciorthid	337
Tarlee rotation (South Aust)	crop-ley-fallow	1977	Rhodoxeralf	475
Wagga Wagga CSU (NSW)	crop	1977	Paleustalf	520
Condobolin (NSW)	crop	1978	Rhodoxeralf	420
Avon (South Aust)	crop-ley	1979	Calciorthid	350
Wagga Wagga SATWAGL	crop-ley	1979	Palustalf	520
Hamilton (Victoria)	ley	1979	Natrustalf	700
Walpeup MR10	crop-ley-fallow	1980	Calciorthid	330
Walpeup MR9	crop-ley	1980	Calciorthid	330
Rutherglen RR1	crop	1981	Natrixeralf	593
Breeza (NSW)	crop	1981	Pellustert	550
Warialda (NSW)	crop	1981	Pellustert	700
Croppa Creek (NSW)	crop	1981	Torrert	600
Walpeup MC14	crop-ley-fallow	1982	Calciorthid	330
Merredin	crop-ley	1982	Paleustalf	302
Billa Billa (Queensland)	crop	1983	Natrixeralf	550
Biloela (Queensland)	crop	1983	Pellustert	698
Kapunda (South Aust)	crop-ley	1984	Rhodoxeralf	500
Narrabri (NSW)	crop	1984	Torrert	600
Tarlee tillage	crop	1984	Rhodoxeralf	475
Warra (Queensland)	crop-ley	1986	Chromustert	685

[1]Soil Taxonomy approximation only (Soil Survey Staff, 1975)

For example, cumulative effects of a particular long-term management practice may result in unforeseen disaster, as identified by the sudden failure of the fallow-wheat treatment at the Mallee Research Station (Figure 3) or on the other hand some form of benefit or stability (Figure 4). In the latter, after a period of 6-10 years the herbage yield from the rock phosphate plots exceeded the yield from the superphosphate plots. Russell (1960) subsequently identified the important role of rock phosphate in maintaining long-term pasture productivity as opposed to pasture establishment. This conclusion could not have been drawn from short-term experimentation.

Long-term trials also are considered essential in identifying real trends in treatments by superimposing the effect of annual climate variability on production, an analysis which requires the completion of a number of climatic cycles. Ecological succession and changes in soil and plant community structure may also occur over time, e.g. pesticide or herbicide resistance, and these effects can only be seen with adequate long-term monitoring. The real value of these trials, however, is as comprehensive databases of knowledge, and even though they may be site and soil specific, they provide essential information for developing and testing simulation models for predictive purposes. It is this recommendation that has provided the impetus for funding bodies to utilize the network as a valuable source of data for two national soil C projects.

Soil carbon studies

Historical soil samples from trials which are over 25 years in duration have been the primary source of data for a study investigating the potential for carbon sequestration in Australian agricultural soils. Funded in mid-1994 for two years by the Rural Industries Research and Development Corporation, the objective is to estimate the size and turnover rates of the slow and inert carbon pools in a range of Australian agricultural soils. The management systems and trials available through the Australian site network offer the full spectrum of carbon dynamics in agricultural systems and contrast the role of soil organic matter as a source or sink for carbon dioxide.

The analytical procedures used in this study enable the soil organic matter pools in the Rothamsted (Jenkinson *et al.*, 1987) and CENTURY (Parton *et al.*, 1987) carbon cycling models to be quantified. The light fraction (LF) is removed by flotation and the particulate organic matter (POM) fraction is retained on a 50 μm sieve. The <50 μm fraction is exposed to a UV source for varying periods of time, with four hours considered sufficient to give an estimate of the physically protected pool (Skjemstad *et al.*, 1993). Infra-red spectroscopy has identified the materials external to both clay- and silt-sized aggregates as proteinaceous in nature and the materials within the aggregates as humic acids. These humic acids appear to be physically protected from photo-oxidation rather than chemically resistant.

This high energy UV photo-oxidation procedure shows a consistent amount of carbon is physically protected at any one site (Table 2), although variation between sites and treatments may be large (0.09-0.61%). The three oldest cropping trials in the network (Dooen, Glen Innes and Urrbrae) range from 70 to 83 years in duration and are found on markedly different soil types (ranging from grey clay to sandy-loam). However, the physically protected organic matter fraction estimated by UV photo-oxidation has been found to be a consistent proportion of the total soil organic carbon at 31% for a wide variety of samples taken over the course of each trial. For forty-nine samples assayed across 8 trials at 7 different locations, this proportion was approximately 26%. Solid state ^{13}C NMR studies have shown that most of this protected carbon is in the form of charcoal (inert carbon) which appears to be a relic of fires prior to European settlement. In particular, over 80% of the carbon in the Glen Innes and Longeronong protected pools is in the form of charcoal (Skjemstad, unpublished). Infra-red spectroscopy is also being used to qualitatively identify the presence of charcoal and reduce the number of expensive NMR analyses.

Data from the half-century trials (Rutherglen, Dooen, Urrbrae, Glen Innes and Walpeup) will be used to calibrate each of the soil carbon models. The carbon data from the quarter-century trials will then be used in the initial validation stage with data from some of the more recent trials being utilized once model selection is finalized. Preliminary modelling studies using the Waite Permanent Rotation trial dataset (Grace, this volume) indicate the protected pool measured by UV photo-oxidation may overestimate the size of the inert pool as required by the Rothamsted model, but is a better measure of the passive pool of the CENTURY model. The charcoal value, however, appears to give a good estimate of the inert pool. This unique database is also being used in the development of the SOCRATES model which is based on concepts in both Rothamsted and CENTURY models and the concepts of microbial and substrate protection outlined in the model of Ladd *et al.* (1995). It is envisaged that this dataset there will also provide enough basic information to establish relationships for initializing the soil pools for a wide range of Australian soils and environments. Data generated in this study is also available to the National Greenhouse Gas Inventory Committee to improve estimates of the total store and form of carbon in Australian soils.

The common arguments against the use of long-term trials are that land managers tend not to adopt the results from these trials as many have little or no relevance to current practice. Data is usually site specific for the wider audience and plot sizes tend to be small thereby increasing the risk of soil and water movement between plots. In some of the older trials the experimental design may also be flawed (compared to modern statistical thinking).

Table 2. *The contribution of physically protected organic carbon as measured by UV photo-oxidation to total soil organic carbon at selected long-term trials in Australia.*

Trial	Samples Assayed (No.)	Organic carbon (%)	'Protected' carbon (%)	'Protected' /Organic carbon
Urrbrae	12	1.91 +/- 0.50	0.66 +/- 0.12	0.32
Glen Innes	9	2.05 +/- 0.39	0.60 +/- 0.05	0.29
Dooen	6	1.02 +/- 0.06	0.35 +/- 0.04	0.34
Walpeup (MM1)	4	0.43 +/- 0.07	0.09 +/- 0.02	0.21
Rutherglen (ley)	3	2.29 +/- 0.44	0.41 +/- 0.03	0.18
Rutherglen (RR1)	6	1.34 +/- 0.34	0.39 +/- 0.03	0.29
Tarlee (rotation)	4	1.42 +/- 0.24	0.30 +/- 0.03	0.21
Wagga (SATWAGL)	5	1.92 +/- 0.30	0.53 +/- 0.05	0.28

Whilst on-farm monitoring or paired sites are seen as an alternative to formally designed trials, accurate record keeping and quality control in maintenance, sampling and analysis becomes a definite concern in many of these programs. On-farm monitoring usually poses more questions than it can answer because of these problems but the continuation of established trials is vital for cause and effect research.

Network maintenance

In what may be considered the first steps in a nation wide rationalisation of long-term experimentation in Australia, rural funding bodies have sponsored (in the short-term) the administrative infrastructure of the site network. However, this support is dependent on a cost-benefit analysis of the value of these trials and a peer review to be completed in mid 1996. The review also includes site evaluations by independent farming system specialists, modellers, statisticians, economists and industry representatives. Guidelines for the management, design, data collection and funding of prioritized core trials will be formalized in the review process as well as the final structure of the network.

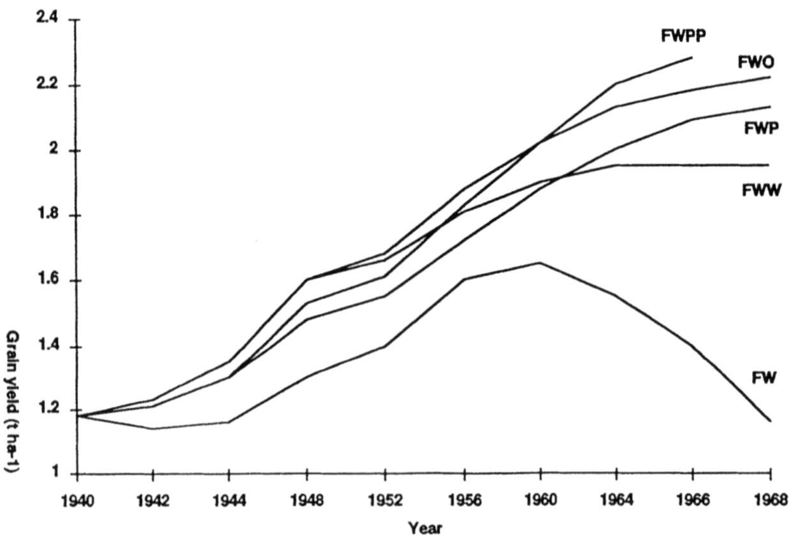

Figure 3. *Wheat grain yield in long-term fallow-based rotations at the Mallee Research Station, Walpeup as reported by Elliott & Jardine (1972). (F-fallow; W-wheat; O-oats; P-pasture).*

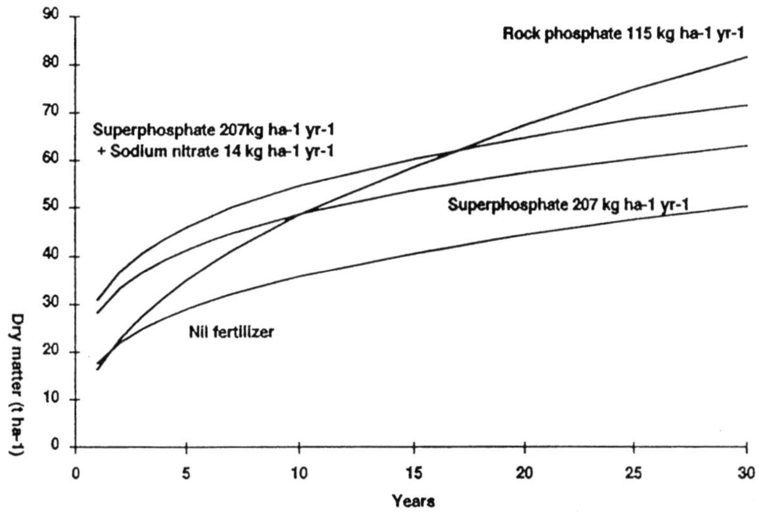

Figure 4. *Trends in herbage yields at the Waite Agricultural Research Institute P1 trial (1925-1955 as reported by Russell (1960).*

Emphasis will be placed on quality control in sample analysis and data collection as well as effective communicationstrategies to disseminate information to the producer and policy maker. The review team is also required to identify potential locations and designs for new trials where farming systems data, specific to an agro-climatic region, is not being collected.

Researchers are now also realizing that as funding sources become less institutional and more industry-derived there is a definite need to be more flexible in treatment selection so as to meet the needs of their clients. This does not mean long-term agronomic experiments need to be terminated but some outdated treatments may be discontinued, re-focused or even replaced with new or innovative practices. An example of this is the decision in 1995 to split all cereal plots at the Waite Agricultural Research Institute Permanent Rotation trial. Since commencing in 1925, crop residue removal and the grazing of the pasture plots throughout the growing season has resulted in topsoil organic carbon being reduced to about half its initial value. The aim of the trial has been to identify productive rotation sequences, with the focus of the soil researchers very much on changes in organic carbon and its implications on both fertility and structure. The trial's usefulness as a sustainability benchmark is unquestioned, however residue retention and minimum tillage has been widely practiced in Australia for the past 15 years and the trial's spectrum of fertility and physical properties is now ideal for looking at the posible rehabilitative effects of residue retention on soil and plant productivity. Residues will now be retained on one-half of all cereal plots, which will make the trial more relevant and continue to address its principal aims. Evolution is a necessity in today's economic climate and long-term experiments are no exception.

Internationally, Australia has one of the largest collections of medium to long-term trials which are directly relevant to current management practices. African and Asia countries have relied more on short-term experimentation, whilst a number of long-term museum pieces have been maintained in Europe, North America and Australia. The latter however can be considered sustainability benchmarks, particularly in establishing threshold or critical levels of soil organic matter, pH and soil structure and their impact on productivity. They now form an extremely valuable source of information for climate change studies and environmental research, e.g. the impact on long-term applications of fertilizers and pesticides on ground water quality. In Australia, the continued existence of the site network in its current (or modified) form is largely driven by industry backed federal agencies (i.e. the Grains, and Land and Water Resources Research and Development Corporations') wishing to use them as indicators of sustainability.

References

Clarke & Russell, J.S. (1977) Crop sequential practices. In: *Soil factors in crop production in a semi-arid environment.* (ed: J.S. Russell & E.L Greacen), University of Queensland Press, St Lucia: 279-300.

Donald, C.M. (1964) Phosphorus in Australian agriculture. *Journal of the Australian Institue of Agricultural Science* **30:** 75-105.

Elliott, B.R. & Jardine, R. (1972) The influence of rotation systems on long-term trends in wheat yield. *Australian Journal of Agricultural Research* **23**: 935-944.

Grace, P.R. & Ladd, J.N. (1995) *SOCRATES (Soil Organic Carbon Reserves And Transformations in agro-EcoSystesm): A decision support systems for sustainable framing systems in southern Australia.* Cooperative Research Centre for Soil and Land Management, Adelaide, 23pp.

Grace, P.R. & Martin, R. (1994) *National workshop on long-term agronomic experiments.* Presentation and poster abstracts. Cooperative Research Centre for Soil and Land Management, Adelaide, 28pp.

Grace, P.R. & Neill, J. (1994) *Australian directory of long-term agronomic trials.* Cooperative Research Centre for Soil and Land Management, Adelaide, 100pp.

Jenkinson, D.S., Hart, P.B.S., Rayner, J.H. & Parry, L.C. (1987) Modelling the turnover of organic matter in long-term experiments at Rothamsted. In: *Soil organic matter dynamics and soil productivity.* (ed: J.H. Cooley), Intecol Bulletin: 1-15.

Ladd, J.N., Amato, M., Grace, P.R. & van Veen, J.A. (1995) Simulation of ^{14}C turnover through the microbial biomass in soils incubated with ^{14}C-labelled plant residues. *Soil Biology and Biochemistry* **27:** 777-783.

Parton, W.J., Schimel, D.S., Cole, C.V. & Ojima, D.S. (1987) Analysis of factors controlling soil organic matter levels in Great Plains grasslands. *Soil Science Society of America Journal* **51**: 1173-1179.

Russell, J.S. (1960) Soil fertility changes in the long-term experimental plots at Kybybolite, South Australia. I. Changes in soil phosphorus. *Australian Journal of Agricultural Research* **11:** 927-947.

Skjemstad, J.O., Janik, L.J., Head, M.J. & McClure, S.G. (1993) High energy ultraviolet photo-oxidation: a novel technique for studying physically protected organic matter in clay and silt-sized aggregates. *Journal of Soil Science* **44:** 485-499.

Soil Survey Staff (1975) *Soil Taxonomy,* Handbook No. 436, USDA, Washington.

Acknowledgements

The author would like to thank NATO for funding his attendance at an extremely fruitful and well organized Advanced Research Workshop.

Long-term data sets from Germany and Eastern Europe

Martin Körschens

Centre for Environmental Research Ltd.

Hallesche Straße 44

D-06246 Bad Lauchstädt

Germany

Summary

Long-term experiments are immensely expensive. It is necessary to use all information available from them and to minimise the costs relative to the gain of scientific findings. This paper gives an overview of 50 selected long-term fertilisation experiments all of which are older than 30 years in Germany and Eastern Europe. Examples of evaluation are given. From the combined results from 13 long-term field experiments, the range of optimum mineralisable carbon contents can be estimated at 0.2 to 0.6 %. The N input from the atmosphere can indirectly be derived from the N uptake from the nil plots of long-term field experiments and amounts to 50 kg ha $^{-1}$.

KEYWORDS: long-term experiments, directory of long-term experiments,
examples of evaluation, carbon

Introduction

Long-term field experiments are an indispensable source of information in soil science. For example, the long-term effect of differentiated crop management can, under field conditions,

NATO ASI Series, Vol. I 38
Evaluation of Soil Organic Matter Models
Edited by David S. Powlson, Pete Smith and Jo U. Smith

only be measured on experiments of long standing. The value of these experiments increases with their age. The increased value is a result of the accumulation of knowledge over decades with a time factor that cannot be reproduced, and the immense input of manpower and money.

Long-term experiments are useful for the following reasons:

- long-term effects are frequently recognizable only after decades
- the state of knowledge on long-term effects is continuously increasing due to new developments in methods of investigation
- additional demands for experimental investigation of soil processes emerge over a long period of time
- sustainable soil management can only be demonstrated by long-term experiments
- long-term changes can be monitored by means of stored soil samples
- the results from long-term experiments can be used as the basis for validation of computer models
- combined appraisal of a number of long-term experiments running on different soils and under different climatic conditions contribute to the interpretation of climatic changes

Agricultural and environmental sciences are currently confronted with urgent problems which can only be solved with the assistance of long-term experiments. These problems include:

1. sustainability of land-use systems
2. effect of potential climate changes on soil fertility and yield
3. quantification of carbon and nitrogen transformations and of the N cycle

Table 1 *Directory of long-term fertilization experiments in Germany with a duration of more that 30 years (Körschens, 1994).*

site	initial year	crops	soil	average temp. °C	annual rainfall mm	plot size m^2	size of experim. ha
Hohenschulen	1951	wheat, barley, ...	loam	7.8	716	22	0.2
Bonn-Dikopshof	1906	wheat, sugar beet, ...	loess	10	637	130	3
Bonn-Poppelsdorf	1906	wheat, oats, ...	loam	10.1	670	100	0.7
Bonn-Meckenheim	1958	wheat, barley, ...	loess	9	600	25	0.1
Bonn-Rengen Daun	1942	grasland	loess	7.1	833	15	0.1
Gießen	1954	sugar beets, wheat, ...	loam	8.7	630	30	0.2
Rauisch-Holzhausen	1954	sugar beets, wheat, ...	loam	8	578	25	0.3
Rauisch-Holzhausen	1962	wheat, barley	loam	8	578	20	0.4
Limburgerhof	1938	cereals, potatoes, ...	l. sand	9.9	550	42	0.5
Limburgerhof	1957	cereals, potatoes, ...	l. clay	9.8	560	38	0.4
Dülmen	1958	potatoes, rye, ...	sand	9.1	725	40	0.3
Speyer	1959	cereals, sugar beets, ...	sand	9.8	583	-	-
München-Freising	1922	wheat, barley, ...	loam	7.7	810	50	0.14
Thyrow	1937	potatoes, corn, ...	sand	8.6	520	72	0.5
Thyrow	1938	potatoes, cereals, ...	sand	8.6	520	360	0.6
Groß Kreutz	1959	potatoes, rye, ...	sand	8.9	537	40	0.3
Müncheberg	1963	potatoes, wheat, ...	sand	8.2	521	30	0.8
Berlin-Dahlem	1923	wheat, potatoes, ...	sand	9.2	549	20	0.4
Bad Lauchstädt	1902	wheat, sugar beets, ...	loess	8.6	480	63	4
Halle	1878	rye, corn, potatoes	loam	9.2	501	290	0.6
Halle	1949	potatoes, oats, ...	loam	9.2	501	30	0.9
Seehausen	1957	potatoes, wheat, ...	s. loam	9	556	26	4.1
Seehausen	1959	wheat, sugar beets, ...	s.loam	9	556	48	1.3

Table 2 *Directory of long-term fetilization experiments in Eastern Europe with a duration of more than 30 years (Körschens, 1994).*

Country/ site	initial year	crops	soil	average temp. °C	annual rainfall mm
Bulgaria					
Sofia-Bozhouvishte	1962	corn, wheat, ...	loess		
Czech Republik					
Prag-Ruzyně	1955	cereals, rootcrops, ...	clay-loam	7.7	487
Lucavec	1956	cereals, rootcrops, clover	s. loam	6.8	686
Caslav	1956	cereals, rootcrops, clover	loam	8.1	590
Ivanovice	1956	cereals, rootcrops, clover	loam	8.4	556
Viglas	1956	cereals, rootcrops, clover		7.7	669
Pohorelice	1956	cereals, rootcrops, clover		9.0	502
Hungary					
Martonvasar	1958	corn, wheat	loess	10.6	514
"	1958	corn	loess	10.6	514
"	1958	corn, wheat, ...	loess	10.6	514
"	1961	wheat, corn, ...	loess	10.6	514
"	1961	corn	loess	10.6	514
Keszthely	1960	corn, wheat, clover	clay	10.8	700
"	1963	wheat, corn, ...	clay	10.8	700
Poland					
Skierniewice	1921	barley, wheat, ...	l. sand	7.8	529
"	1922	vegetable	l. sand	7.8	529
Chilice	1955	potatoes, barley, ...	loess		
Baborwokow	1957	rye, oats, ...	sand		
Biebrza	1957	gras, fodder crops	peat		
Lyczyn	1960	barley, potatoes, ...	sand		500
Slovakia					
Banska Bystrica	1961	meadow	loam	8.2	757
Pincina Lucenev	1961	meadow	loam	9.1	619
Former Soviet Union					
Moskau	1912	rye, potatoes, ...	loam		
Mironovka	1912	sugar beets, ...	loam		
Daos	1931	rye, potatoes, ...	clay-loam		
Primorskoje	1941	corn, wheat, ...	clay-loam		
Perm	1946	rye, potatoes, ..	loam		

Directory of long-term experiments in Germany and Eastern Europe

The number of long-term experiments around the world is very difficult to estimate and probably cannot be listed comprehensively. We did a preliminary summary in 1994 which contained 240 long-term experiments from all over the world. For Germany we included only experiments older than 20 years, for all other countries only experiments older than 30 years. We included experiments with many different scientific objectives. If long-term experiments older than 10 years had been included this would have encompassed over 1000 experiments. Swift et al. (1994) for example reported more than 139 experiments in Southern and Central Africa older than 5 years. However, less than 20 of these experiments were older than 20 years.

Tables 1 and 2 provide a summary of 50 selected long-term fertilisation experiments all of which are older than 30 years. 23 of them are located in Germany, 27 in Eastern Europe. Compiling such an overview raises large problems as some experiments have been reported in scientific journals only sporadically and may have been terminated in the meantime. This summary therefore, whilst not at all complete, nevertheless illustrates the vast number of long-term experiments - especially fertilisation experiments - under very different soil and climatic conditions. This provides an outstanding experimental basis for solution of the questions mentioned at the beginning of this paper. Our experience in data collection shows that the funding may have gaps due to changes in experimental interests, in some cases restricted feasibilities to monitor the relevant factors and treatments and, finally, due to different methods of investigation. Complete documentation of all important results, such as that of the Static Fertilisation Experiment Bad Lauchstädt which includes complete results from the pre-computer era is hardly ever found. It is often very difficult then to reconstruct the data of former years. Frequently no information is available about site conditions which is imperative for the interpretation of experimental results. For this reason it is difficult to do a serious evaluation using only single experiments.

The classical long-term fertilisation experiments differ distinctly in their scientific objectives and experimental treatments from the more recent ones. Before and around the turn of the century and until the first half of this century, the object of investigation was primarily the effect of single plant nutrients and of their combinations. This was frequently either compared to the effect of farmyard manure or tested in combination with farmyard manure, producing a qualitative ranking of treatments (Table 3). With increasing application of mineral fertilisers, mainly nitrogen fertilisers, the experimental investigations focused on the effect of different application rates of nutrients, producing a quantitative grading of the treatments. At the same time the classical experimental design, which had been mostly systematic, was substituted by randomised multifactorial experiments. Especially after the Second World War, experiments were initiated with quantitative equidistant gradings of N fertilisation together with different manure dressings as multifactorial experiments in orthogonal combination. These experiments contain comparable treatments or combinations of treatments and are thus more suitable, especially for secondary evaluation. An example of a planned series are the international nitrogen fertiliser experiments which Professor von Boguslawski initiated 12 years ago. 22 experiments using the same design were started in 12 European countries to test the influence of different organic fertilisers and different mineral N fertiliser rates on yield and soil properties with respect to site conditions.

Table 3 *Treatments of selected long term experiments initiated in the first half of this century*

Country	Germany	Germany	Russia	Poland
Site	Bad Lauchstädt[1)]	Thyrow	Moskow	Skierniewice
Initial year	1962	1937	1912	1927
Soil	loess	sand	sandy loam	sand
Treatments	NPK	Ca NPK	NPK	Ca NPK
	NP	Ca NP	NP	Ca NP
(N = nitrogen	NK	Ca NK	NK	Ca KN
P = Phosphorus	PK	Ca PK	PK	Ca PK
K = Potasium	N	-	N	-
Ca = Calcium	Zero	Zero	Zero	Ca Zero
FYM = Farm	-	FYM	-	-
yard manure)	-	NPK	-	NPK
	-	FYM+NPK	FYM+NPK	-
	-	-	P	-
	-	-	K	-

[1)] in orthoganal combination with 3 levels of farmyard manure

Examples of Evaluation

A question of equal importance to both agriculture and the environment which can only be solved by use of results from long-term experiments is the quantitative assessment of the effect on yield of soil organic matter and, consequently, the optimal humus (or carbon) content of the soil. Some examples of evaluating long-term experiments are given below.

An extended evaluation of numerous national and international long-term experiments which comprised a total of some 1100 experimental years showed that an optimal combination of organic and mineral fertilisers brought about an average additional yield of 7% as compared to the yield obtained when mineral fertiliser were used alone. The additional yield amounts to 10% on sandy soils and to 6% on loamy soils. This means that if the optimal amount of mineral fertiliser is applied the primary organic matter input by root and crop residues alone is sufficient to build up a humus level in the soil which allows an exploitation of at least 90% of the yield potential.

The next example, quantifying the content of mineralisable carbon and nitrogen in the soil, is to a large extent based on data from the literature. For this application it is necessary to divide the soil organic matter into at least two fractions. One of them is "inert", i.e. not participating in mineralisation processes, the other fraction is mineralisable and depends on management. The inert fraction is dominated mainly by site conditions (texture, weather) and therefore cannot be changed, while the mineralisable part depends on

- the amount and composition of root and crop residues remaining on and in the soil
- the amount and composition of organic fertiliser applied
- the intensity of tillage.

Only the mineralisable carbon content can be influenced by plant production and is thus responsible for nitrogen release. This fraction, however, is relatively small, as are the differences in mineralisable carbon in different locations.

If for practical reasons the inert carbon content is defined as the minimum content under field conditions with zero fertiliser application and either, cultivation of humus-consuming plants or bare fallow, the mineralisable fraction can be derived from long-term experiment results as the difference between nil plot and the respective variant with the highest application of organic and mineral fertiliser.

Figure 1 shows the range of carbon and nitrogen contents which were reached after decades of differentiated fertilisation in thirteen long-term experiments. They amount to:

0.21 to 0.66% for total carbon (C_t) with a mean of 0.40 %

0.018 to 0.062% for total nitrogen (N_t) with a mean of 0.034 %.

In the Static Fertilisation Experiment Bad Lauchstädt, a combined application of 15 t ha^{-1} farmyard manure plus mineral fertiliser over 90 years resulted in an additional 0.66% C_t as compared to the nil plot. According to the recent results the optimum for this location, which is estimated at 0.5 %, has already been exceeded.

From the combined results of 13 long-term field experiments, the range of optimum mineralisable carbon contents can be estimated at 0.2 to 0.6% under the prevailing conditions.

The results from soil investigations depend on the date of sampling, the cultivated crop, the preparation of samples and the method of investigation. For specific questions it is necessary to take all samples on the same day and to prepare and investigate them in the same way. The following investigation to quantify the effect of clay and C_t content on physical soil properties was done in this way. Soil samples from the extremes of the range of carbon contents were taken from five long-term experiments in Germany at the same time to determine a number of parameters. Some of the results are shown in Figure 2.

Figure 2 shows a very close correlation between carbon content and water capacity (coefficient of correlation 0.87) and the bulk density (coefficient of correlation 0.93). Different effects of organic matter are discernible in different locations. On the ideally textured loess

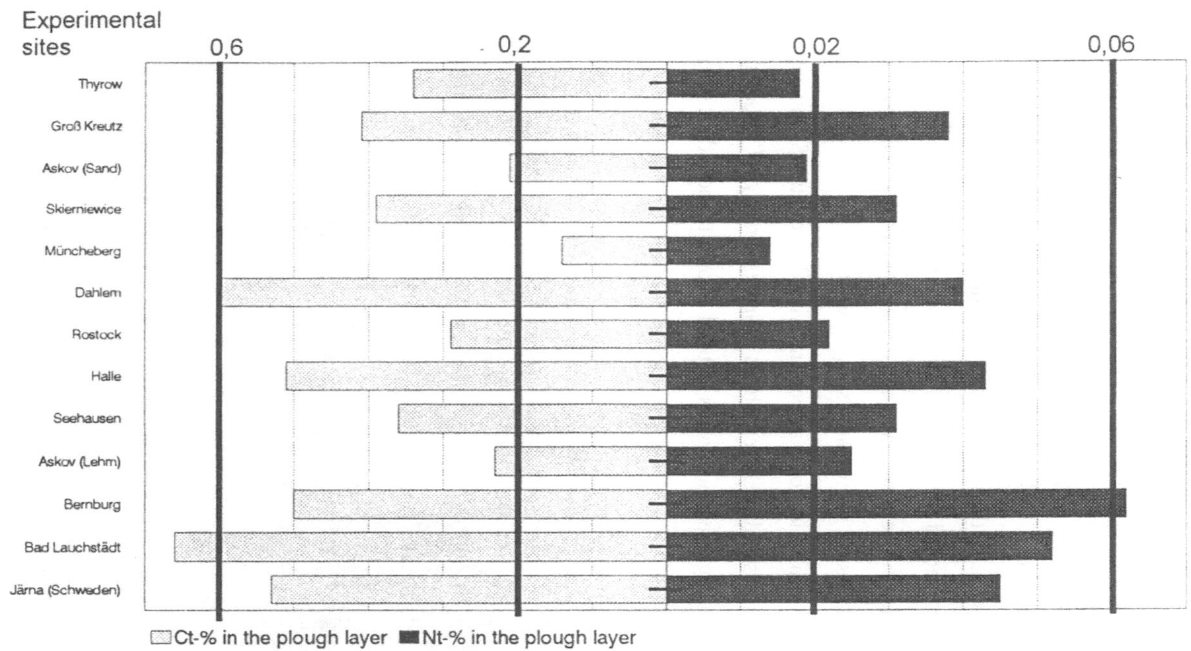

Figure 1 *Differences in the Ct- and Nt-content between "without fertilization" and "organic manure plus NPK" of selected long term experiments.*

black earth at Bad Lauchstädt the increase in water capacity from 31 to 34 mass per cent is of less importance than from 15 to 17 mass per cent on the light sandy soil at Thyrow.

Another example is shown in Table 4. The N input from the atmosphere can indirectly be derived from the N uptake of the nil plots of long-term field experiments, if we assume that the experiments have reached steady-state conditions and thus input equals output (see Lauchstädt, Halle, Prague, Skierniewice) or that the N amount released from this soil N pool is negligibly small. The results from loamy soils show an overall atmospheric N input larger than 50 kg ha^{-1} y^{-1} which is in good agreement with the results of Goulding (1990) and Russow (1995) who measured the atmospheric N inputs directly. N uptakes are lower on light sandy soils due to larger leaching losses, but still amount to more than 30 kg ha^{-1} y^{-1}.

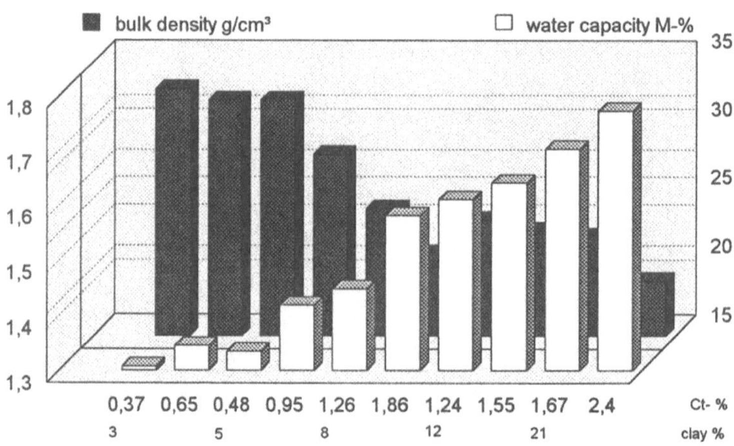

Fig. 2: Influence of different Ct-contents on bulk density and water capacity in the plough layer of long term experiments with varying texture

Table 4 *Nitrogen uptake by unfertilized plots of selected long term fertilizer experiments*

Country	Site	Soil	Initial year	Period	N-uptake kg ha^{-1}	Reference
Germany	Bad Lauchstädt	loam	1902	1983-1994	56	Körschens & Müller, 1995
	Halle	loam	1949	1980-1985	69	Stumpe, *et al.*, 1990
	Seehausen	s.loam	1967	1967-1990	78	Hülsbergen, 1992
	Groß Kreutz (M4)	sand	1967	1967-1986	47	Asmus, 1990
	Groß Kreutz (P60)	sand	1959	1959-1986	34	Asmus, 1990
	Müncheberg	sand	1967	1983-1988	33	Smukalski, *et al.*, 1990
Czech. Republic	Prag	clay loam	1955	1974-1994	53	Klir *et al.*, 1995
Poland	Skierniewice	sand	1923	1985-1994	37	Mercik, 1995

Conclusions

A number of problems can only be solved by using long-term experiments, especially those which have reached steady state after a correspondingly long duration so that balances can be calculated on the basis of input minus output. For comprehensive evaluation of the meaning of an unplanned series of experiments it is necessary, however, to make the most important measurements frequently and thus to determine whether constant factors are sufficiently precise. It becomes clear that these valuable and costly experiments must be planned and carried out in a way that they can also be used for extensive secondary evaluations.

From conception, a long-term experiment should be planned to allow
- the experiment to become part of an unplanned series or be used for secondary evaluation
- the experiment to be used to investigate other questions without changing the treatments.

Long-term experiments are immensely expensive. They should therefore be used comprehensively and made more accessible to a wider group of interested people. When initiated, potential users and future problems should be considered. In this way it is possible to use all the information available from them and to minimise the costs relative to the gain of scientific findings.

References

Asmus, F. (1990) *Versuch M 4 Groß Kreutz - Wirkung organischer und mineralischer Düngung und ihrer Kombination auf Pflanzenertrag und Bodeneigenschaften.* Dauerfeldversuche, Berlin: 245-250

Goulding, K.W.T., 1990. Nitrogen deposition to land from the atmosphere. *Soil Use and Management* **6**: 61-63

Hülsbergen, K.-J., 1992. Erträge, Stickstoff- und Kohlenstoffbilanz im Kombinationsversuch Seehausen. *Tag.-Ber. zum Symposium Dauerfeldversuche und Nährstoffdynamik, 9-12 June 1992 in Bad Lauchstädt, Leipzig*: 108-113

Klir, J.; Kubat, J.; Pora, D., 1995. Stickstoffbilanzen der Dauerfeldversuche in Prag. *Mitteilgn. Dtsch. Bodenkundl. Gesellsch.* **76**: 831-834

Körschens, M. (ed.) (1994) *Der Statische Düngungversuch Bad Lauchstädt nach 90 Jahren.* Teubner-Verlag Stuttgart, Leipzig, 179pp.

Körschens, M. & Müller, A. (1995) The Quantification of Carbon and Nitrogen Dynamics and their Management, (in press).

Mercik, St. (1995) unpublished.

Russow, R., Faust, H., Dittrich, P., Schmidt, G., Mehlert, S., Segschneider, H.-J. & Sich, I. (1995) *Untersuchungen zur N-Transformation und zum N-Transfer in ausgewählten Agrarökosystemen mittels der Stabilisotopentechnik.* Teubner-Verlag Stuttgart-Leipzig, (in press)

Swift, M.J., Seward, P.D., Frost, PGH., Qureshi, J.N. & Muchena, F.N. (1994) *Long-term Experiments in Africa: Developing a Database for Sustainable Land Use under Global Change.* CAB International, Wallingford, 229-251

Smukalski, M., Kundler, P. & Rogasik, J. (1990) *Der Müncheberger Nährstoffsteigerungs-versuch.* Dauerfeldversuche, Berlin: 251-259

Stumpe, H.; Garz, J.; Hagedorn, E. (1990) *Die Dauerdüngungsversuche auf dem Versuchsfeld in Halle.* Dauerfeldversuche, Berlin, 25-71

Establishing a European GCTE Soil Organic Matter Network (SOMNET)

Pete Smith, David Powlson and Margaret Glendining
Soil Science Department
IACR-Rothamsted
Harpenden
Herts, AL5 2JQ, U.K.

Summary

Soil organic matter (SOM) is recognised as being of critical importance as a source and sink of carbon in the biosphere. As a result, research into predicting the effects of global environmental change on soil organic matter has been identified as a high priority within the Global Change and Terrestrial Ecosystems (GCTE) programme of the International Geosphere-Biosphere Programme (IGBP). The objectives of GCTE Task 3.3.1., "Soil Organic Matter" require that a global network of SOM modellers and experimenters be established. This global soil organic matter network (SOMNET) will comprise a number of regional networks, one of which will be the European GCTE SOMNET.

The European GCTE SOMNET will provide a forum for soil organic matter researchers to share models and datasets, and will establish a database system holding metadata, models and long-term datasets from Europe. The database system will comprise a central metadata database linked via the World-Wide-Web to many other European databases containing quality-controlled data in standard format.

The data from long-term experiments that exists in Europe that could be made available by the European GCTE SOMNET is estimated to have cost, in 1989 terms, nearly $140,000,000 to collect.

NATO ASI Series, Vol. I 38
Evaluation of Soil Organic Matter Models
Edited by David S. Powlson, Pete Smith and Jo U. Smith

In this paper we describe the progress so far made in establishing the European GCTE SOMNET and outline some of the activities planned for the next few years to build the European SOM network and its associated database system.

Introduction

Why is soil organic matter important ?

Soil organic matter (SOM) represents a major pool of carbon within the biosphere, estimated at about 1400 x 10^{15} g globally, roughly twice that in atmospheric CO_2 (Post *et al.*, 1982). Changes in climate are likely to influence the rates of accumulation and decomposition of SOM through changes in temperature, moisture and rhizodeposition (Jenkinson *et al.*, 1991). Other changes, especially in land use and management, will have even greater effects, whether caused by climate change or by environmental, agricultural or economic / social policies.

Among the changes of significance are the encouragement of set-aside, afforestation, extensification, growth of biomass crops and the wider application of organic wastes to soils under agriculture or forestry. In addition to the direct role of SOM as a source and sink of carbon, it plays a central role in other key environmental processes including nutrient (N, P) availability and loss (and, hence, movement to aquifers), soil stability and erosion risk, and the flux of trace greenhouse gases (especially nitrous oxide and methane) between land surface and the atmosphere.

Given the importance of changes in SOM in response to global environmental change, it is perhaps not surprising that the International Geosphere-Biosphere Programme (IGBP) has identified research on the interactive effects of various components of global change on SOM as of high priority within its Global Change and Terrestrial Ecosystems (GCTE) programme (see Gregory & Ingram, 1995). The objectives of GCTE Task 3.3.1., "Soil Organic Matter" require that a global network of SOM modellers and experimenters be established. The GCTE Global SOMNET will comprise a number of regional networks since this structure has advantages over a single, global network (see Elliott & Paustian, 1995).

One of the regional SOM networks will be the European GCTE SOMNET. Europe contains many of the most thoroughly analyzed long-term field experiments in the world which are now being used for new and innovative purposes; this resource will contribute a central component of the global network (GCTE SOMNET), officially launched in May 1995. In this paper we describe the progress so far made in establishing the European GCTE SOMNET and outline some of the activities planned over the coming months and years to build the European Network and its associated database system.

A Network of models and long-term experiments

What functions will the network perform ?

The European GCTE SOMNET will perform two functions. Firstly, it will provide a forum for soil organic matter researchers to share models and datasets. An electronic bulletin board and list-server will be established to facilitate rapid communication between network members. Secondly, it will establish a database system holding metadata (metadata is detailed descriptive information *about* the models and datasets; not the actual data or model source code), and in some cases actual models and actual long-term datasets. Modellers and dataholders will sign a data and model sharing agreement similar to that used in other GCTE networks (e.g. GCTE, 1994). Collaboration between different researchers will be encouraged through the information provided in the form of metadata whilst preserving the intellectual property rights of dataholders and model developers. The database system will comprise a central metadata database linked via the World-Wide-Web to many other European databases containing quality-controlled data in standard format. The database system is described in greater detail below. A schematic representation of the functions performed by the European GCTE SOMNET is given in figure 1.

Why are SOM models required in the Network ?

Mathematical models of SOM dynamics are essential for predicting the effects of changes in climate, atmospheric composition and land use, testing specific scenarios and developing management strategies to both minimise adverse impacts and capitalize on beneficial impacts. At the simplest level, they can be used to organise data in a rational way and provide the

Figure 1 *A Schematic representation of the functions performed by the European GCTE SOM Network.*

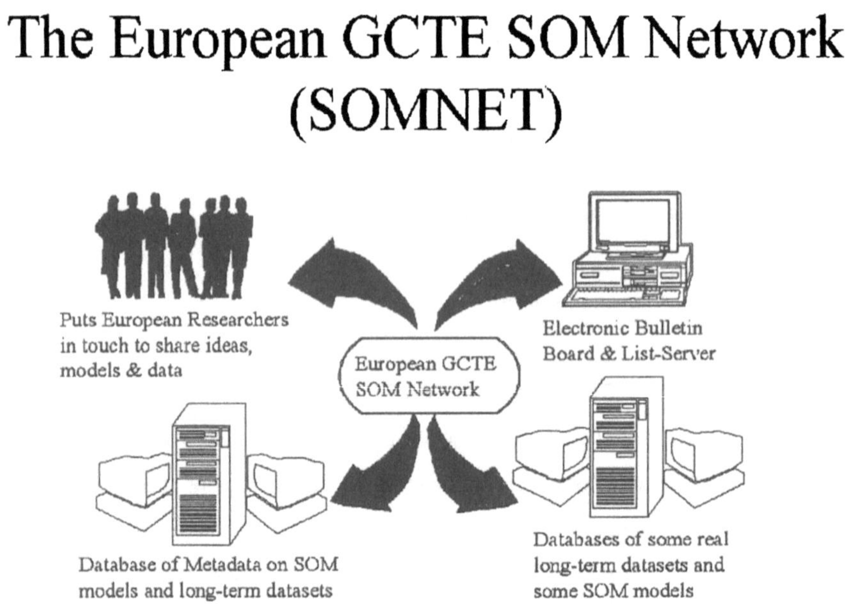

most robust way of extrapolation from trends detected in data at individual sites. Any network aiming to facilitate research into the effects of global environmental change on soil organic matter therefore needs to include SOM models. A large number of models exist but these have mostly been developed using data from one specific land-use type, e.g. arable agriculture, grassland or forest. There is a need to test them more widely to evaluate their suitability for use in considering changes in land use or climate. A pre-requisite for model testing is the identification of relevant datasets together with an evaluation of their suitability with respect to data quality, compatibility with model input requirements and availability of associated data on soil properties and climate. It is essential therefore, that long-term experiments with SOM measurements are also included in the Network.

Why are datasets from long-term experiments required in the Network ?

Data on SOM changes from well-monitored long-term experimental sites are invaluable for testing the validity of models in a wider range of environments from those for which they were originally developed (e.g. Jenkinson *et al.*, 1994).

The value of such data was high-lighted at the workshop on SOM modelling held as part of the EC COST action on "Decomposition and Accumulation of Organic Matter in Terrestrial Ecosystems" (Buurman, 1994). Concurrently, the same conclusion was drawn by other groups including a project on agricultural sustainability organized by the Rockefeller Foundation (which has compiled a directory of long-term experiments in areas other than Europe; Steiner and Herdt, 1993) and a workshop on "Long-term soil management experiments in the tropics" held in June 1993 at Ohio State University (Lal & Stewart, 1995).

There are a number of reasons why long-term, rather than short-term, experiments are needed for SOM model evaluation in the context of global environmental change. The first reason is that small changes in SOM content are difficult to measure accurately over short periods, because of year-to-year variation in crop growth (and thus inputs of SOM), and because any changes must be measured against the large background of SOM already present in the soil. Arable topsoils commonly contain 1-3% organic C (approximately 20-50 t ha^{-1} in the top 25cm), with grassland and forest soils often containing more, particularly if poorly drained (Jenkinson, 1988).

The most important reason for using long-term experiments for SOM model evaluation is that changes in SOM occur slowly. There are many examples of this (e.g. Balesdent *et al.*, 1988; Gregorich *et al.*, 1995; Jenkinson, 1988; Johnston, 1991; Juma, 1994; Körschens, 1994; Schjønning *et al.*, 1994; Vanotti *et al.*, 1995). Figure 2 shows the accumulation of total nitrogen in silty clay loam soils at Rothamsted when old arable soil is sown to permanent grass (Johnston, 1991). The 200 and 300 year values are from the Park Grass Experiment, which had been in permanent grass for at least 200 years before the experiment began in 1856. It had taken at least 100 years for the soil organic nitrogen content to change from that characteristic of an old arable soil to that of a grassland soil, and about 25 years to reach the half-way stage.

Figure 2 *Why do we need long-term experiments? The accumulation of total nitrogen in silty clay loam soils at Rothamsted when old arable land is sown to permanent grass. Adapted from figure 18.10 of Jenkinson (1988). For further details, see text.*

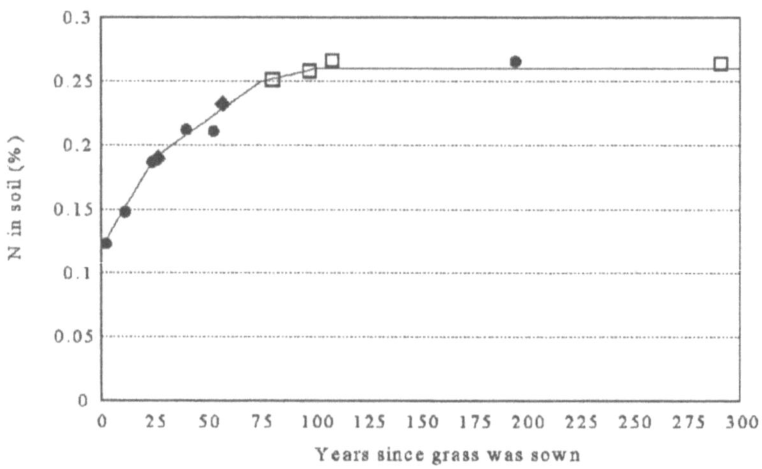

After data from A.E. Johnston, □, ◆; and J.B. Lawes & J.H. Gilbert, ●

Figure 3 shows the effects of long-continued applications of organic fertilizers on soil organic carbon content of the Hoosfield Continuous Spring Barley Experiment at Rothamsted (Jenkinson & Johnston, 1977). One plot received 35 t ha^{-1} farmyard manure (FYM) each year from 1852 until the present, another received no fertilizer from 1852 until the present, and another received 35 t ha^{-1} FYM from 1852-1871 and nothing thereafter. Over one hundred years later, the plot that received FYM between 1852 and 1871 only, still contains more SOM than the plot that has never received fertilizer. The annual application of FYM since 1852 has more than trebled soil organic C, yet equilibrium has still not been reached.

Figure 3 *Why do we need long-term experiments? The effects of long-continued applications of organic fertilizers on soil organic carbon content of the Hoosfield Continuous Spring Barley Experiment at Rothamsted. Adapted from figure 18.11 of Jenkinson (1988). For further details, see text.*

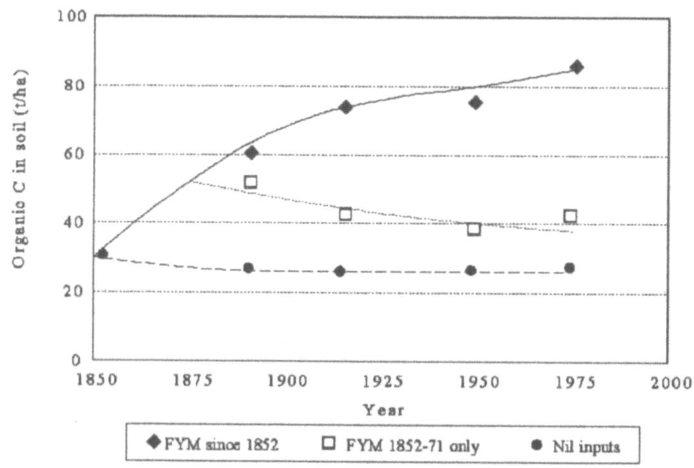

It is clear from these examples (of which there are many more) that changes in SOM occur very slowly and long-term experiments are therefore invaluable for evaluating the performance of SOM models over years, decades and centuries.

How will the European GCTE SOMNET interact with other regional GCTE SOM networks ?

Networks of long-term sites exist in North America (Paustian *et al.*, 1995), Australia (Grace, 1995) as well as in other regions of the World (e.g. the tropical long-term experiment network - Prof. M. Swift at TSBF, Nairobi). It is envisaged that these site networks will become GCTE regional SOM networks, similar to that described here. Collaboration and cooperation between the regional networks will be maintained and strengthened by frequent communication, and through the international activities of GCTE.

The European GCTE SOMNET database system

Structure of the database system

The database system will comprise a central metadata database that will be linked via the Internet to other databases that contain quality controlled data in standard format (see below) at the sites responsible for the data. The central database will contain all metadata, any datasets and models that network participants want to make available centrally, and database tools (such as data manipulation routines, statistical tools and model input/output routines) developed by Network participants. The electronic bulletin board and list server will also be administered as part of the central database. Access to all site databases will be available via menu options from the central database.

By retaining datasets at the sites where the data originated, access to data can be controlled by dataholders through the use of passwords, thus protecting intellectual property rights. It also ensures that the data remains close to the experts who can advise on its use and answer any queries that may arise. For those sites not accessible through the Internet, datasets can be retained at sites and distributed as appropriate, either by file-transfer-protocols (FTP), or by distribution on disk. Figure 4 gives a schematic representation of the database system.

The construction of the database system has begun. The metadata database has been constructed as an ORACLE database on the Sun Server at IACR-Rothamsted. The database can be accessed via the World-Wide-Web (Internet) using NCSA Mosaic for MS Windows, Netscape or similar programs at remote sites, and will be made accessible in late 1995. The current status of the metadata database is described below.

The metadata database

With the help of our collaborators in the USA (Dr Ted Elliott & Dr Keith Paustian), Australia (Dr Peter Grace) and in Germany (Professor Martin Körschens & colleagues), we have collected metadata on SOM models and long-term experiments with relevant SOM measurements from around the world. We have currently received metadata on 58 long-

Figure 4 *A Schematic representation of how databases will be linked together in the database system of the European GCTE SOM Network.*

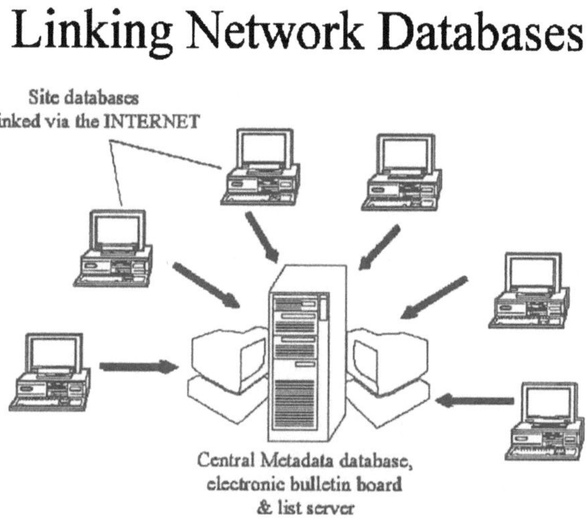

term experiments and 23 SOM models. Datasets from non-arable land-use systems and from non cool-temperate climatic zones are rare; over 90% are from arable land-use and 90% are from cool-temperate climatic zones.

Of these global returns, 34 of the long term experiments and 14 of the SOM models are from Europe. Figure 5 shows the country of origin of the European long-term experiments for which we have metadata.

This is clearly a good start, but we must put this in the context of the total number of long-term experiments in Europe. Figure 6 shows the country of origin of European long-term experiments that have been identified by Steiner & Herdt (1993), Körschens (1994) and Glendining & Powlson (1995) for which we do not yet have metadata. There are 163 long-term experiments in this category. Whilst not all of these experiments will have SOM measurements, this vast number demonstrates two things; firstly Europe is a very rich source of long-term data, and secondly, there is much work yet to be done in establishing the European GCTE SOMNET.

Figure 5 *A map of Europe showing the country of origin of datasets for which we currently have metadata in the GCTE SOMNET metadata database. Of the 34 European datasets so far included in SOMNET, over 90% are arable or grassland experiments. Numbers denote number of entries from a country, un-numbered markers denote a single entry.*

European Long-Term Experiments with SOM measurements that are part of SOMNET

Figure 6 *A map of Europe showing the country of origin of long-term experiments known to exist for which we do not yet have metadata in the GCTE SOMNET metadata database. Numbers denote number of entries from a country, un-numbered markers denote a single entry.*

European Long-term experiments that are not yet part of SOMNET

Site databases containing actual datasets

Site databases will contain the actual datasets. Datasets will consist of a series of files similar to those used at the Workshop for which this paper was prepared, i.e. meteorological data in daily, weekly and monthly formats, site information, and land-management data (e.g. fertilizer applications, tillage, irrigation, vegetation cover, harvest and planting dates etc.). For this data to be readily usable, it will need to be in a standard format and will need to be quality controlled. The following sections briefly describe network activities that will design standard data formats, and define quality control criteria and protocols.

Designing standard data formats

Data held at site databases will need to be in a standard data format. Standard data formats provide a number of advantages. Firstly, for a given model, only one input/output routine needs be written to use any dataset in the network. Secondly, data manipulation routines can be written that operate on datasets in standard format to allow, for example, weekly and monthly meteorological data to be generated from daily meteorological records. The third advantage is that routines written by network participants can be shared with other network partners and can be made available on the central database via the Internet.

Designing a standard data format is a complex process. The GCTE wheat network have adopted an existing data format (Goudriaan *et al.*, 1994) and whilst other standard data formats exist (e.g. Kragt & Jansen, 1991), there is no existing format suitable for use with all SOM models for which we so far have metadata. For this reason, a standard needs to be developed that is tailored to the needs of network participants.

A prototype standard data format was developed for use at the Workshop for which this paper was prepared, and was defined by examining the input requirements for all SOM models for which there is metadata. This prototype will however need to be refined considerably and to do this effectively, it is desirable to gain the input of as many modellers and dataholders as possible. It will therefore be an important network activity to define an agreed standard data format.

Defining quality control criteria and protocols

Datasets to be included in the European GCTE SOMNET database will need to be reviewed for quality in a similar way to datasets included in other GCTE Networks (e.g. GCTE, 1994). It is therefore important that quality criteria are defined and agreed, and that standard quality control protocols are designed.

As many network participants as possible should contribute to defining quality control criteria. Once defined, a quality control checklist or standard protocol should be applied to all datasets (and models) in the Network. This will enable indicators of various aspects of data and model quality to be attached to the metadata database entries and to the actual datasets (and models). Some of the methods and procedures used in long-term field experiments that can lead to quality problems are discussed by Glendining & Poulton (1995).

It is anticipated that many of the quality indicators will be of a minor nature (e.g. no daily meteorological data is available before 1935; only weekly averages present etc.) but these quality indicators provide invaluable information for others using the data, especially when assessing the uncertainty associated with SOM model predictions. There are, however, likely to be more fundamental quality issues associated with some long-term datasets, for example problems relating to the missing data, experimental design, available documentation, management practice (e.g. soil movement between plots) and analytical inconsistencies. These, and other quality issues, are discussed in detail in Glendining & Poulton (1995).

Actual datasets currently available within the European GCTE SOMNET

A number of European datasets were prepared for use at the Workshop for which this paper was prepared. These datasets, although not subjected to review using agreed quality control protocols and although not yet available in the finalised standard network data format, do exist in the prototype standard data format. Datasets in this category are given below:

- Park Grass - IACR-Rothamsted, U.K. (Thurston *et al.*, 1976; Jenkinson *et al.*, 1994)
- Geescroft Wilderness - IACR-Rothamsted, U.K. (Jenkinson, 1971)

- Essai Permanent - Station de Phytotechnie, Gembloux, Belgium (Frankinet *et al.*, 1993)
- Askov Long Term Experiments on mineral fertilisers and animal manures - Research Centre Foulum, Denmark (Christensen *et al.*, 1994)
- Statischer Düngungsversuch - UFZ-Bad Lauchstädt, Germany (Körschens, 1994)
- Dry Meadow Experiment - Russian Academy of Sciences, Moscow, Russia (Yakimenko, 1987)
- Praha - Ruzyně - Research Institute of Crop Production, Praha-Ruzyně, Czech Republic (Klír *et al.*, 1992)
- Keszthely - Pannon University, Keszthely, Deak, Hungary (Debreczeni & Debreczeni, 1994)

These European datasets will form the core of the European GCTE SOMNET.

Other Possible Network Activities

A key activity for the network will be establishing common sampling protocols and analysis methods to be used at all network sites (similar to those used within the Environmental Change Network; see Tinker, 1994). Having established these protocols and methods, a common sampling and analysis of soils at all network sites could be undertaken at one time (to give directly comparable measurements), and archived samples from each site could also be analyzed using common methods. There are many other examples of possible Network activities but such discussion is beyond the scope of this paper.

Conclusions: Why is the European GCTE SOMNET valuable ?

The European GCTE SOMNET will provide a forum for data-sharing, model- sharing and collaboration among European SOM researchers and will provide a database system containing metadata, actual datasets and models that will be accessible via the Internet. The Network will form one regional component of the GCTE Global SOMNET and will help to meet the objectives of GCTE Task 3.3.1., "Soil Organic Matter". The value of such a Network is also recognised within the EC. The recently published EC TERI (Terrestrial Ecosystem Research Initiative) document identifies SOM as a special focus area, particularly the analysis of existing datasets and the development of SOM models. Furthermore, there is an emphasis in the document on network building and the on the construction of databases.

As well as providing a forum for European SOM researchers to evaluate the suitability of SOM models for predicting the effects of global environmental change on SOM, it will also allow network partners to begin to explore the impact of potential future climate/land-use change scenarios. Predictions can then be made, using the clearly defined assumptions inherent in the models, and based on measured changes at individual sites, of the effects of global environmental change on SOM as a source and sink of carbon, as a driving force controlling nutrient (N, P) availability and loss (and, hence, movement to aquifers), as an important factor affecting soil stability and erosion risk, and as a factor contributing to the flux of trace greenhouse gases (especially nitrous oxide and methane) between land surface and the atmosphere.

As well as fulfilling a recognised scientific need, the European GCTE SOM Network will also make more available valuable data collected during long-term field experiments. A rough estimate of the economic value of this data can be obtained by estimating the cost of collecting the data, a calculation that we have performed below.

The Rothamsted Classical Experiments (of which there are eight) were estimated to cost $100,000 per year to run in 1989 (Jenkinson, 1991). If we assume that each long term experiment costs $12,500 per year (which is a conservative estimate since experiments run in isolation are more expensive than those conducted as part of a large field experimental programme), data collected during the thirty-four long-term experiments so far included as metadata in the European SOMNET (which range in age between 27 and 152 years) cost an estimated $23,625,000 (in 1989 terms) to collect. A similar distribution of experiment durations among the remaining European long-term experiments for which we do not have metadata would bring the cost of datasets available within Europe to $ 137,000,000 in 1989 terms.

Based on these simple calculations, then, the European GCTE SOMNET could potentially make data valued at around $140,000,000 available to network participants. Coupled with the scientific value of the data and models harnessed for use by network participants, the value of the European GCTE SOMNET in facilitating research into the effects of global environmental change on soil organic matter, is clear.

References

Balesdent, J., Wagner, G.H. & Mariotti, A. (1988) Soil organic matter turnover in long-term field experiments as revealed by Carbon-13 natural abundance. *Soil Science Society of America Journal* **52**: 118-124.

Buurman, P., (1994) *Decomposition and accumulation of organic matter in terrestrial ecosystems.* Commission of the European Communities. Report on the EC Cost Action, 26-92 Environment, 108pp.

Christensen, B.T., Peterson, J., Kjellerup, V. & Trentemøller, U. (1994) The Askov long-term experiments on animal manure and mineral fertilizers: 1894-1994, Danish Institute of Plant and Soil Science, SP Report 43, volume 2, 85pp.

Debreczeni, B. & Debreczeni, Bné (Eds.) (1994) *Fertilization research 1960-1990.* Akadémiai Kiadó, Hungary, 411pp.

Paustian, K., Elliott, E.T., Paul. E.A., Collins, H.P., Cole, C.V. & Frey, S. (1995) The North American Site Network. This volume.

Frankinet, M., Raimond, Y., Destain, J.P., Roisin, C. & Grevy, L. (1993) Organic matter management and calcic amendments in order to maintain or improve soil fertility. In: *Soil biota, nutrient cycling, and farming systems.* (ed: Paoletti, M.G., W. Foissner & D. Coleman), Lewis Publishers, Boca Raton, USA: 27-39.

GCTE (1994) *Global Change and Terrestrial Ecosystems Report No. 2. GCTE Focus 3 Wheat Network: 1993 Model and Experimental Meta Data.* GCTE, Canberra, Australia, 147pp.

Glendining, M. J. & Poulton, P. (1995) Interpretation difficulties with long-term experiments. This volume.

Glendining, M.J. & Powlson, D.S. (1994) The effect of long-continued applications of inorganic nitrogen fertilizer on soil organic nitrogen - a review. In: *Soil Management. Experimental basis for sustainability and environmental quality.* (ed: R. Lal & B.A. Stewart), *Advances in Soil Science,* CRC Lewis Publishers, Boca Raton, USA: 385-446.

Goudriaan, J., van de Geijn, S.C. & Ingram, J.S.I.(Eds.) (1994) GCTE Focus 3 Wheat Modelling and Experimental Data Comparison Workshop Report, Lunteren, Netherlands; 2-4 November, 1993, 16pp.

Grace, P. (1995) The Australian Site Network. This volume.

Gregorich, E.D., Ellert, B.H., Drury, C.F. & Laing, B.C. (1995) Fertilization effects on soil organic matter turnover and corn residue C storage. *Soil Science Society of America Journal* (in press).

Gregory, P.J. & Ingram, J.S.I. (1995) The contribution of the Soil Organic Matter Network (SOMNET) to GCTE. This volume.

Jenkinson, D.S. (1971) The accumulation of organic matter in soil left uncultivated. *Rothamsted Annual Report for 1970, Part 2*: 113-137.

Jenkinson, D.S. (1988) Soil organic matter and its dynamics. In: *Russell's soil conditions and plant growth, 11th Edition.* (ed: A. Wild), Longman, London: 564-607.

Jenkinson, D.S. (1991) The Rothamsted long-term experiments: are they still of use? *Agronomy Journal* **83**: 2-10.

Jenkinson, D.S. & Johnston, A.E. (1977) Soil organic matter in the Hoosfield continuous barley experiment. *Rothamsted Annual Report for 1976, Part 2*: 87-101.

Jenkinson, D.S., Adams, D.E. & Wild, A. (1991) Model estimates of CO_2 emissions from soil in response to global warming. *Nature* **351**: 304-306.

Jenkinson, D.S., Bradbury, N.J. & Coleman, K. (1994) How the Rothamsted Classical Experiments have been used to develop and test models for the turnover of carbon and nitrogen in soil. In: *Long-term experiments in agricultural and ecological sciences*. (ed: R.A. Leigh & A.E. Johnston), CAB International, Wallingford, UK: 117-138.

Jenkinson, D.S., Harkness, D.D., Vance, E.D., Adams, D.E. & Harrison, A.F. (1992) Calculating net primary production and annual input of organic matter to soil from the amount and radiocarbon content of soil organic matter. *Soil Biol. Biochem.* **24**: 295-308.

Jenkinson, D.S., Potts, J.M., Perry, J.N., Barnett, V., Coleman, K. & Johnston, A.E. (1994) Trends in herbage yields over the last century on the Rothamsted long-term continuous hay experiment. *Journal of Agricultural Science, Cambridge* **122**: 365-374.

Johnston, A.E. (1991) Soil fertility and soil organic matter. In: *Advances in soil organic matter research*. (ed: W.S. Wilson), Royal Society of Chemistry: 299-314.

Juma, N.G. (1994) Dynamics of soil C and N in a typic cryoboroll and a typic cryoboralf located in the cryoboreal regions of Alberta. *Advances in soil science* (in press).

Klír, J., Škarda, M., Marková, J. & Vojtíšková, I. (1992) Effect of organic fertilizing on the nutrients balances in long term field experiments. In: *Proceedings of Symposium "Long-term field experiments and nutrient dynamics, 1992, Bad Lauchstädt, Germany*.

Kragt, J.F. & Jansen, E.J. (1991) *Standardisation of datasets for the testing of simulation models. A contribution to the EC-Project, "Nitrate in Soils"*. The Winand Staring Centre for Integrated Land, Soil and Water Research Report 25, Wageningen, The Netherlands, 75pp.

Körschens, M. (Ed.) (1994) *Der Statische Düngungsversuch Bad Lauchstädt nach 90 Jahren*. B.G.Teubner Verlagsgesellschaft, Leipzig, Germany, 179pp.

Lal, R. & Stewart, B.A. (Eds.) (1995) *Soil Management. Experimental basis for sustainability and environmental quality. Advances in Soil Science,* CRC Lewis Publishers, Boca Raton, USA, 555pp.

Elliott, T.E. & Paustian, K. (1995) Why regional networks ? This volume.

Post, W.M., Emanuel, W.R., Zinke, P.J. & Stangenberger, A.G. (1982) Soil carbon pools and world life zones. *Nature* **289**: 156-159

Schjønning, P., Christensen, B.T. & Cartensen, B. (1994) Physical and chemical properties of a sandy loam receiving animal manure, mineral fertilizer or no fertilizer for 90 years. *European Journal of Soil Science* **45**: 257-268.

Steiner, R.A. & Herdt, R.W. (1993) *A Global Directory of Long-Term Agronomic Experiments,* Volume 1, Rockefeller Foundation.

Tinker, P.B. (1994) Monitoring environmental change through networks. In: *Long-term experiments in agricultural and ecological sciences*. (ed: R.A. Leigh & A.E. Johnston), CAB International, Wallingford, UK: 407-421.

Thurston, J.M., Williams, E.D. & Johnston, A.E. (1976) Modern developments in an experiment on permanent grassland started in 1856: Effects of fertilizers and lime on botanical composition and crop and soil analyses. *Ann. Agron.* **27**: 1043-1082.

Vanotti, M.B., Bundy, L.G. & Peterson, A.E. (1995) Nitrogen fertilizer and legume-cereal rotation effects on soil productivity and organic matter dynamics in Wisconsin. In: *Soil organic matter in temperate agroecosystems: long-term experiments in North America*. (ed: Paul, E.A., Paustian, K. & Elliot, E.T.), Lewis Publishers, Boca Raton. (in press).

Yakimenko, E. (1987) Pedogenesis under dry meadows. *Pochvovedenie (Soviet Journal of Soil Science)* No. 5.

Acknowledgements

We acknowledge funding for P.S. from the Terrestrial Initiative in Global Environmental Research (TIGER) programme of the U.K. Natural Environment Research Council (NERC), and NATO for funding the Advanced Research Workshop for which this paper was prepared. We thank Dr Jo Smith for her hard work in helping to secure the TIGER and NATO funding and Andy Caiger for his expert role in establishing the Internet GCTE SOMNET metadata database. IACR receives grant-aided support from the Biotechnology and Biological Sciences Research Council of the United Kingdom.

Interpretation Difficulties with Long-Term Experiments

M. J. Glendining and P. R. Poulton
Soil Science Department
IACR-Rothamsted
Harpenden, Herts
AL5 2JQ, UK

Summary

Many long-term experiments were not originally established to measure changes in soil organic matter (SOM) content. Thus, there are inevitably difficulties associated with interpreting SOM data from these sites. The general difficulties associated with long-term experiments mainly arise from the experimental design, sampling methods and record keeping. These include: little or no replication or randomisation; lack of time-zero samples; inappropriate control treatments; incomplete description of the experiment and sampling protocols; modifications to the experiment with time; soil movement between the plots. In addition, there are some difficulties of interpretation specifically associated with measuring long-term changes in SOM. These include: modifications to soil sampling protocol (depth, method, number and distribution of samples); changes in soil bulk density; methods of measuring soil C. In this paper we describe these potential problems, and identify the main sources of error when interpreting SOM data from long-term experiments.

Introduction

Changes in SOM can occur very slowly (see examples in Smith *et al.*, 1995), and long-term experiments (or more precisely, data from long-term experimental sites) are the only satisfactory means of obtaining information on the long-term effects any physical, chemical and biological factors may have on SOM content. Most long-term experiments were not

NATO ASI Series, Vol. I 38
Evaluation of Soil Organic Matter Models
Edited by David S. Powlson, Pete Smith and Jo U. Smith

originally established to measure changes in SOM content, or even specifically as long-term experiments (e.g. Johnston, 1994). Thus, there are inevitably difficulties associated with interpreting data on changes in SOM from these sites.

General interpretation difficulties

There are a number of difficulties commonly associated with interpreting data from long-term field experiments. These mainly arise from experimental design and sampling methods, and include the following (taken from Glendining & Powlson, 1995):

1) Lack of time-zero samples. Soil samples may not have been taken until the experiment was well under way. Thus, the initial SOM content, before the treatments were imposed, may have to be estimated by extrapolation (e.g. Jenkinson, 1977; Jenkinson & Johnston, 1977). However, it may still be possible to follow changes in SOM due to different treatments, or due to other treatments started later on in the experiments (e.g. on the Broadbalk Wheat Experiment, Glendining & Powlson, 1991; Jenkinson, 1977). Lack of time-zero samples may also lead to problems in estimating the rate of change of SOM.

2) Lack of replication and poor randomisation. Particularly common in the oldest long-term experiments, this may be partially overcome if the plots are very large, as in most of the Rothamsted 'Classical' Experiments. Observed differences in SOM content may have been present *before* the treatments were imposed, due to spatial variability in soil properties. No true measure of error can be made, except by using years as replicates. Statistical approaches relevant to long-term unreplicated experiments are discussed by Barnett (1994).

3) Inappropriate treatments and controls. Treatments and controls which are appropriate for SOM research may not have been chosen when the experiment was originally started. The control treatments were often designed to be relevant to the interpretation of treatment effects on crop yields. Other treatments may have been changed or stopped, if they were no longer thought to be appropriate for current agricultural or forestry practice. For example, at the Dehérain Experiment in France (started in 1875), the different fertilizer treatments were started at different times: NPK in 1902, PK in 1929, etc (Morel *et al*, 1984). Thus it is difficult to determine the effects of N fertilizer on SOM in isolation from PK fertilizers.

4) Modifications to the experiment over time. This includes changes to fertilizer application rates, and improved pest and disease control. There has always been a conflict between updating the management of an experiment, so that it remains relevant to current agricultural, forestry and environmental research, and maintaining the original treatments, which may soon become obsolete, but do provide an uninterupted set of data. If the original experimental area is large enough, the best solution is to do both, as has occurred at some of the long-term sites at Rothamsted (Jenkinson, 1991; Leigh *et al.*, 1994), but if the original plots were small this is impossible.

5) Incomplete description of the experiments and the sampling protocols. Examples include whether lime was applied to counteract the acidifying effects of ammonium-based fertilizers, exact depth of soil sampling, number and distribution of samples within each plot and whether they are bulked, whether crop residues are removed or retained, direction of ploughing/cultivation, etc. There may also be interpretation difficulties specific to each site, (e.g. slope, shading from adjacent buildings or trees, etc).

6) Soil movement between plots. This is principally due to cultivation, but also due to water, wind and soil fauna. At some long-term sites, barriers to prevent soil movement by cultivation have been installed. For example, concrete partitions are sunk into the ground at the 'Ewiger Roggenbau' experiment at Halle (Garz & Hagedorn, 1990), permanent grass strips separate the plots at Sanborn Field (Buyanovsky & Brown, 1995) and at the Jordan experiment (Holben, 1955). Cultivation practices should be designed to minimize soil movement. There should be no cross cultivation between plots, i.e cultivations should always be along the length of the plots. In addition, the direction of ploughing should be reversed each year to prevent unidirectional movement of soil across plots (Leigh *et al*, 1994).

Sibbesen (1986) simulated soil movement due to cultivation at 21 long-term sites (50 years or more), based on a model of changes in soil P at the long-term experiment at Askov in Denmark. Assuming the same rate of soil movement as at Askov, he calculated that on average less than a third of the topsoil now present in the central quarter of each plot originated from topsoil present at the start of the experiment. The extent of soil movement will depend on many factors, including size of plot, soil type, number, method and direction of cultivation. If the plots are small and not bounded by some form of barrier, the spread of soil during cultivation will eventually destroy the integrity of the plot (Leigh *et al*, 1994; Powlson, 1994).

Soil erosion, whether by wind or water, will be important at some sites. It not only blurs plot edges, but may lead to a decrease in measured SOM content which is not due to SOM decomposition. In addition to direct removal of topsoil, which can lead to dilution with subsoil with a lower SOM content, both wind and water erosion is selective. The SOM content of the eroded material may be much higher than that of the soil that remains (Allison, 1973).

Soil movement is generally confined to *cultivated* long-term experiments. This is illustrated in two examples from Rothamsted. Figure 1 shows soil pH in two adjacent plots of the Park Grass experiment (see Poulton, 1995a for further details). The application of 144 kg N ha^{-1} as $(NH_4)_2SO_4$ annually since 1856 to plot 11/2d has decreased soil pH to around 3.5. In contrast plot 12d, with no fertilizer inputs, has a pH of around 5.0. Figure 1 shows the very distinct boundaries between the plots, demonstrating that there is little sideways movement of soil or nutrients at this flat, undisturbed site.

Figure 1 *Soil pH in two adjacent plots in the Park Grass Experiment at Rothamsted, 1993. See text for details of fertilizer treatments.*

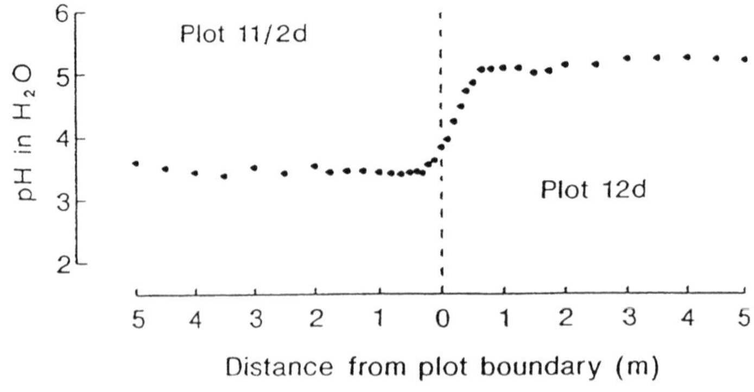

In contrast, Figure 2 shows soil zinc content in the Woburn Market Garden Experiment, in the plot given sewage sludge for 20 years, and the surrounding plots which did not receive sewage sludge (see McGrath & Lane, 1989, for further details). After 40 years of intensive cultivations on this sandy soil, it is no longer possible to clearly differentiate between adjacent plots, because there has been so much soil movement. However, McGrath and Lane (1989) showed that nearly all the heavy metals added to these plots in sewage sludge were retained in the topsoil if the sideways movement of the soil was taken into account.

Figure 2 *Soil zinc content in the Woburn Market Garden Experiment. See text for details of sewage sludge treatments.*

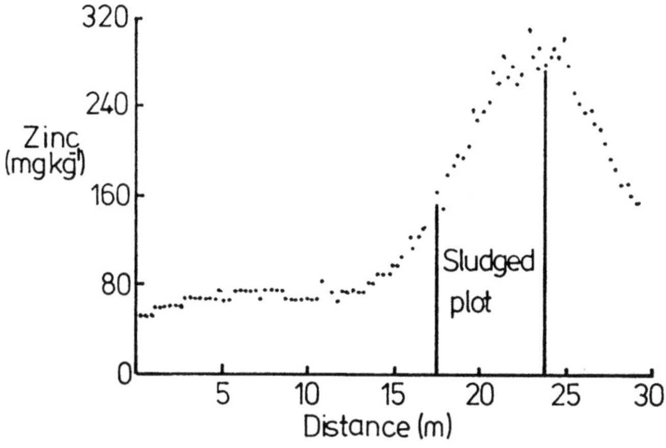

Specific interpretation difficulties associated with SOM measurements

In addition to the above general difficulties of interpretation, there are several difficulties specifically associated with measuring changes in SOM at long-term sites.

Soil sampling techniques.

At each long-term site, soil samples taken at different times should be directly comparable. Modifications to sampling protocol (sampling depth, method, number and distribution of samples within each plot) may mean that this is not so. Some authors have reverted to original sampling methods and depths, to ensure comparibility (e.g. Jenkinson, 1971). Thus it is essential to be aware of and *record* full details of soil sampling protocol. Archived soil samples taken from many soil layers may enable corrections or comparisons to be made with later soil samples.

Alterations to sampling depth, particulary in the topsoil, where most of the changes in SOM occur, present a considerable problem, as the samples are no longer directly comparable. It is often preferable therefore to keep to the original sampling depth, but this may not always be sensible. For example, increasing the ploughing depth will dilute the original topsoil. If samples were originally taken from a shallower depth than that achieved by the deeper ploughing, then subsequent samples taken to that same depth will be meaningless. However it may be possible to accommodate such modifications if deeper soil layers were also sampled, both before and after any changes.

Modifications to sampling method, e.g. from spade to auger, have occured in most long-term experiments, and are usually associated with modifications to the number and distribution of samples taken from each plot (e.g. in the Park Grass Experiment, Poulton, 1995a). Early samples often consisted of a few large samples, which may or may not have been bulked together. These gave good estimates of soil bulk density, but little information on soil variability, and, depending on that variability, possibly a poor estimate of the mean SOM content of the plot. Modern samples tend to consist of many small cores, bulked together, taken randomly or systematically across the plot, but avoiding the edges. These provide a better estimate of mean SOM content, a poor estimate of bulk density and, if the samples were bulked together, no information on soil variability. For a detailed discussion of the design of sampling schemes and the application of spatial statistics, see Webster & Oliver (1990).

Some common changes in soil sampling protocol are illustrated in Figure 3, which shows the Broadbalk Wheat Experiment at Rothamsted. Originally the plots extended the whole length of the field (300m). Six soil samples were taken with an open ended metal box, 15cm square and 23cm deep, from the centre of the plot. When the experiment was divided into five Sections in 1926, samples were first taken by spade, two per plot, and then by auger, when many more small samples were taken and bulked together. In 1968 the experiment was further divided into the present 10 Sections, with only four remaining in continuous wheat, and thus directly comparable with the earlier samples. As the early box samples were archived separately, rather than being bulked together, it is possible to get some indication of soil spatial variability early in the experiment. Around 16 auger samples are now taken from each Section, avoiding a 1m discard around the plot edges. Throughout the experiment the soil has been sampled to 23cm, even though the ploughing depth was originally only to 12 or 15cm. This consistency has greatly added to the value of this long-term site. Deeper soil samples have also been taken on numerous occasions.

Figure 3 *Changes in soil sampling protocol on the Broadbalk Wheat Experiment at Rothamsted. See text for details.*

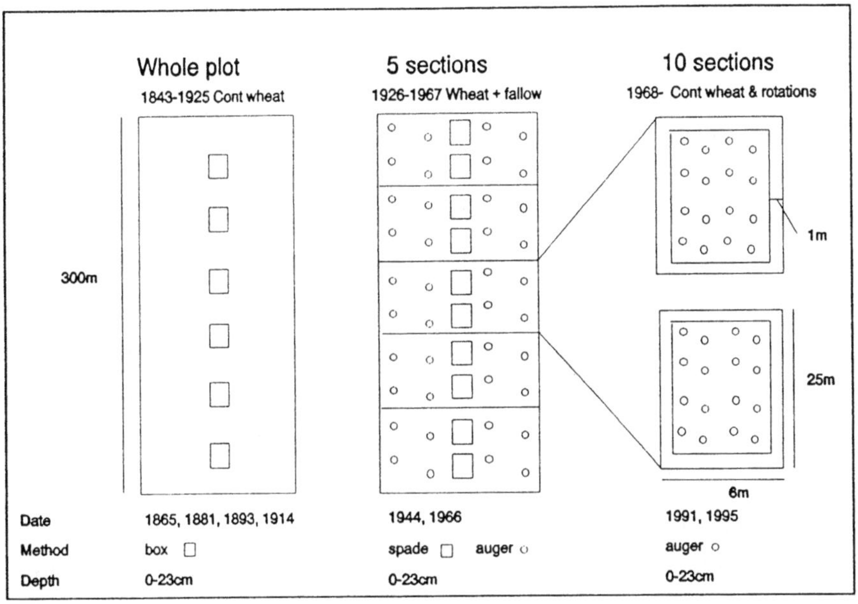

Changes in soil bulk density.

Soil bulk density (the weight of sieved, organic matter-free soil per unit volume, for a specified soil layer) may change. It may increase, due to compaction from heavy machinery, or a reduction in organic inputs, e.g. when grassland is ploughed up. It may decrease, due to greater organic inputs, e.g. when using organic manures, or when converting arable land to grassland or woodland.

Changes in soil bulk density can have substantial effects when calculating SOM content on an area basis for a given soil layer. If deeper soil layers have also been sampled, it may be possible to correct for changes in bulk density by calculating the 'equivalent depth' to which the soil should have been sampled to contain the same weight of soil. Poulton (1995b) discusses this for the Geescroft Wilderness. If no allowance is made for the change in soil bulk density at this site, the increase in SOM between 1883 and 1985 due to reversion to natural woodland is underestimated by at least 20%.

Method of measuring soil C.

Nelson & Sommers (1982) provide a comprehensive review of the subject. Total C in soil consists of organic and inorganic C. Most organic C is present in the SOM fraction, whereas inorganic C is largely found in carbonate materials. In some soils, inorganic C concentration may exceed organic C concentration.

Soil organic matter is difficult to estimate quantitatively. Past procedures involved determining weight change after destroying organic compounds by H_2O_2 treatment or by ignition at high temperature. Both techniques are subject to error. Alternatively, SOM content has been estimated by multipling the organic C content by a constant factor based on the percentage of C in organic matter. Traditionally the Van Bemmelen factor of 1.724 was used, based on the assumption that organic matter contains 58% organic C. However, many studies have shown that the proportion of C in SOM is highly variable for a range of soils and depths. Thus it is preferable to determine and report organic C concentration in a soil rather than to convert to SOM content through the use of an approximate conversion factor (Nelson & Sommers, 1982).

Many different methods have been used to measure total or organic C in the data-sets used at this workshop: dry combustion (using various CHN analysers based on the Dumas method, e.g. LECO, Carlo Erba); wet combustion; Tinsley; Walkley-Black; hot water-soluble C; Tyurin method and others. See Nelson & Sommers (1982) and Tiessen & Moir (1993) for details of these and other methods. It is important to record exactly what has been measured, (total C, organic C, etc). It may be necessary to devise conversion factors to ensure that the different analytical methods are providing comparable data. There have been changes in the analytical methods used to measure soil C in nearly all the long-term experiments (e.g. a change from a dichromate oxidation-titration technique, such as Tinsley, to a commercially available automated total C analyser). Preferably, changes in analytical techniques should be kept to a minimum and archived soil samples, if available, should be re-analysed when a new method is introduced.

Sources of error

Most analytical techniques used to measure soil C give consistent results, if performed carefully, and if the analysis is repeated when discrepancies between replicates exceed accepted limits. Larger errors will be associated with soil sampling techniques - number, size, accuracy and randomisation of samples per plot, and from handling the soil prior to analysis, rather than the analysis itself. As already discussed, modifications to sampling protocol, and changes in soil bulk density are potentially large sources of error.

Soil variability (both spatial and temporal) is another potential source of error, and may be greater when the site is under perennial crops or natural vegatation, rather than when it is regularly cultivated. For example, total soil C was measured at two sites at Rothamsted, (plot size 6 x 28m), taking 18 cores from the 0-15cm depth (P. Hargreaves, *pers comm*). The coefficient of variation was around 25% at the woodland site (Geescroft Wilderness) but only 8% at the arable site (Broadbalk plot 14). The presence of trees will undoubtedly be a major factor contributing to SOM variability, in addition to the lack of cultivation. If time-zero samples are available for all plots in a replicated long-term experiment, it is possible to decrease the standard errors of later measurements, by taking into account the pre-existing spatial variability of SOM (Saffigna *et al*, 1989).

Conclusions

There are inevitably difficulties with interpreting SOM changes at long-term sites. These can often be at least partially overcome, especially if a good range of archived soil samples is available. Full details of experimental techniques and of any changes in methodology are vital to the interpretation of results. In spite of their limitations, well-monitored long-term experiments are the only satisfactory means of measuring long-term changes in SOM content.

Data from such experiments are invaluable for testing the validity of SOM models, but should always be interpreted in the light of any known limitations.

References

Allison, F.E. (1973) Soil organic matter and its role in crop prroduction. *Developments in Soil Science 3,* Elsevier, Amsterdam.

Barnett, V. (1994) Statistics and the Long-term Experiments: Past Achievements and Future Challenges. In: *Long-term Experiments in Agricultural and Ecological Sciences.* (eds: R.A. Leigh & A.E. Johnston), CAB International, Wallingford: 165-183.

Buyanovsky, G.A. & Brown, J.R. (1995) Sanborn Field. This Volume.

Garz, J. & Hagedorn, E. (1990) Der Versuch 'Ewiger Roggenbau' nach 110 Jahren. In: *110 Jahre Ewiger Roggenbau.* (ed: H.-J. Liste), Kong. Tag.-Ber. Martin-Luther-Univ., Halle-Wittenberg, Halle: 9-30.

Glendining, M.J. & Powlson, D.S. (1995) The Effects of Long Continued Applications of Inorganic Nitrogen Fertilizer on Soil Organic Nitrogen - A Review In: *Soil Management: Experimental Basis for Sustainability and Environmental Quality*, (eds: R. Lal & B.A. Stewart), Advances In Soil Science, Lewis Publishers, Boca Raton: 385-446.

Glendining, M.J. & Powlson, D.S. (1991) The effects of long-term applications of inorganic nitrogen fertililzer on soil organic nitrogen. In: *Advances in Soil Organic Matter Research: the Impact on Agriculture and the Environment.* (ed: W.S. Wilson), The Royal Society of Chemistry, Cambridge: 329-338.

Holben, F.J. (1955) History of the Jordan Plots; Practises, Changes and People. In: *75th Anniversary of the Jordan Soil Fertility Plots.* Bulletin 613, The Pennsylvania State University, College of Agriculture, Agricultural Experiment Station, Pennsylvania: 5-11.

Jenkinson, D.S. (1971) The Accumulation of Organic Matter in Soil Left Uncultivated. *Rothamsted Annual Report for 1970, Part 2*: 113-137.

Jenkinson, D.S. (1977) The Nitrogen Economy of the Broadbalk Experiments. I. Nitrogen Balance in the Experiments. *Rothamsted Annual Report for 1976, Part 2*: 103-109.

Jenkinson, D.S. (1991) The Rothamsted Long-term Experiments: Are they still of use? *Agronomy Journal* **83**: 2-10.

Jenkinson, D.S. & Johnston, A.E. (1977) Soil organic matter in the Hoosfield Continuous Barley Experiment. *Rothamsted Annual Report for 1976, Part 2*: 87-101.

Johnston, A.E. (1994) The Rothamsted Classical Experiments. In: *Long-term Experiments in Agricultural and Ecological Sciences* (eds: R.A. Leigh & A. E. Johnston), CAB International, Wallingford: 9-37.

Leigh, R.A., Prew, R.D. & Johnston, A.E. (1994) The Management of Long-term Agricultural Field Experiments : Procedures and Policies Evolved from the Rothamsted Classical Experiments. In: *Long-term Experiments in Agricultural and Ecological Sciences.* (eds: R.A. Leigh & A. E. Johnston), CAB International, Wallingford: 253-263.

Morel, R., Lasnier, T. & Bourgeois, S. (1984) Les essais de fertilisation de longue durée de la station agronomique de Grignon. Dispositif Dehérain et des 36 Parcelles. Resultats expérimentaux (1938-1982). INA-PG, INRA, Paris

McGrath, S.P. & Lane, P.W. (1989) An explanation for the apparent losses of metals in a long-term field experiment with sewage sludge. *Environmental Pollution* **60:** 235-256

Nelson, D.W. & Sommers, L.E. (1982) Total Carbon, Organic Carbon and Organic Matter In: *Methods of Soil Analysis, Part 2.* (eds: A.L. Page *et al.*), ASA/SSSA, Madison, Wisconsin: 539-579.

Poulton, P.R. (1995a) The Park Grass Experiment, 1856-1995. This volume.

Poulton, P.R. (1995b) Geescroft Wilderness, 1883-1995. This volume.

Powlson, D.S. (1994) Quantification of Nutrient Cycles using Long-term Experiments. In: *Long-term Experiments in Agricultural and Ecological Sciences,* (eds: R.A. Leigh & A.E. Johnston), CAB International, Wallingford: 97-115.

Saffigna, P.G., Powlson, D.S., Brookes, P.C. & Thomas, G.A. (1989) Influence of sorghum residues and tillage on soil organic matter and soil microbial biomass in an Australian vertisol. *Soil Biology and Biochemistry* **21**: 759-765.

Sibbesen, E. (1986) Soil movement in long-term field experiments. *Plant and Soil* **91**: 73-85.

Smith, P., Powlson, D.S. & Glendining, M.J. (1995) Establishing a European GCTE Soil Organic Matter Network (SOMNET). This volume.

Tiessen, H. & Moir, J.O. (1993) Total and Organic Carbon In: *Soil Sampling and Methods of Analysis*, (ed: M.R. Carter), Canadian Society of Soil Science, Lewis Publishers, Boca Raton: 187-199.

Webster, R. & Oliver, M. A. (1990) *Statistical methods in soil and land resource survey.* Oxford University Press, New York, 316pp

Acknowledgements

IACR-Rothamsted receives grant-aided support from the Biotechnology and Biological Sciences Research Council of the United Kingdom. We thank NATO for funding the Advanced Research Workshop for which this paper was prepared.

Review and Classification of Ten Soil Organic Matter (SOM) Models

William B McGill

Department of Renewable Resources

4-42 Earth Sciences Building, University of Alberta

Edmonton, Alberta, T6G 2E3

Canada

Summary

This paper describes ten SOM models briefly, presents a scheme for classifying information about them and summarizes relationships among model attributes and among models. All are process-oriented multicompartment models. Few attributes are needed to distinguish among them and with few exceptions they are empirical in nature. There appears to be a trend to include inert organic matter but not to specify its nature nor rate of formation. Soil texture is increasingly included in SOM models both to protect or slow decomposition of soil organic components, and to regulate partitioning of C among compartments. About half the models represent a homogeneous soil unit. Concepts pertaining to the nature of litter are different from those pertaining to SOM, as indicated by inclusion of biomass as a SOM component but not a litter component.. Many models in the past have specified biochemical fractions to follow C and N through litter and SOM components, but in no case were biochemical separations applied to litter or SOM.

I conclude that a simple classification scheme to organize information about models provides a focus for concepts within them and comparisons among them. The scheme revealed convergence in the kinetic compartmentalization, growing use of soil clay content, and inclusion of an IOM component. Future developments of SOM models could benefit from: i) ways such as physical fractionation to characterize kinetic components of soils, if possible, and to relate soil structure through organism-substrate accessibility to turnover rates, ii)

NATO ASI Series, Vol. I 38
Evaluation of Soil Organic Matter Models
Edited by David S. Powlson, Pete Smith and Jo U. Smith

cross-scale comparisons and GIS links and evaluations to determine the level of detail needed as a function of scale, and the extent to which these models can move from site-specific simulations to landscape simulations; and, iii) a more mechanistic treatment of the role of soil organisms.

Introduction

Soils connect the atmosphere, hydrosphere, geosphere, and biosphere. Understanding transfers of CO_2, between them and storage of C within the pedosphere requires understanding of soil organic matter turnover. Increasingly, such understanding is embodied in models which are then used to make projections about the condition of the planet given diverse scenarios. As scenarios become more complex, models are required that have a high degree of generality and that may also be complex.

SOM models represent the transformations of C and / or N in soil-plant systems with the quantity of soil organic components as a state variable. They may or may not be mechanistic in their treatment of organic C addition to and transformation within soils. They subsume on one hand multicompartment simulation models that aim to represent the flows of mass between compartments, and on the other analytical mathematical models that represent the size of a few state variables over time. Hybrids of these two poles have also been published.

Models of SOM change have been published at least since the 1940s. Jenny (1941) used

$$\frac{dX}{dt} = -kX + A$$

where dX = change in a state variable (e.g. soil C or N),

k = first order rate constant (t^{-1}); and,

A = addition rate (mass t^{-1}) which is independent of loss and amount present

to describe C accumulation or loss from soils. Similarly Hénin & Dupuis (1945), Newton *et al.*, (1945), Woodruff (1949), Bartholomew & Kirkham (1960) and Russell (1960) used such models to analyze SOM (C or N) under a variety of rotations and climates.

It has long been recognized, however, that SOM has many components varying in stability, and if the proportions of the constituents in SOM change over time, then the value of k changes over time (see review by Van Veen *et al*, 1981). Strategies to accommodate changing values of k range from adding more components (as reviewed by Jenkinson, 1990), to developing time-dependent ways to modify k (e.g. Russell, 1975), using decay series in which residues are treated as homogeneous but each year the residual portion becomes more stable (Powers *et al* 1975) or decay series in which k is an exponential function of time (Mathers & Goss, 1979).

The large number of models published that include SOM turnover make it important to develop ways to compare and organize information about them. Jenkinson (1990) distinguished four categories of SOM turnover models: i) single homogeneous compartment, ii) two-compartment, iii) non-compartmental decay models; and, iv) multicompartmental models. Paustian (1994) reviewed several multicompartmental models of soil biology and biogeochemical processes and grouped them into either organism-oriented, or process-oriented models. Organism-oriented models focus on flow of energy and matter through food-webs; process-oriented models focus on processes controlling energy and matter transformations. Organisms are represented in process models as a form of generic biomass or active organic matter. Paustian (1994) grouped information about models under four headings: i) non-living organic matter pools, ii) organism components, iii) rate kinetics; and, iv) special attributes. He concluded that process-oriented models were the most commonly used to describe soil biology in agricultural systems, and that they shared some common elements: i) dominance of first-order kinetics, ii) discrete SOM components with characteristic rates of decomposition; and, iii) interconnected dynamics of C and N. The dominance of first-order kinetics in such models further implies that the metabolic potential of the biomass or soil organism community exceeds the substrate supply; consequently, changes in biomass would not be associated with changes in rates of decomposition. He identified three areas for future development: i) ways to obtain empirical data about kinetic pools now included in SOM models, ii) connections between soil organism community structure and processes; and, iii) connections between soil structure and organic matter dynamics. If SOM models are to meet the need for broad generality, then it may be essential to include sensitivity to soil conditions at the scale in which they operate, and find ways to measure the quantity of soil C in kinetic compartments.

This paper describes ten SOM models briefly, proposes a scheme for classifying information about them and then examines each in the context of the proposed classification scheme.

The Models

Nine of the models examined were part of the NATO Advanced Research Workshop: *Evaluation of soil organic matter models using long-term data sets*. Eight of these: CANDY, CENTURY, DAISY, DNDC, ITE, NCSOIL, ROTHC and VVV were developed for agricultural or grassland ecosystems; and one, SOMM was developed for forested ecosystems. The tenth, QSOIL did not participate in the workshop; it is a noncompartmental cohort decay model that was developed for heterogeneous substrates with few ecosystem-specific attributes or embedded assumptions.

CANDY (CArbon- Nitrogen-DYnamics) is a modular system of simulation models and a data base system for model parameters, measurement values, initial values, weather data and soil management data (Franko, 1995; Franko *et al.*, 1995). It simulates dynamics of soil N, temperature and water in order to provide information about N uptake by crops, leaching and water quality. CANDY uses a semi-cohort system to track litter decay, and calculates a biologically active time to allow comparisons among sites. The inert organic matter (IOM) component is calculated from the proportion of soil particles < 6 m.

CENTURY was developed to simulate long-term (decades to centuries) SOM dynamics, plant growth and cycling of N, P and S. It uses a monthly time step with monthly average maximum and minimum temperatures and monthly precipitation data (Parton *et al.*, 1987b; Parton & Rasmussen, 1994). It comprises two forms of litter: Metabolic and Structural, and three SOM compartments: Active, Slow, and Passive. C leaving the Active organic matter compartment is partitioned into either CO_2 or Slow forms with the split determined by soil texture. Soil texture also regulates the rate of transfer between Slow and Passive forms. CENTURY has been used to simulate C accumulation during soil formation (Parton *et al.*, 1987a), and changes in soil C storage following climate change scenarios (Schimel *et al.*, 1994).

DAISY simulates crop production, and dynamics of soil water and nitrogen under diverse agricultural management systems (Hansen, *et al.*, 1991). It was developed as a field management tool as well as for regional administrative purposes, and has been applied to catchment areas (Styczen & Storm, 1993), farmland areas (Jensen *et al.*, 1994a) and specific sites (Jensen *et al* 1994b). DAISY contains a hydrological model with a soil water submodel, a soil nitrogen model with a soil organic matter submodel, and a crop model with a nitrogen uptake submodel. Rate constants are modified by soil clay content, a semi-cohort accounting system is used for litter decay and soil microbial biomass is a dependent variable although concepts of zymogenous and autochthonous microorganisms are included.

DNDC (DeNitrification and DeComposition) couples denitrification and decomposition processes as influenced by the soil environment to predict emissions of CO_2, N_2O and N_2 from agricultural soils (Li *et al.*, 1992a, 1992b). It has been supplemented with a plant growth submodel including subroutines for cropping practices such as fertilization, irrigation, tillage, crop rotation and manure addition to simulate SOM turnover in arable lands (Li *et al.*, 1994). Clay adsorption of humads allows some soil-specificity; decomposition is first order, such that biomass formed during decomposition is a dependent variable.

ITE is the soil submodel of the Hurley Pasture Model and the ITE (Institute of Terrestrial Ecology - Edinburgh) Forest Model. The pasture model aims to simulate N cycling in a grazed soil-plant system, and comprises three submodels (Thornley & Verberne, 1989). They are: i) a grazing animal-intake model in which the C fluxes and N content of faeces and urine are sensitive to the N content of ingested plant material, ii) a vegetative grass growth model which responds to light, temperature and N; and, iii) a soil organic matter submodel that responds to faeces, urine, and decaying plant residues. All decomposition rates are a function of the quantity of microbial biomass. Although some components are mineralized to CO_2 or NH_3 without passing through the biomass, the rate of such mineralization is a function of quantity of biomass. This model has also been combined with a transport-resistance model to describe C and N compartments and fluxes in a plantation forest soil system (Thornley & Cannell, 1992).

NCSOIL simulates N and C flow through soil microbes and organic moieties. It comprises four organic compartments: plant residues, microbial biomass (Pool I), Humads (Pool II) and Stable organic matter (Pool III) (Nicolardot *et al.*, 1994). The original version didn't include stable organic matter (Molina *et al.*, 1983). Flows of C and N are interconnected, and increasing stability of organic matter results from metabolism and not from sorption mechanisms that would be sensitive to clay content. Microbial succession is simulated on residues, and although decomposition rate is independent of microbial biomass, microbial succession leads to more stable materials. In addition to ^{12}C and ^{14}N it simulates ^{14}C and ^{15}N dynamics. NCSOIL has been incorporated into a deterministic model (NCSWAP) of the soil-plant system that simulates interactions of N and C dynamics with crop growth and soil water (Molina & Richards, 1984; Clay *et al.*,, 1985; Lengnick & Fox, 1994).

QSOIL is an example of a noncompartmental cohort decay model. It embodies concepts described in the theoretical analyses of Bosatta & Ågren (1985, 1994) which characterize the quality of each fresh litter cohort by a continuous real variable, q, which determines its decay rate. For each cohort, q varies over time according to a dispersion function representing the reduction in substrate quality (accessibility and ability to promote microbial growth) as C is transformed by soil microorganisms.

ROTHC is the Rothamsted C model in which the turnover of C in aerobic soil is sensitive to soil type, temperature, moisture and plant cover (Jenkinson *et al.*, 1987; Jenkinson, 1990; Coleman & Jenkinson, 1995). Nitrogen and C dynamics are not interconnected in ROTHC, the IOM component is quantified using C-dating, and starting values are obtained by running the model to equilibrium. Unlike other models examined, ROTHC has been used for calculating organic matter inputs to soil (Jenkinson & Coleman, 1994) and net primary productivity (Jenkinson *et al.*, 1992) using soil organic matter and radiocarbon measurements. It is formulated as a discrete sums-of-exponentials that can be transformed into a continuous form which can yield analytical results that may provide additional useful insights into soil organic matter turnover (Parshotam, 1995).

SOMM is described in essence as the raw humus submodel of a single plant ecosystem model (SPECOM) developed for forested ecosystems (Chertov, 1990). It treats the soil litter layers: L, F, and H explicitly. C flow from L to H are governed by biological activity and progressive humification of the residual material. Activities of soil animals are implicit regulators of C flow; distinctions among humus forms such as Mull in contrast to Mor reflect proportions of earthworms in contrast to micro-arthropods etc. This is the only model of the group examined that simulates C accumulation in soil organic horizons explicitly.

VVV (Van Veen/Verberne) aims to simulate N and water balance in a grassland soil-plant system in order to predict yield, N uptake, N leaching, N mineralization and accumulation of soil organic N (Verberne, 1992). The grass growth submodel is based on the model described by Thornley & Verberne (1989) (See ITE above). The soil is treated as a multilayer mineral soil system, and is simulated using three submodels: i) soil water submodel, ii) soil organic matter submodel; and, iii) soil N submodel. The soil organic matter submodel is based on Verberne *et al.*, (1990). Transformations follow first order kinetics, with physical protection caused by soil clay resulting in protected and nonprotected biomass and organic matter. Specific decomposition rates are modified by soil moisture and temperature, but microbial biomass doesn't influence decay rate. Clay protection reduces microbial biomass turnover by a factor of 100. Plant residues are partitioned into three compartments: Decomposable, Structural, and Resistant; there is a Stabilized organic matter component in addition to Protected and Nonprotected active organic matter. In contrast to the other models, this model emphasizes the influence of clay on protection of microorganisms and soil organic components.

Classification Scheme.

A scheme consisting of a partially catenary sequence of attributes was prepared and used to organize information about SOM models (Figure 1). The components of it, and rationale for them is presented below. The scheme and component designations are presented to stimulate discussion about ways to classify such models.

Type of Environmental Conditions	
Dynamic	Static

Scale: Temporal as Surrogate for Spatial		
Hours	Months	Centuries

Vertical Discontinuities		
Homogeneous	Layered: Mineral Only	Layered: Organic Over Mineral

Regulation by Soil Properties	
None	Soil texture / Clay Content

Location of Biotic Component			
None	Above Ground	Below Ground	Both

Litter / SOM Distinction	
Combined	Separate

Litter Compartmentalization					
Biochemical	Biom / Kinetic	Kinetic	Funct./ Morph.	Cohort	None

SOM or SOM + Litter Compartmentalization					
Biochemical	Biom / Kinetic	Kinetic	Funct./ Morph.	Cohort	None
Control					
Process			Agent		

Figure 1 *Scheme for describing SOM models with a view to their classification.*

Model Type

The highest, or most general, level pertains to the nature of the environmental conditions that drive the model. They may vary, such as daily air temperature, incoming radiation, and precipitation, which I designated Dynamic. Alternatively the driving environmental variables may be constant, designated as *Static*. The fundamental assertion in this scheme is that the variability in environmental conditions that an SOM model subsumes should be its most basic attribute. Most models that would fall into the categories described by Jenkinson (1990) as "single homogeneous compartment", or "two-compartment models" and some "non-compartmental decay models" would operate under *Static* environmental conditions. Most "multicompartmental models" accommodate *Dynamic* environmental conditions. Consequently, this level in the classification scheme maps onto the categories recognized by Jenkinson (1990) and groups them into two rather than four types.

Scale

Biotic and abiotic processes structure ecosystems across space and time (Holling, 1992). Consequently both process- and organism-oriented models must consider temporal and spatial scales. I used temporal scale as a surrogate for both temporal and spatial scales. There is some agreement that short time intervals are most appropriate for fine levels of spatial resolution, and that longer time intervals may be valid over coarser spatial scales. SOM models are concerned with change over time, so temporal scale rather than spatial scale was deemed to be the most appropriate single scale to use. An explicit spatial dimension could be added to this analysis, but I doubt it would alter the outcome.

Although division of a continuous variable such as time into discrete units may appear arbitrary, it is perhaps consistent with the discontinuities in frequency of processes at specified temporal and spatial scales that Holling (1992) demonstrated. I chose three units: *Hours, Months*, and *Centuries*. With 720 hours per month, the first and second division are separated by 2 orders of magnitude; and with 1200 months per century, the second two are separated by 3 orders of magnitude. Although, this scheme covers 5 orders of magnitude, restricting each model to one category was difficult, and I fear more arbitrary than I had expected. No model fits any category ideally, and model developers often report that their model will function over at least two and perhaps all three of these scale categories. Holling (1992) distinguished three classes of processes operating over three ranges of scale. The size of the first is centimeters to tens of meters operating at time scales of days to decades. The mesoscale group was hundreds of meters to hundreds of kilometers in size and years to decades in time. The macroscale extreme was hundreds to thousands of kilometers in size and centuries to millennia in time. Post *et al.*, (1995) analyze modelling requirements for global estimations of soil carbon dynamics and use the ROTHC model to calculate transient responses of global SOM to climate change. They conclude that a major challenge remains to capture small scale (< 50 km) variations in topography, drainage, land use, and parent material that regulate soil C dynamics. In addition, one wonders if scale-specific processes regulate SOM dynamics? How can cross-scale analyses be performed rigorously? Are landscape processes additive, such that models of small patches can be summed over space? Further inquiry into the scale-dependence and cross-scale adaptability of models is warranted.

Soil Horizons

Spatial scale usually implies the horizontal plane. Soils, however, exhibit characteristic vertical horizonation. The pedon is an important integrative unit for characterizing agricultural ecosystems (Anderson *et al.*, 1983), and the variation in soil properties with depth influences environmental conditions of soil microorganisms (e.g. moisture content Zhang *et al.*, 1988). Consequently, vertical discontinuities in, or distribution of, soil properties was considered next. This is a spatial attribute, but was independent of the time scale above it. Three categories covered the range of SOM models: *Homogeneous*, no layering; *Layered Mineral Only*, mineral horizons over mineral horizons; or *Layered Organic Over Mineral*, soils with organic horizons overlying one or more mineral horizon.

Regulation by Soil Properties

Abiotic soil properties, regulate biotic behavior. Inclusion of soil properties might increase the universality of a model by accommodating regional or site-specific soil properties. The range in properties used to regulate models remains small, so two categories were recognized: None, or *Soil Texture* (clay content or sand content were used as surrogates for texture). Is soil texture a fundamental property, or is it a surrogate for soil structure?

Paustian (1994) emphasized the need to relate soil structure to turnover of SOM in models. Soil structure influences organism-substrate contact, habitable space, water relations, air relations and hence the environmental driving variables in many SOM models. Explicit treatment of soil structure either as a static or dynamic variable was absent from all models, although the VVV model includes clay protection of biomass and SOM. Inclusion of physical fractions (Christensen, 1992, 1995b) may connect pools in SOM models with empirical evidence (Christensen, 1995a). It isn't sufficient, however, to obtain experimentally measurable fractions; quantitative relationships between measurable components and rates of turnover must also be present consistently and be measurable.

Another aspect of long term SOM models that was ignored, but which might require further inquiry is feedback between soil properties and model behavior. Soil development means change. Models now recognize the quantitative changes in soil over time, but soils also change qualitatively. These changes may involve vertical redistribution of materials, soil fabric reorganization and structural change all of which help regulate soil C dynamics.

Biotic Component

A SOM model may be first order with organisms treated as catalysts, it may designate organisms but retain a first order treatment with organisms as generic biomass or living organic matter that doesn't influence the dynamics of other organic moieties, or organisms may be treated as responsible for transformations of SOM. The categories are: None, no biotic component; *Above Ground*, only higher plants; *Below Ground*, soil microbial biomass; or *Both*, in which higher plants and soil biomass were recognized.

Although some SOM models separated soil microbial biomass into two or more functional groups, none included explicit food web or soil faunal components. Consequently none would be considered organism-oriented by the criteria presented by Paustian (1994). The model SOMM, however, bases some distinctions on implied activities of soil fauna. Further, although biomass was a frequent "pool" in these models, its treatment was often indistinguishable from active forms of SOM. One might consider inclusion of soil biomass in this way to be tokenism.

From the perspective of SOM dynamics, however, treatment of soil biomass as a form of SOM may be justified. Consider the problem of identifying the kinetically homogeneous components in soil. Microbial biomass may be one such component. It can be measured and its accumulation or depletion related to simulated results. From this perspective it adds an identifiable soil component in the continuum from active to very resistant or inert SOM and serves the purpose of many models well. To restrict microbial biomass to a measurable SOM component, however, renders models nonmechanistic and removes the possibility that a model might simulate changes in SOM dynamics as a result of changes in activity or characteristics of soil organisms. Contaminated soils come to mind here.

Interactions between plant roots and soil are important for nutrient and energy exchange and for soil structural reorganization. The models examined don't simulate these interactions. Although the location of plant residue addition to soil is important, it could be accommodated only in models that treat soil as a layered system. This attribute of models was not given a separate category, although future classifications might do so.

Litter / SOM Distinction

Litter and SOM are qualitatively different with distinct dynamics. As litter decomposes and becomes transformed in soil its turnover rate slows. Litter and SOM often occupy unique locations and in some cases litter is a habitat as well as a substrate (McGill *et al.*, 1981; McGill & Myers, 1987). But just as soil organisms may be viewed as an active form of SOM, litter may also be seen as a recent and active version of SOM. Consequently, the diverse conceptualizations of SOM dynamics lead to possibilities to treat litter and SOM as discrete or as part of the same continuum. The two categories are: Combined; or *Separate*.

Litter and SOM Compartmentalization

The same range of categories was used to describe approaches to compartmentalizing litter and SOM. Using the same suite of categories allows differences in perceptions of litter and SOM to be discerned. Although diversity characterizes compartmentalization criteria, there is evidence of some convergence. The categories selected represent the complete spectrum of approaches. Six categories are used. Kinetic, to recognize the diversity of turnover rates found in soils. Kinetically homogeneous units may be biologically and chemically dissimilar but have similar accessibility in soil, or chemically diverse components which are associated with each other may not be metabolized independently. Components that are distinguished by their turnover rate (*Kinetic*) are difficult to separate analytically and such compartmentalization leads to potential ambiguity in the empirical evidence about a soil. *Biochemical* separations recognize the discrete chemical components of litter and to some degree SOM. They can be analytically separated and their dynamics measured unambiguously. Biochemical compartmentalization, however is valid only of the discrete chemicals behave independently in soil. Mixed systems are common, and were called *Biomass / Kinetic*, because they include biomass as a unit of compartmentalization, but the biomass is defined on the basis of its turnover rate. They are intermediate between the kinetic separations above and *Functional / Morphological* separations that are also used. Functional separations recognize a connection between the properties that define the biological function of a plant or microbial component (e.g. cell wall) and the properties governing its fate in soil. Recognition of litter layers was also considered to fit into this category. In recognition of the variety of ways SOM and litter may be compartmentalized, the diverse kinetics associated with them, and the inability to determine which is best, it may be appropriate to consider litter and SOM as a continuum and treat each addition of litter as a cohort that decomposes to progressively more stable products over time. I called this the *Cohort* system. Finally, no compartmentalization may have been employed, in which case the designation was *None*.

Control

Environmental driving variables control SOM dynamics either at the Process level by altering first order rate constants for the turnover of each component, or by altering the biological activity of *Agents* responsible for the pertinent processes. Usually the agents are organisms. Using these two designations leads to segregation of models along mechanistic lines, such that direct regulation of a process may require fewer parameters and assumptions about habitats and activities of soil organisms, but may be less mechanistic and potentially less universal.

Overview of Model Questionnaire Data and Model Attributes

Questionnaires were returned by model developers to organizers of the NATO workshop. The questionnaires sought information about model scale, input values, output, structure, environmental controls, evaluation and sensitivity testing, model use and accessibility. The characteristics recorded for each model are presented in Tables 1 through 4, and reflect my assessment of information in the questionnaires and in publications about each model. Every effort was made to avoid misrepresenting any model.

Table 1 *Basic characteristics of ten SOM models.*

	Type[1]		Scale			Reg'n	Vertical Distribution		
Model	Dynamic	Static	Hours	Months	Centuries	Texture	Homo.	Min/Min	Org/Min
CANDY	X			X		X		X	
CENTURY	X				X	X		X	
DAISY	X			X		X		X	
DNDC	X			X		X	X		
ITE	X			X			X		
NCSOIL		X	X				X		
Q-SOIL		X			X		X		
ROTHC	X				X	X	X		
SOMM	X			X					X
VVV	X			X		X		X	
Total =	8	2	1	6	3	6	5	4	1

1 Terms are defined under Classification Scheme

Relationships Among Attributes

The distribution of basic properties among models is displayed in Figure 2. Most are of the dynamic type with scales of months (6) or centuries (3); only one model appeared to have been used much at hourly time scales. Inclusion of soil texture as a control appeared in 6 of the 10 models. Perhaps there is a trend among model developers to accommodate soil and regional variation in models. Such a development is critical if SOM models are to be universal and are to make a contribution to understanding of soils in a global environment. It will be interesting to discover how successful is the use of clay or texture in this regard, and how well soil structural dynamics can be incorporated into future models. Soil texture was used both to protect or slow decomposition of soil organic components, and to regulate partitioning of C among pools. These contrasting approaches warrant some discussion. Most models (5) still represent a homogeneous soil unit, but 4 represent soils as layered mineral and one included organic horizons over a mineral horizon. In only 2 cases did litter and SOM appear to be treated as a combined unit.

Table 2 *Biological attributes and relation of litter to SOM in ten models of SOM dynamics.*

	Biotic Component[1]				Litter/SOM Distinction	
Model	None	Above	Below	Both	Combined	Separate
CANDY		X				X
CENTURY				X		X
DAISY				X		X
DNDC				X		X
ITE				X		X
NCSOIL			X			X
Q-SOIL			X		X	
ROTHC			X			X
SOMM				X	X	
VVV				X		X
Total =	0	1	3	6	2	8

[1] Terms are defined under Classification Scheme

From Figure 3 it appears that concepts pertaining to the nature of litter are different from those pertaining to SOM. Litter was treated as kinetically separable units in 4 cases, and functional or morphological separations were made in 4 models. In no case did a biomass component enter the compartmentalization of litter. In contrast, biomass/kinetic mixtures appeared in 6 models and purely kinetic compartmentalization in only 1. Many models in the past have relied on biochemical fractions to distribute C and N among litter and SOM components. Although biochemical concepts are implicit in these models, in no case were biochemical separations applied to litter or SOM. In most cases C/N ratio was used to define fractions.

An inert organic matter component with no turnover rate was included in 3 models, and there may be a trend towards the use of such a strategy to deal with very persistent organic material. Although less intellectually satisfying than using a specific low turnover rate, it appears to work and makes no assumptions about the quality of estimates of such slow rates. Most controls were on the process (9), reflecting the empirical nature of the SOM models examined.

Table 3 *Litter and SOM compartmentalization in ten SOM models.*

Model	Litter comparments	SOM or SOM + Litter Compartments
CANDY	Semi-Cohort	Kinetic (EF)
CENTURY	Functional (EF-LN)	Biomass/Kinetic (DM-Cl+S)
DAISY	Functional & Kinetic (EF-CN)	Biomass/Kinetic (EF-CN)
DNDC	Kinetic (EF-CN)	Biomass/Kinetic (EF)
ITE	Functional (EF-CN)	Biomass/Kinetic (EF-CN)
NCSOIL	Kinetic (EF-CN)	Biomass/Kinetic (EF-CN & No)
QSOIL	Cohort (EF)	
ROTHC	Kinetic (EF)	Biomass/Kinetic (EF)
SOMM	Functional (EF-N & Ash)	Morphological (EF-N & Ash)
VVV	Functional (EF-CN)	Biomass/Kinetic (EF-Clay)

Ash = ash, CN = C/N ratio, EF = Empirical fitting, LN = Lignin / N ratio, No = N mineralization potential (Stanford & Smith), DM = Direct Measurement, Cl+S = Clay & Sand

Relationships Among Models

An effort was made to seek the connections among models and to represent them visually (Figure 4 and 5). It appears that few attributes are needed to distinguish among models. Many have shared elements, and there is evidence of convergence among some. NCSOIL and QSOIL fall together because of the static environmental conditions (Figure 4), but with NCSOIL incorporated into NCSWAP, it falls into the larger group. Of the rest, CANDY and DAISY appear to have the greatest number of shared components including an IOM pool (Figure 5), and there is considerable concordance among all models on the right hand side of Figure 5. ITE shares many conceptual similarities with VVV. DNDC and ROTHC fall close together, but they differ in the most appropriate time scale in my estimation.

Table 4 *Inert organic matter (IOM) and controls on C and N flows.*

	IOM	Control	
Model	Pool	Process	Agent
CANDY	X	X	
CENTURY		X	
DAISY	X	X	
DNDC		X	
ITE			X
NCSOIL		X	
Q-SOIL		X	
ROTHC	X	X	
SOMM		X	
VVV		X	
Total =	3	9	1

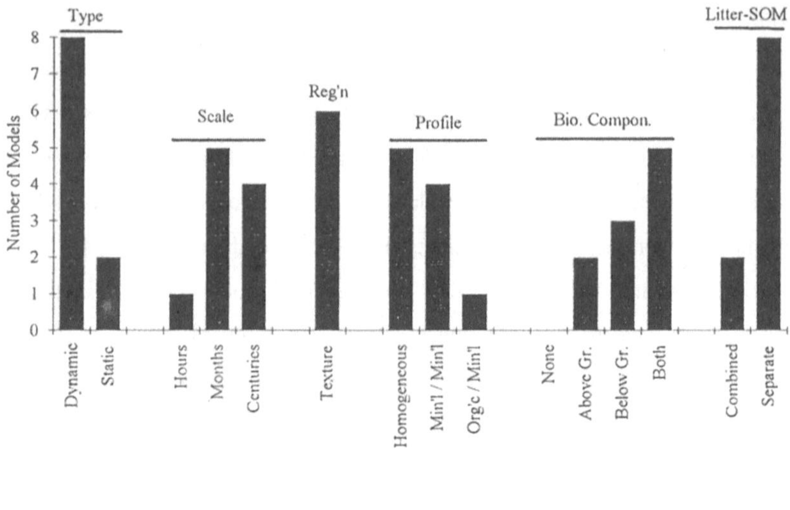

Figure 2 *Distribution of basic characteristics among ten SOM models.*

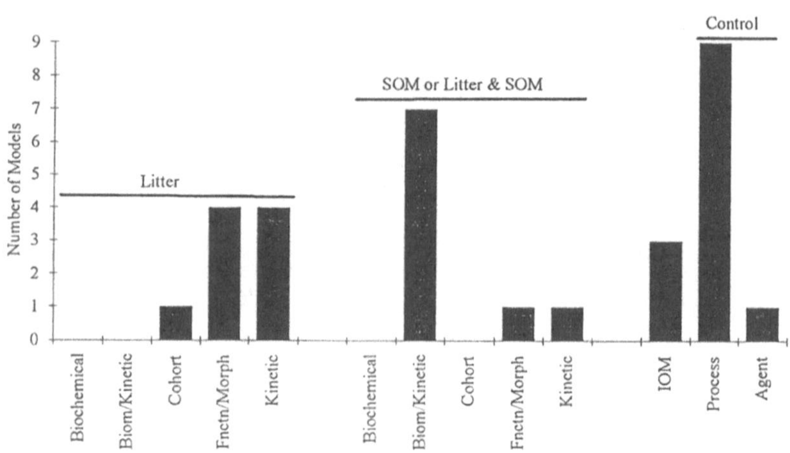

Figure 3 *Litter and SOM compartmentalization, and control on organic matter dynamics.*

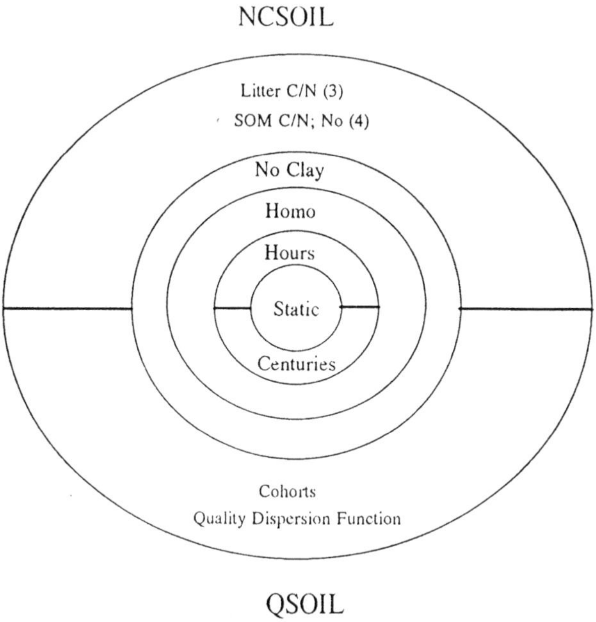

Figure 4 *Schematic representation of relationships among two SOM models operating under constant environmental conditions.*

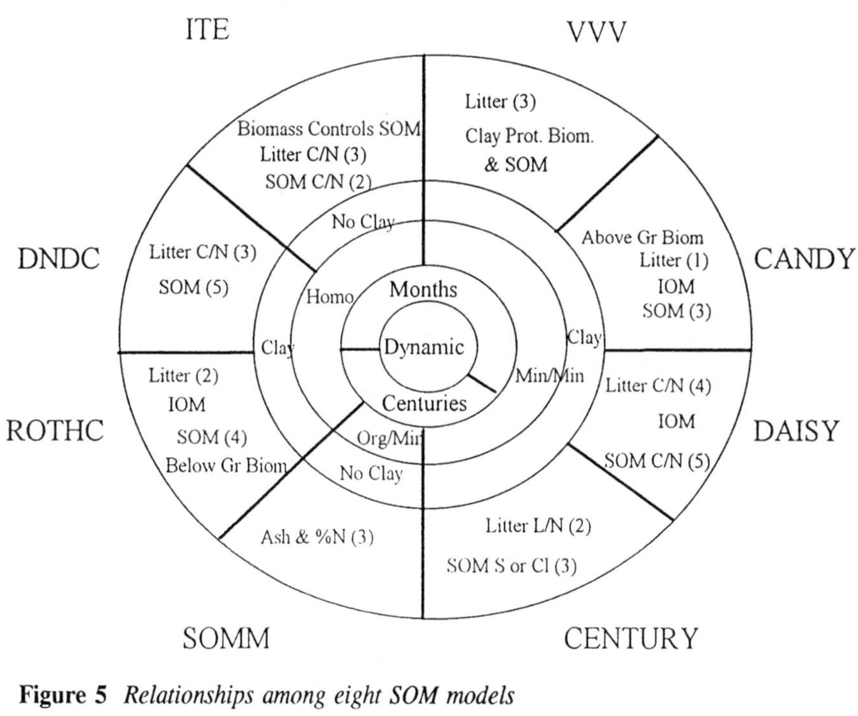

Figure 5 *Relationships among eight SOM models*

References

Anderson, D.W., Heil, R.D., Cole, C.V., & Deutsch, P.C (1983) Identification and characterization of agricultural ecosystems at different integrative levels. p 517-531 In: *Nutrient cycling in agricultural ecosystems.* R. Lowrance, R. Todd, L. Ausmussen, & R Leonard, (eds.) Special Publication 23, University of Georgia, College of Agriculture experiment Station, Athens, GA.

Bartholomew, W.V & Kirkham, D (1960) Mathematical descriptions and interpretation of culture induced soil nitrogen changes. *Trans. VII International Congress of Soil Science,* **2**:471-477.

Bosatta, E & Ågren, G.L (1994) Theoretical analysis of microbial biomass dynamics in soils. *Soil Biol. & Biochem.* **26**:143-148.

Bosatta, E & Ågren, G.L (1985) Theoretical analysis of decomposition of heterogeneous substrates. *Soil Biol. & Biochem.* **16**:63-67.

Chertov, O.G (1990) SPECOM - A single tree model of pine stand / raw humus soil ecosystem. *Ecological Modelling* **50**:107-132.

Christensen, B.T (1992) Physical fractionation of soil and organic matter in primary particle size and density separates. *Adv. Soil Sci.* **20**:1-90.

Christensen, B.T (1995a) Matching experimentally verifiable SOM fractions with SOM pools in simulation models of SOM turnover. Proceeding, NATO Advanced Research Workshop: *Evaluation of soil organic matter models using long-term data sets.* (This Volume).

Christensen, B.T (1995b) Carbon in primary and secondary organomineral complexes. *Adv. Soil Sci.* (In Press).

Clay, D.E., Clapp, C.E., Molina J.A.E., & Linden, D.R (1985) Nitrogen-tillage-residue management: I. Simulating soil and plant behavior by the model NCSWAP. *Plant & Soil* **84**:67-77.

Coleman, K. & Jenkinson, D.S (1995) ROTHC-26.3: *A model for the turnover of carbon in soil.* IACR Rothamsted, Harpenden, Herts.

Franko, U. (1995) Simulation of nitrogen dynamics in agroecosystems. p 153-163 In: *Modelling the fate of agrochemicals and fertilizers in the environment.* C Giupponi, A. Marani, & F. Morari (eds.) European Society for Agronomy.

Franko, U., Oelschlägel, B., & Schenk, S (1995) Simulation of temperature-, water-, and nitrogen-dynamics using the model CANDY. *Ecological Modelling,* **81**:213-222

Hansen, S., Jensen, H.E., Nielsen, N.E. & Svendsen, H (1991) Simulation of nitrogen dynamics and biomass production in winter wheat using the Danish simulation model DAISY. *Fert. Res.* **27**:245-259.

Hansen, S., Jensen, H.E., Nielsen, N.E. & Svendsen, H (1990) DAISY: *A soil plant system model. Danish simulation model for transformation and transport of energy and matter in the soil plant atmosphere system.* 360 pp. The National Agency for Environmental Protection, Copenhagen.

Hénin, S., & Dupuis, M (1945) Essai de bilan de la matiere organique de sol. *Ann Agron.* **15**: 17-29.

Holling, C.S (1992) Cross-scale morphology, geometry, and dynamics of ecosystems. *Ecological Monographs* **62**:447-502.

Jenkinson, D.S. & Coleman, K (1994) Calculating the annual input of organic matter to soil from measurements of total organic carbon and radiocarbon. *European J. of Soil Sci.* **45**:167-174.

Jenkinson, D.S., Harkness, D.D., Vance, E.D., Adams, D.E. & Harrison, A.F (1992) calculating net primary production and annual input of organic matter to soil from the amount and radiocarbon content of soil organic matter. *Soil Biol. & Biochem.* **24**:295-308.

Jenkinson D.S., Hart P.B.S., Rayner J.H. & Parry L.C (1987) Modelling the turnover of organic matter in long-term experiments. *INTECOL Bulletin* **15**:1-8

Jenkinson, D.S (1990) The turnover of organic carbon and nitrogen in soil. *Phil. Trans. Royal Soc. London B* **329**:361-368.

Jenny, H (1941) *Factors of soil formation.* McGraw-Hill, New York.

Jensen, C., Stougaard, B. & Østergaard, H.S (1994) Simulation of nitrogen dynamics in farmland areas of Denmark (1989 - 1993). *Soil Use & Management* **10**:111-118.

Jensen, C., Stougaard, B. & Olsen, P (1994b) Simulation of water and nitrogen dynamics at three Danish locations by use of the DAISY model. *Acta Agric. Scand., Sect. B* **44**:75-83.

Lengnick, L.L. & Fox, R.H (1994) Simulation by NCSWAP of seasonal nitrogen dynamics in corn: I. Soil nitrate. *Agronomy J.* **86**:167-175.

Li, C., Folkring, S., & Folkring, T.A (1992a) A model of nitrous oxide evolution from soil driven by rainfall events: 1. Model structure and sensitivity. *J. of Geophysical Res.* **97**:9759-9776.

Li, C., Folkring, S., & Folkring, T.A (1992b) A model of nitrous oxide evolution from soil driven by rainfall events: 2. Model applications. *J. of Geophysical Res.* **97**:9777-9783

Li, C., Folkring, S., & Harriss, R (1994) Modelling carbon biogeochemistry in agricultural soils. *Global Biogeochemical Cycles* **8**:237-254.

Mathers, A.S & Goss, D.W (1979) Estimating animal waste application to supply crop nitrogen requirements. *Soil Sci. Soc. Amer. J.* **43**:364-366.

McGill, W.B., Hunt, H.W., Woodmansee, R.G., & Reuss, J.O (1981) PHOENIX: A model of the dynamics of carbon and nitrogen in grassland soils. In FE Clark and T Rosswall, eds, *Terrestrial nitrogen cycles: processes, ecosystem, strategy and management impacts. Ecol Bull (Stockholm)* **33**: 49-115.

McGill, W.B. & Myers, R.J.K (1987) Controls on dynamics of soil and fertilizer nitrogen p 73-99, In: *Soil fertility and organic matter as critical components of production systems.* Soil Sci. Soc. Amer. Spec. Pub. no. 19. Madison WI

Molina, J.A.E. & Richards, K (1984) *Simulation model of the nitrogen and carbon cycle in soil-water-plant system, NCSWAP; guide for the preparation of input data files and execution of NCSWAP.* Soil Ser. 116. Dept. Soil Sci. University of Minnesota, St. Paul.

Molina, J.A.E., Clapp, C.E., Shaffer, M.J., Chichester, F.W. & Larson, W.E (1983) NCSOIL, a model of nitrogen and carbon transformations in soil: Description, calibration, and behavior. *Soil Sci. Soc. Amer. J.* **47**:85-91.

Newton, J.D., Wyatt, F.A., & Brown, A.L (1945) Effects of cultivation and cropping on the chemical composition of some western Canada prairie province soils. Part III. *Scientific Agriculture* **25**:718-737.

Nicolardot, B., Molina, J.A.E. & Allard, M.R (1994) C and N fluxes between pools of soil organic matter: Model calibration with long-term incubation data. *Soil Biol. & Biochem.* **26**:235-243.

Parshotam, A (1995) The Rothamsted soil-carbon turnover model - discrete to continuous form. *Ecological Modelling* (In Press).

Parton, W.J., Schimel, D.S., Cole, C.V., & Ojima, D.S (1987a) Analysis of factors controlling soil organic matter levels in great plains grasslands. *Soil Sci. Soc. Amer. J.* **51**:1173-1179.

Parton, W.J., Stewart, J.W.B., & Cole, C.V (1987b) Dynamics of C, N, S, and P in grassland soils: A model. *Biogeochemistry* **5**:109-131.

Parton, W.J. & Rasmussen, P.E (1994) Long-term effects of crop management in wheat-fallow: II. CENTURY model simulations. *Soil Sci. Soc. Amer. J.* **58**:530-536.

Paustian, K (1994) Modelling soil biology and biochemical processes for sustainable agriculture research. p 182-193, In: *Soil biota management in sustainable farming systems.* ZE Pankhurst, BM Doube, VVSR Gupta, & PR Grace (eds.) CSIRO Information Services, Melbourne.

Post, W.M., King, A.W., & Wullschleger, S.D (1995) Soil organic matter models and global estimates of soil organic carbon. Proceeding, NATO Advanced Research Workshop: *Evaluation of soil organic matter models using long-term data sets.* (This Volume).

Powers, W.L., Wallingford, G.W., & Murphy, L.S (1975) Formulas for applying organic wastes to land. *J Soil & Water Conservation* **30**:286-289.

Russell, J.S (1960) Soil fertility changes in long-term experimental plots at Kybybolite, South Australia. I. Changes in pH, total nitrogen, organic carbon and bulk density. *Australian J. Agric. Res.* **11**:902-626

Russell, J.S (1975) A mathematical treatment of the effect of cropping system on soil organic nitrogen in two long term sequential experiments. *Soil Science*, **120**:37-44.

Schimel, D.S., Braswell, B.H., Holland, E.A., McKeown, R., Ojima, D.S., Painter, T.H., Parton, W.J., & Townsend, A.R (1994) Climatic, edaphic, and biotic controls over storage and turnover of carbon in soils. *Global Biogeochemical Cycles* **8**:279-293.

Styczen, M. & Storm, B (1993) Modelling of N-movement on catchment scale - a tool for analysis and decision making. I. Model description. *Fert. Res.* **36**:1-6.

Thornley, J.H.M & Cannell, M.G.R (1992) Nitrogen relations in a forest plantation - Soil organic matter ecosystem model. *Annals of Botany* **70**:137-151.

Thornley, J.H.M & Verberne E.L.J (1989) A model of nitrogen flows in grassland. *Plant Cell and Environment* **12**:863-886.

Verberne, E.L.J., Hassink, J., De Willigan, P., Groot, J.J.R. & Van Veen, J.A (1990) Modelling organic matter dynamics in different soils. *Netherlands J. of Agric. Sci.* **38**:221-238.

Verberne, Els (1992) *Simulation of nitrogen and water balance in a system of grassland and soil.* DLO-Institut voor Bodemvruchtbaarheid, Osterweg 92, Postbus 30003, 9750 RA Haren, Netherlands 56 pp + Appendices.

Woodruff, C.M (1949) Estimating the nitrogen delivery of soil form the organic matter determination as reflected by Sanborn field. *Soil Sci. Soc. Amer. Proc.* **14**:208-212.

Zhang, T. & Berndtsson, R (1988) Temporal patterns and spatial scale of soil water variability in a small humid catchment. *J. of Hydrology.* **104**:111-128.

Acknowledgements

I am indebted to the model developers who provided detailed information about their models; every effort has been made to be faithful to the information provided. NATO provided funding to attend this Advanced Research Workshop. Reflection on these ideas has been informed by discussions with colleagues in the Department of Renewable Resources and in the Soil Science community at large. Although I accept responsibility for what I have written, I am indebted to Dr RC Izaurralde and Dr NG Juma for suggestions and valuable comments on an earlier version of this manuscript; to colleagues who participated in the Advanced Research Workshop in Rothamsted, May 1995, for their insights; and to: VL Bailey, CC Gomez, SB González, DC Jans-Hammermeister, RA Janzen, SM Luther, JL Offord, KM Ower, JL Roy PM Rutherford, & MJ Salloum for their honest comments, thoughtful suggestions and lively supportive environment.

Ecosystem model comparisons: science or fantasy world?

William J. Parton
Natural Resource Ecology Laboratory
Colorado State University
Fort Collins, CO 80523
USA

Summary

Ecosystem models have been used extensively to evaluate the impact of environmental change and management practices on ecosystem dynamics. Recent results from model comparison activities suggest there are substantial differences among models in predicted ecosystem impacts of altered climatic change scenarios in spite of the fact that the simulated model results are similar for current climatic conditions. These results suggest that there is substantial uncertainty about the ability of ecosystem models to simulate the effect of environmental changes on ecosystem dynamics and that it is risky to use results from only one ecosystem model. The results also suggest that there is substantial uncertainty about the combined impact of changing climate and atmospheric CO_2 level on ecosystem dynamics, and about the interactive impact of nutrient availability and environmental stress on plant growth. The existing model comparisons have been very effective at highlighting scientific uncertainty and suggest critical research needed to enhance our ability to simulate ecosystem response to environmental changes.

A major difficulty with current model comparisons is the lack of a common modelling framework for coding ecosystem models. Models need to be written using a modular format that would allow the substitution of submodels from one model into another model. A common modular framework for developing models would promote analysis of conceptual differences among models and determine the consequences of these differences. A

NATO ASI Series, Vol. I 38
Evaluation of Soil Organic Matter Models
Edited by David S. Powlson, Pete Smith and Jo U. Smith

conceptual framework for modularization of models needs to be formulated by the ecosystem modelling community.

Introduction

A large number of ecosystem models have been developed during the last 25 years for grasslands, savannas, tundra and forest systems. The initial ecosystem modelling work was promoted by the International Biological Program (IBP) and resulted in the development of models for the Grassland Biome, Tundra Biome, Coniferous Forest Biome, Desert Biome, and Deciduous Forest Biome. These models were primarily developed as tools for improving our scientific understanding about how the ecosystems work. Ecosystem models have primarily been used for a limited number of sites, however, with recent advances in modern computers and software, ecosystem models are now being used at regional and global scales. Many of the current ecosystem models are being used as management tools for evaluating the impact of different agroecosystem management practices (i.e., grazing levels, fertilizer and water additions, forest cutting practices, and management of cultivated land) and evaluating the potential impact of environmental and climatic change scenarios. The use of ecosystem models for environmental problem solving reflects the improvement in our understanding of how ecosystems operate and advances in the speed of computers and interactive computer graphics. A general review of the use of ecosystem models is presented by Ågren *et al.* (1991), Hansen *et al.* (1985) and Parton *et al.* (1994).

Until recently there has been very little interest in comparison of different ecosystem models and most new ecosystem models were developed without using components from existing models. The modelling community was more interested in developing new models than comparing and evaluating existing models. There was concern that model comparisons would stifle scientific originality and was unnecessary since most models were primarily used by the model developer. Funding agencies showed very little interest in supporting formal model comparisons. However, during the last five years there has been substantial interest in ecosystem model comparison. This is a result of recent advances in computer technology which makes it easier for non-modelers to use ecosystem models, an interest by funding agencies to evaluate whether new models need to be developed, and an expanded interest in using ecosystem models to evaluate and solve environmental problems. Most current ecosystem model review articles generally compared the structure of the different models, driving variables for the models, objectives of the models, and the theoretical basis for the models. This type of comparison is useful, however, it does not prove a basis for quantifying the relative strengths and weakness of models, provide any quantitative comparison of simulated results from different models or compare the ability of models to simulate observed data from common sites.

Recently there have been a number of research activities started to compare ecosystem models. The goals of these model comparison activities are quite different and include: (1) calculating error bars for climatic change assessment, (2) selection of a specific model for use in environmental assessment, and (3) advancing scientific understanding about ecosystems. These model comparisons were specifically designed to quantify differences in the models, and attempt to understand what caused differences in the model behavior. This paper will review the results from model comparison activities and suggest how model comparison can be improved. The discussion will be divided into three sections which include: (1) model comparisons of climatic change assessment models, (2) model comparisons for improving scientific understanding, and (3) selection of specific models for impact assessment.

Climatic change assessment model comparisons

During the last ten years there has been substantial interest in using ecosystem models to simulate the impact of climatic change scenarios. Most of the work has focused on using a specific ecosystem model to evaluate the effect of climatic change scenarios on ecosystem properties (Ågren *et al.*, 1991). This section will present results from two model comparison activities where different ecosystem models used the same climatic change scenarios and results from the models were compared. These comparisons were attempting to identify the common responses of the models to environmental change and potential differences between the models. Differences reveal lack of understanding of fundamental ecological processes and limits in our ability to predict response of ecosystems to potential environmental changes (e.g., climatic change, land use changes and increased atmospheric CO_2). The two major recent ecosystem model comparisons include the VEMAP (VEMAP, in press) and forest model (Ryan *et al.*, in press a, b) comparisons. Both of these model comparison activities simulated the response of ecosystem models to changes in the climate, atmospheric CO_2 levels and the combined impact of altered climate and atmospheric CO_2 levels.

The forest model comparison (Ryan *et al.*, in press b) was designed to compare the response of eight forest growth models to environmental change scenarios. The models were set up to simulate forest dynamics for coniferous forest sites in Sweden and Australia. The models were parameterized for both sites using observed soils and plant growth data from control, irrigated, fertilized and irrigated and fertilized treatments at the two sites. A comparison of the simulation model results with observed biomass and plant production data showed that the model did a reasonable job simulating the impact of the different treatments on ecosystem dynamics (Ryan *et al.*, in press a) however, there were substantial differences between observed data and model results for some of the treatments. Note that this comparison is a model verification since the observed data was used to parameterize the models. The climatic change scenarios that were evaluated include increasing air temperature by $+4^0$C, increasing the atmospheric CO_2 level (increased to 720 ppm), and the combined impact of increasing atmospheric CO_2 and air temperature.

Results from model comparisons showed that large uncertainties exist on how the models responded to changes in temperature and atmospheric CO_2. The largest uncertainties exist for the interactions between plant and soil systems and the long-term response to increased atmospheric CO_2. Most of the models suggested that increasing atmospheric CO_2 increased net primary production and net carbon storage. The major disagreement between model predictions for the increase in air temperature resulted from the fact that some of the models suggested that increased temperatures would result in increased nutrient mineralization and that the increased nutrient availability would result in increased plant production, while other models suggested that increased air temperatures would increase drought stress and reduce plant production and storage of C in the tree boles. The disagreement in model predictions about the impact of atmospheric CO_2 was a result of uncertainties about the interactions with nutrient supply and long-term effect of increased CO_2. The results suggest the need to consider both carbon and nutrient cycling when simulating ecosystem response to environmental changes and that results will be affected by the age of the forest stand (e.g., aggrading *vs.* mature stand).

The VEMAP model comparison (VEMAP, in press) considered three biogeography models (MAPSS, BIOME2, and DOLY) and three biogeochemistry models (CENTURY, TEM, and BIOM-BGC). The models were all compared under contemporary conditions for sensitivity to increased CO_2 and climatic change. The model comparisons also simulated the linked response of the biogeochemistry and biogeography models by using the biogeography model outputs as input into the biogeochemistry models. The models simulated dynamics of all of the ecosystems in the United States at 0.5 degree resolution. The models used the same data bases for current climate, soils, vegetation and climatic change scenarios. The inputs required by the models were different and ranged from daily to monthly weather data, different climatic and soil input variables and different representations of the driving variables. Daily and monthly climatic data sets were generated for the current climate and three atmospheric general circulation models (GCM). All of the models used the same vegetation classification which allows the biogeochemistry models to be linked to vegetation redistribution models. The major soil input variables required by the models include soil texture and depth. The climatic change scenarios were based on three GCM coupled-CO_2 equilibrium climate experiments. The three climatic change scenarios were selected to represent the range of climate sensitivity over the United States. The output variables that were considered for the biogeochemistry models include net primary production, carbon in living vegetation, soil organic C, actual evapotranspiration and net N mineralization.

The comparison of the biogeography models for the contemporary climate shows that all of the models agree for the geographic distribution of the major vegetation types. The major differences between models are the boundary between the C3 and C4 grasslands, the location of ecotone between the central grassland and eastern forests, and the location of tropical forest biome in Florida. The sensitivity of the models to climatic change differ between models and, in general, the models diverged most for the extreme climatic change scenarios. All of the models showed some response to increased CO_2 and the combined impact of changing the climate and increasing CO_2 was to reduce the divergence between the models by reducing the drought effects. The major fundamental differences among the biogeography models is the sensitivity to drought stress with BIOME2 being less sensitive than MAPSS and DOLY.

The contemporary climate comparison of the biogeochemistry models for the ecosystem level predictions of NPP and total storage of C showed that there was substantial agreement among the models. A comparison of the NPP and carbon storage at the individual grid cells also show that the models are highly correlated. Comparison of the models results for the climatic change scenarios showed that for CENTURY and TEM, increasing temperature resulted in increased decomposition, N availability and higher NPP. For BIOME-BGC, increasing temperature resulted in increased drought stress, decreased NPP and decreased total C storage. All three of the models suggested that increasing atmospheric CO_2 would result in increased NPP and total C storage. The results for BIOME-BGC and CENTURY showed that the combined effect of CO_2 and climate change impacts tended to additive, while the TEM results showed a significant iterative term. The model results for the coupled biogeochemistry and biogeography models showed that the biogeochemistry models were sensitive to changes in the biome distributions. The most dramatic changes occurred in the southeastern U.S.

The paper describing the VEMAP model comparison (VEMAP, in press) identified four areas of research which require attention: modularization of models; reduction of uncertainties regarding key processes; validation of models; and development of models of transient ecological responses. Modularization of models was identified because of difficulty in pinpointing the important conceptual differences among the models and exploring the consequences of these differences. If the models were coded using a more modular structure, then it would be possible to replace computer code of key ecological processes with common submodels and evaluate the importance of different model formulations. The VEMAP model comparison identified major differences in the way models represent the effect of atmospheric CO_2 on nutrient and water use, allocation of N and C to different plant parts, and coupling of carbon and nitrogen storage over the long-term. Research in these areas needs to be promoted if we are going to improve our ability to simulate the impact of environmental change on natural ecosystems. Validation of the models was identified because of the difficulty in comparing model outputs at the regional scale and being able to differentiate which model behavior was most reasonable. The VEMAP paper also concluded that it was necessary to develop models that incorporate transient dynamics. These types of models will allow the simulation of real time predictions of changes in vegetation distribution and biogeochemical output variables.

Model comparison for scientific understanding

This section will review results from two model comparison activities where the goal was to compare models for specific ecosystem processes. The decomposition model comparison (DECO) was set up to evaluate litter decomposition models, while the RICE/PILPS (Shao and Henderson-Sellers, in press) model comparison was designed to compare soil moisture models used in land-surface schemes. The results from the ecosystem model comparisons (see previous section) show that, in general, it is very difficult to determine why ecosystem models differ in their response to environmental changes. The ecosystem models are quite complex and use different submodels for the abiotic (heat and water flows), plant production and nutrient cycling submodels, and thus make it very difficult to identify what features of the ecosystem models are responsible for differences among models. Process-oriented model comparisons are designed to focus attention on specific processes and evaluate the advantage and disadvantage of different model formulations.

The RICE/PILPS model comparison (Shao and Henderson-Sellers, in press) was designed to evaluate soil moisture models used in ecological, hydrological and atmospheric models, and to improve our understanding of the processes included in hydrological models. The model comparison included 14 land-surface schemes. The models used the same land-surface parameters and atmospheric driving variables, ran the same numerical experiments, and compared the simulated results with observed soil water, latent and sensible heat flux data for the HAPEX site (Goutorbe and Tarrieu, 1991). The observed data was used to parameterize the models for the HAPEX site. The model comparison shows that there are substantial differences in the simulated seasonal patterns of soil water, annual cumulative evapotranspiration and total runoff plus drainage. Comparison of the simulated results with the observed data show that the land-surface schemes can be tuned to the observed data but no single scheme simulated all of the land-surface hydrology variables well. The greatest differences in the simulated patterns of soil water were during the summer months when transpiration occurred. The allocation of precipitation into drainage, soil storage and evaporation were substantially different for the land-surface schemes. Model experiments also suggested that the simulated differences in soil water and temperature among the models would greatly impact ecosystem models (Parton *et al.*, in press).

One of the major limitations of the RICE/PILPS model comparison is that many of the important biological land-surface parameters (e.g., leaf area index, seasonal plant biomass, vegetation cover, canopy roughness length, etc.) were not measured for the HAPEX experiment and had to be estimated. Errors in some of the these estimated parameters could contribute to some of the differences between the observed and simulated data. The model comparisons would have been much more definitive if sensible and latent heat fluxes had been measured for the whole year (measured values were only available for a 40-day period). A more detailed process-oriented hydrological data set is needed to determine the relative strength and weakness of the different land-surface schemes.

The DECO model comparison was specifically designed to compare the processes that control decomposition of surface litter. Four different litter decomposition models were considered. The models used a common observed data set to test and parameterize these models and an independent data set was used to validate and compare model predictions. All of the models used the same abiotic decomposition submodel to simulate the effect of soil water and temperature on decomposition. Observed litter decomposition data from different sites and litter types in the LIDET litter decomposition experiment (LIDET, submitted) were used for model validation and comparison. The sites included a desert grassland, a tropical forest, an arctic grassland and a temperate deciduous forest. The models predictions of carbon and nitrogen remaining in litter bags with time were compared with the LIDET data. Preliminary results from model comparison showed that all of the models simulated the observed carbon dynamics reasonably well, however, none of the models did well at simulating the nitrogen dynamics. The model results were sensitive to the soil N availability which was not measured in the LIDET experiment and was assumed to range from low to high N availability in the model runs.

Model selection process

The U.S. Environmental Protection Agency (EPA) has funded the use of ecosystem models to evaluate the impact of environmental change on natural and managed ecosystems. They also are interested in evaluating how changes in agroecosystem management practices would impact terrestrial carbon stocks. A recent paper by Barnwell *et al.* (1992) documents the process that EPA used to evaluate methods, models and databases for soil carbon sequestration in agroecosystem. They reviewed the literature pertaining to soil carbon modelling and selected a list of models that have the potential to be used for simulating management impacts on agricultural soil carbon levels. Some of the criteria for comparing the models include degree of field testing and validation, documentation of the computer code, ability to use the model for regional and national scales, ability to respond to the dominant agricultural management practices, and compatibility of the model to available national level databases. As a result of this process, they selected a specific model to evaluate the impact of alternative management practice on soil carbon sequestration in the U.S. corn belt during the next 50 years (Donigian *et al.*, 1994). The major weakness of this approach is that they did not formally compare the candidate models with different observed data sets, and used a qualitative approach for selecting a specific model.

This approach contrasts the VEMAP model comparison where different models were used to evaluate the impact of various climate change scenarios. The results from the VEMAP model comparison suggest that it is potentially misleading to use results from only one model since there were substantial differences in the climate change ecosystem impacts predicted by the VEMAP models. With the present state-of-the-art in ecosystem model development, it is risky to use only one model to evaluate potential impacts of environmental and management change on ecosystem dynamics.

References

Ågren, G.I., McMurtrie, R.E., Parton, W.J., Pastor, J. & Shugart, H.H.(1991) State-of-the-art of models of production decomposition linkages in conifer and grassland ecosystems. Ecological Applications 1: 118-138.

Barnwell, Jr., T.O., Jackson IV, R.B., Elliott, E.T., Burke, I.C., Cole, C.V., Paustian, K., Paul, E.A., Donigian, A.S., Patwardhan, A.S., Rowell, A. & Weinrich, K. (1992) An approach to assessment of management impacts on agricultural soil carbon. Water, Air & Soil Pollution 64: 423-435.

Donigian, Jr., A.S., Patwardhan, A.S., Jackson IV, R.B., Barnwell, Jr., T.O., Weinrich, K.B. & Rowell, A.L. (1995) Modeling the impacts of agricultural management practices on soil carbon in the central U.S. In: Soil Management and Greenhouse Effect (ed: R. Lal, J. Kimble, E. Levine, B.A. Stewart), Chapter 11, CRC Lewis Publishers, Boca Raton, FL.

Goutorbe, J.P. (1991) A critical assessment of the SAMER network accuracy. In: Land Surface Evaporation (ed: Schmugge and Andre), Springer-Verlag: 171-182.

Goutorbe, J.P. & Tarrieu C. (1991) HAPE-MOBILHY data base. In: Land Surface Evaporation (ed: Schmugge and Andre), Springer-Verlag: 403-410.

Hanson, J.D., Skiles, J.W. & Parton, W.J. (1988) A multi-species model of rangeland plant communities. Ecological Modelling 44: 89-123.

LIDET (submitted) Meeting the challenge of long-term, large-scale ecological experiments. Environmental Management (submitted).

Parton, W.J. & Ojima, D.S. (1994) Environmental change in grasslands: assessment using models. Climatic Change 28: 111-141.

Parton, W.K., Haxeltine, A., Thornton, P. & Anne, R. (1995) Ecosystem sensitivity to land-surface models and leaf area index. Global and Planetary Change (in press).

Ryan, M.G., Hunt, Jr., E.R., Agren, G.I., Friend, A.D., Pulliam, W.M., Linder, S., McMurtrie, R.E., Aber, J.D., Rastetter, E.B. & Raison, R.J. (in press, a) Comparing models of ecosystem function for temperate conifer forests. I. Model description and validation. For SCOPE Volume Effects of Climate Change on Forests and Grasslands (in press).

Ryan, M.G., Hunt, Jr., E.R., Agren, G.I., Friend, A.D., Pulliam, W.M., Linder, S., McMurtrie, R.E., Aber, J.D., Rastetter, E.B. & Raison, R.J. (in press, b) Comparing models of ecosystem function for temperate conifer forests. II. Simulations of the effect of climate change. For SCOPE Volume Effects of Climate Change on Forests and Grasslands (in press).

Shao Y. & Henderson-Sellers, A. (in press) Soil Moisture Simulation Workshop Review. Global and Planetary Change (in press).

VEMAP. (1995) Vegetation/ecosystem modeling and analysis project (VEMAP): Assessing biogeography and biogeochemistry models in a regional study of terrestrial ecosystem responses to climate change and CO_2 doubling. Global Biogeochemical Cycles (in press).

Acknowledgements

I would like to acknowledge support for this research from the CPER "Long-Term Ecological Research Program: Shortgrass Steppe", NSF Grant No. BSR-9011659, and the NASA supported project titled "Using multi-sensor data to model factors limiting carbon balance in global grasslands", NAGW-2662.

Matching Measurable Soil Organic Matter Fractions with Conceptual Pools in Simulation Models of Carbon Turnover: Revision of Model Structure

Bent T. Christensen
Danish Institute of Plant and Soil Science
Research Centre Foulum
P.O. Box 23
DK-8830 Tjele, Denmark

Summary

A major limitation of current compartment models of soil organic matter (SOM) turnover is that most of the conceptual pools they contain do not correspond to experimentally verifiable fractions. Exceptions are some of the input pools and the microbial biomass. Available information suggests that physical fractionation of soil may delineate SOM fractions that relate more directly to SOM structure and function, and thereby also to biological turnover. Such fractionations may therefore define SOM model pools that are functionally meaningful, leading to a redefinition of the structure of current models and the factors that regulate the material fluxes between compartments. A first approximation to a revised model structure is ventured. It is based on soluble and non-soluble input fractions and separate above- and below-ground input routes. It includes two dynamic pools of particulate organic matter (free and intraaggregate light fraction OM), one pool of inert light fraction OM, and two pools of organomineral associated SOM (silt- and clay-SOM). The need to incorporate soil fauna and soil structural features is stressed.

NATO ASI Series, Vol. I 38
Evaluation of Soil Organic Matter Models
Edited by David S. Powlson, Pete Smith and Jo U. Smith

Introduction: Importance of SOM turnover

Soil organic matter (SOM) constitutes a basic component of any terrestrial ecosystem, SOM being structurally and functionally integrated into fundamental ecosystem processes. Besides being closely associated with physical and chemical soil properties vital to plant growth, SOM also provides carbon and energy to the saprotrophic system responsible for nutrient cycling. During biological turnover, plant nutrients held in SOM may become mineralized and available for root uptake or loss via leaching from the root zone or gaseous losses to the atmosphere. Leaching of nitrate and emissions of radiatively active trace gases are of direct environmental concern.

The CO_2 released during SOM turnover is a major contributor to the overall carbon exchange between terrestrial ecosystems and the atmosphere. Globally, the stock of carbon in SOM is about twice that present as CO_2 in the atmosphere, and the annual flow of carbon through terrestrial ecosystems is about 20 times greater than the present net increase in atmospheric CO_2 (IPCC, 1995). Consequently, even relatively small changes in SOM dynamics may significantly affect atmospheric CO_2 levels. CO_2 is responsible for 50-60% of the increase in radiative forcing arising from anthropogenic emissions of gases to the atmosphere (IPCC, 1995).

The SOM encompasses plant, animal and microbial residues in all stages of decay and a diversity of heteropolymeric organic decomposition products intimately associated with the inorganic soil matrix. The turnover rate of the different SOM components varies continuously due to the complex interactions of biological, chemical and physical processes in soil, with numerous feed-back mechanisms operating in a hierarchical manner. By affecting both plant productivity and the activity of the saprotrophic system, external abiotic factors, such as soil temperature and moisture, exert overall control over SOM dynamics.

Changes in climate influence not only vegetation zones and agricultural production potentials, but also the exchange of CO_2 between soils and the atmosphere. Climate changes affect SOM turnover through changes in soil temperature, moisture and rhizodeposition. The input of plant residues to the soil may change in quantity and in quality, influencing the rate of initial residue decomposition and the turnover of SOM accumulated from previous inputs. Since changes in climate occur slowly, simulation models are probably the only way to predict climate derived changes on SOM storage.

Changes in land use and soil management influence SOM turnover through changes in quantity and quality of plant residues entering the soil, their seasonal and spatial distribution, the ratio between above- and below-ground inputs and through changes in soil disturbance (e.g. tillage) and nutrient inputs (e.g. fertilization). Changes in total SOM induced by changes in the soil management system will be experimentally verifiable only over extended time periods, especially in temperate soils cultivated for centuries (Christensen & Johnston, 1995), whereas the effects of land use changes may be manifest over a shorter time scale. But even when native and other permanently vegetated soils high in SOM are brought into cultivation or when long-term arable soils low in SOM are converted to permanent grass or woodlands, long periods are needed before the steady-state level of SOM characteristic of a particular site is reached. Therefore simulation models combined with long-term experimentation are indispensable in order to evaluate land use and soil management-induced changes in SOM.

About modelling SOM turnover

Although SOM turnover varies continuously and includes numerous hierarchically operating feed-back mechanisms influenced by stochastic events, most dynamic simulation models of SOM turnover can be classified as deterministic multicompartment models based on a mixture of mechanistic and empirically derived relationships. Other approaches exist however, including the continuously varying litter quality model of Bosatta & Ågren (1985; 1991; Ågren & Bosatta, 1987), and the model described by Janssen (1984) involving the concept of "apparent initial age" of organic matter inputs to soil.

Recognizing that mathematically formulated simulation models of SOM turnover are inevitably simplified representations of the part of reality delineated by the system under study, deterministic compartment models are adopted mainly because of their simplicity and requirements for input data. Model validation and verification is also easier using either direct experimental equivalents to state (pool) or rate variables, or values derived indirectly from model outputs and from experimental activity. A true verification of model outputs against experimentally verifiable SOM properties is usually preferable. Together with a high model transparency in terms of an explicitly stated and biologically founded model structure, verification is crucial for the transferability of models from one system to another. Transferability in turn allows a more reliable prediction of long-term changes in SOM to be made, whether these are caused by changes in climate, land use or soil management.

General features of current compartment models

Most of the current compartment models that simulate SOM turnover in arable soils or grasslands show a similar basic structure featuring three main modules: litter inputs, the microbial biomass responsible for transformations of inputs and SOM, and one or more pools of SOM (Table 1). This basic structure was adopted by the now classical five compartment model of Jenkinson & Rayner (1977), in which the litter input was divided into decomposable and resistant plant material and the SOM into biomass and physically and chemically stabilized pools.

The grassland simulation model of Hunt (1977) also includes two separate litter inputs (labile and resistant litter) whereas the NCSOIL model (Molina *et al.*, 1983; Nicolardot *et al.*, 1994) operates with one litter input pool only. Most other models have however expanded the model structure by further divisions of basic modules. Thus the model of van Veen & Paul (1981; van Veen *et al.*, 1984; Verberne *et al.*, 1990), the CENTURY model of Parton *et al.* (1987), and the DAISY model (Hansen *et al.*, 1991) rely on three input pools: a readily decomposable pool (water solubles, sugars, proteins), a pool of structural cell wall materials (cellulose, hemicellulose), and a pool of resistant lignified residues. The last mentioned pool adds directly to one of the SOM pools while the other two inputs pass the microbial biomass before entering the SOM module.

The microbial biomass has also been subdivided and represented by two pools termed non-protected (labile, dynamic) and physically protected (resistant or stable) biomass (Molina *et al.*, 1983; van Veen *et al.*, 1984; Verberne *et al.*, 1990; Hansen *et al.*, 1991). Pool dichotomy was also adopted by Hunt (1977) who for energetical reasons allocated the decomposers to active and inactive biomass pools. The version of the Rothamsted model described by Jenkinson *et al.* (1987) splits the microbial biomass into a zymogenous and an autochthonous component whereas the current version (Jenkinson, 1990) operates with one biomass pool. In contrast, the CENTURY model (Parton *et al.*, 1987, 1988) includes the microbial biomass together with microbial metabolites and other SOM with a short turnover time into one pool termed active SOM.

Table 1 *Examples of compartments included in earlier simulation models of SOM turnover.*

Main module	Ref. *)	Compartments
Input	1	Decomposable & resistant
	2	Labile & resistant (above- & belowground; different depths)
	3	Decomposable, structural & ligniferous (above- & belowground)
	4	Residue
	5	Metabolic, structural & lignin
Decom-posers	1	Microbial biomass
	2	Active & inactive decomposers
	6	Protected & non-protected biomass
	7	Zymogenous & autochthonous biomass
	4	Labile & resistant
SOM	1	Physically & chemically stabilized
	3	Decomposable & recalcitrant (+/- physical protection)
	2	Humic material (different soil depths)
	4	Humads (labile & resistant) & stable humus
	6	Decomposable, recalcitrant, active protected & old
	7	Humified & inert
	5	Active, slow and passive

*) 1: Jenkinson & Rayner (1977); 2: Hunt (1977); 3: van Veen & Paul (1981); 4: Molina *et al.* (1983); 5: Parton *et al.* (1987); 6: van Veen *et al.* (1984); 7: Jenkinson *et al.* (1987).

The identification of SOM pools with different attributes in terms of stabilization mechanism (physical and chemical), bioavailability and thus turnover time is based on qualitative concepts rather than measurable entities. As originally stated by Jenkinson & Rayner (1977), the delineation of different SOM pools with different turnover times is speculative and remains a working hypothesis postulated to explain the data used for model calibration. The current Rothamsted model (Jenkinson, 1990) operates with one dynamic SOM pool only (termed humified organic matter), and a small pool of inert SOM (charcoal etc.) not engaged in biological turnover.

Most contemporary SOM turnover models postulate a pool of passive, old or stable SOM. In contrast to the inert SOM pool of the Rothamsted model, these pools are often quantitatively very significant with as much as half of the SOM being effectively left out of consideration by model dynamics. Stated more bluntly, these pools can be considered parking lots for SOM left over during model calibration and included to improve model performance rather than to satisfy well defined operational concepts.

Experimental equivalents to SOM pools in models

The input pools in models may be approached experimentally by chemical analyses and by using data from decomposition studies of different litters. The analytical equivalent of the readily decomposable fraction of litter input may be a (cold or hot) water soluble fraction (Christensen, 1985). Solubility in water is one attribute of readily metabolizable organic matter, and implies that this fraction can be transported in the litter/soil system by leaching.

The role of lignin and lignin to N ratios in regulating the turnover rate of forest litter has been clearly demonstrated (e.g. Meentemeyer, 1978; Melillo *et al.*, 1982; Berg, 1986), and the resistant lignified residue may be approached experimentally through the van Soest acid detergent fibre method (Anderson & Ingram, 1989). The pool of non-lignified structural cell wall material may similarly be obtained by chemical analyses. In the CENTURY model, the lignin to N ratio controls the split of plant residues into metabolic and structural material and the lignin fraction itself is added directly into the slow SOM pool. Lignin, cellulose and hemicellulose are however often intimately associated in litters, the lignin protecting cellulosic and hemicellulosic material against decomposition. The functional equivalent of chemically defined input fractions may therefore not be straight forward. When considering inputs other than plant residues (e.g. dead faunal components, animal manure and sewage sludge), it becomes even more complicated to define measurable equivalents to conceptual input pools. One approach could be to replace the entire input module of compartment models with the "apparent initial age" concept of Janssen (1984), the apparent initial age being determined from decomposition studies with various categories of organic inputs.

In the model of Jenkinson & Rayner (1977), the microbial biomass compartment corresponded directly to the biomass determined by the chloroform fumigation-incubation method, pioneered by Jenkinson & Powlson (1976). For models that split the microbial biomass into non-protected and physically protected biomass pools, no direct experimental verification is yet possible. In contrast to the protection in soil of non-biomass materials which is considered to be proportional to the total amount present, van Veen *et al.* (1984) envisage that each soil has a specific capacity to protect biomass and that biomass in excess of the protection capacity is decomposed at a much higher rate than protected biomass. The total biomass can be determined and then indirectly allocated to the two pools according to the protective capacity of the soil. The protective capacity relates to soil texture and aggregation, but no independent measurement of this capacity has yet been devised. Since the capacity of soil to protect biomass is a key parameter in several SOM turnover models, experimental equivalents are much needed.

Although incorporating biologically sensible concepts, and chemical and physical mechanisms known to occur in the soil environment, a major limitation of current simulation models is the lack of experimentally verifiable counterparts to the conceptual SOM pools they contain. The original pool definitions drew heavily on information from analytical wet chemical studies of SOM structure and 'humus' formation (e.g. Anderson, 1979), and did not explicitly consider soil structure and the spatial arrangement of SOM in the soil matrix. Thus the original model approach did not allow pools to be compared directly to measurements.

A major impetus for developing SOM turnover models is the need to simulate long-term changes in SOM dynamics. However, since most of the SOM is allocated to model pools that cannot be determined experimentally, there is an urgent need to identify measurable SOM fractions that qualify as candidates for compartments in SOM turnover models. Increased transparency and transferability of models awaits further progress in the identification of fractions that integrate structural and functional properties of SOM, and that can be quantified experimentally.

Feasibility of physical fractionations

Experimental work over the last 10 to 15 years demonstrates that physical soil fractionation methods are able to define and delineate SOM pools that integrate structural and functional properties of SOM. This approach relates much more directly to SOM dynamics *in situ* than the more classical wet chemical SOM fractions (Oades, 1989, 1993; Elliott & Cambardella, 1991; Christensen, 1992, 1995; Cambardella & Elliott, 1993a). Physical fractionation of soil emphasizes the role of soil minerals and soil structure in SOM turnover. The term physical fractionation covers a range of different methods, each designed for a specific purpose, and no generally accepted fractionation scheme has been devised so far. However, most fractionation schemes attempt to avoid chemical changes in SOM during the fractionation step and distinguish between SOM that is not firmly associated with soil minerals, SOM that is incorporated into primary organomineral complexes, and SOM that is trapped within aggregates (secondary organomineral complexes).

Physical fractionation involves density and size separations and various degrees of soil dispersion. Isolation of *primary organomineral complexes* relies on a complete dispersion of soil and sedimentation/sieving in water, and leads to separates equivalent in size to the clay, silt and sand of standard textural analyses. In the context of SOM turnover, primary organomineral size separates may be considered as a functional analogue to soil texture.

Limited soil dispersion and wet sieving techniques are usually applied to isolate *secondary organomineral complexes*. The rationale and experimental procedures behind these fractionations draw on work related to aggregate formation and stability, and on concepts of aggregate hierarchy in soils (Tisdall & Oades, 1982; Oades & Waters, 1991). Fractionation by density is based on submersion of samples into organic liquids or inorganic salt solutions with a specific density of typically 1.6 to 2.0 g cm^{-3} and limited sample dispersion (Christensen, 1992). Although the bonding strength between SOM and soil inorganics varies continuously, the original concept behind density fractionation assumes that SOM can be divided into two discreet pools with different structural and functional characteristics.

Attempts to relate SOM fractions, isolated by physical fractionation methods, directly to pools in existing SOM turnover models have been made (Cambardella & Elliott, 1992, 1994; Buyanovsky *et al.*, 1994), but it is envisaged that the introduction of physically isolated SOM fractions as experimental equivalents to pools in models will lead to a redefinition of pools and the material fluxes between them, including rate modifiers associated with soil structure.

The following sections summarize experimental evidence from studies on physical fractionations of soil according to particle size or density. This summary is based on recent detailed reviews given by Christensen (1992, 1995) and Gregorich & Janzen (1995) and readers are refered to these for primary sources of information.

Light fraction organic matter (LF-OM)

Less than 10% of the SOM isolated with sand sized particles is found as true primary organo-mineral complexes. The major part occurs as LF-OM (free, particulate or macro-OM) encompassing mainly particulate plant and animal residues undergoing decomposition, but occasionally also biologically inert SOM (e.g. charcoal; Skjemstad *et al.*, 1990). A number of studies suggest that LF-OM isolated from sand sized separates and from dispersed whole soil samples may represent identical SOM pools. LF-OM is an important pool of bioavailable carbon. Part of the LF-OM is incorporated and physically stabilized in macroaggregates e.g. as intra-aggregate particulate organic matter (Cambardella & Elliott, 1992, 1993b). Physical protection of LF-OM and its role in aggregate formation and stabilization may be crucial in regulating the proportion of new SOM that is further transformed and stabilized in clay and silt sized organomineral complexes. LF-OM, with its high C/N ratio, may show a significant N immobilisation potential (Adams, 1980).

The amount of LF-OM is sensitive to fluctuations in litter inputs and shows seasonal and spatial variability especially in forest soils (Spycher *et al.*, 1983; Boone, 1994). Seasonal variations are less prominent in arable soils (Adams, 1980; Conti *et al.*, 1992), but LF-OM has been found to reflect differences in the cropping and tillage system (Cambardella & Elliott, 1992; Beare *et al.*, 1994a, b; Biederbeck *et al.*, 1994; Bremer *et al.*, 1994). Accumulation of LF-OM is favoured in cold and dry soils, in soils low in pH and in permanently vegetated soils with significant returns of plant litter (e.g. forests and permanent grasslands). In these soils, LF-OM may account for over 40% of the SOM whereas LF-OM in long-term arable soils often holds less than 10%. LF-OM tends to be readily depleted when native soils are brought under cultivation and when permanent grasslands are converted to all arable rotation. Often the drop in LF-OM accounts for the greater part of the initial SOM loss.

Thus available results show that the LF-OM content of the soil is a dynamic property, reflecting shorter-term shifts in SOM storage and turnover, whether induced by changes in soil management or land use. The turnover of LF-OM is linked to macroaggregation and mechanisms of physical protection.

SOM in organomineral complexes

The highest concentrations of SOM are observed in $<5\ \mu m$ primary organomineral separates. For a number of long-term arable soils, SOM enrichments of the clay and silt sized separates decline following a power function as the content of these separates increase in whole soil. Clay generally accounts for over 50% of the SOM in soils low in LF-OM, clay and silt ($<20\ \mu m$) together may account for over 90%. Concentrations of carbohydrates generally increase with decreasing particle size, and a greater microbial input to clay than to silt is indicated by hexose to pentose ratios. Analyses of CuO oxidation products reveal that lignin degradation is more advanced in clay than in silt organomineral separates (Guggenberger *et al.*, 1994). Studies based on ^{13}C-NMR have demonstrated that clay associated SOM is more aliphatic in nature than SOM in coarser separates, while aromatic structures peak in silt, and that O-alkyl carbon decrease and alkyl carbon increase with decreasing particle size (Guggenberger *et al.*, 1995; Randall *et al.*, 1995).

Generally the SOM mineralization rate decreases in the order: sand, clay, and silt. Following addition of simple substrates, new SOM is found in most separates although clay sized organomineral complexes often show greater accumulations. In the longer term, the proportion of substrate derived SOM in clay declines more rapidly than that in silt indicating a higher stability of silt-SOM. It is indicated that newly formed clay SOM is less stable in sandy than in clayey soils. The loss of LF-OM associated with cultivation of virgin soils, forests and grasslands is due to mineralization and to transfer of SOM to silt and clay sized separates. The chemical composition of SOM in <20 μm separates is not drastically influenced by changes in soil management and land use although permanently vegetated soils and arable rotation may differ somewhat in terms of SOM quality.

Macroaggregates and whole soil appear to hold similar proportions of clay, silt and sand sized primary particles, provided that aggregate classes are corrected for loose sand accumulating during the sieving procedure. Differences in SOM content of clay and silt isolated from dry macroaggregates of different size are relatively small. For soils low in free and intraaggregate LF-OM, the SOM content of true macroaggregates is correlated with their clay + silt content. SOM in microaggregates appears to be more stable than the SOM responsible for binding microaggregates into macroaggregates, substantiating the concepts of aggregate hierarchy and of intraaggregate LF-OM. The mineralization rate of SOM tends to differ between various aggregate size classes, and crushing of macroaggregates generally increases SOM mineralization.

Although macroaggregates seem to contain primary particles in proportions similar to the textural composition of the corresponding fraction of the whole soil, this does not necessarily imply a fortuitous spatial organization of all primary particles engaged in an aggregate. Rather it emphasizes the role of LF-OM in aggregate formation and stabilization. Encrustation of particulate residues is considered a major mechanism in the accumulation of physically protected but otherwise bioavailable LF-OM.

A first approximation to a revised model structure

SOM is extremely heterogenous in terms of its chemical structure, its association with soil minerals, and thus its bioavailability. Individual physical fractions will therefore inevitably include SOM which differs in turnover rate, regardless of size class and density limits. Nevertheless, the experimental evidence obtained so far suggests that differences between LF-OM (free and interaggregate) and SOM in clay and silt sized organomineral complexes are sufficiently distinct to justify regarding these fractions as operationally and experimentally verifiable SOM pools with different structural and functional relationships to SOM turnover.

In most arable and grassland soil models, the turnover of above- and belowground inputs is considered to be similarly influenced by driving variables and rate modifiers. In arable soils aboveground inputs and roots in the tilled layer are mechanically mixed into the top soil, while roots and rhizodeposits within the deeper soil layers may be left to decompose *in situ*. In permanently vegetated soils, however, aboveground residues will be left on the soil surface (or transported into the soil by faunal activity or by leaching) whereas roots and rhizodeposits at all depths will be left *in situ* in the soil. Also, the distribution of residues in the soil profile may be changed. Changing the land use from arable to permanent grassland or woodland may warrant separate handling of the two main input routes if model realism is to be maintained and transferability increased. Furthermore, the above- and below-ground inputs are liable to be different in terms of litter quality (e.g. content of lignin and soluble material), contact with soil mineral, spatial and temporal variability, and exposure to faunal activity and stochastic events such as freeze/thaw and dry/wet cycles.

Figure 1 presents a first approximation to a revised structure of a carbon turnover model that incorporates pools matching physically separated SOM fractions. In line with Hunt (1977) and van Veen & Paul (1981), the plant litter is allocated to above- and below-ground input routes, but here each input is subdivided into soluble and non-soluble material. The soluble fraction includes litter components that are easily leached (e.g. by rain) and are readily decomposable. The decomposition rate of the non-soluble litter fraction is regulated primarily by its lignin content.

Only one pool represents the microbial biomass as in the current Rothamsted model. The main part of the SOM is found in silt and clay sized organomineral associations. These pools replace the various SOM pools of current simulation models. A main new feature of the proposed structure is the inclusion of two pools representing free and intraaggregate LF-OM respectively. One pool is allocated to inert SOM, which is considered to be charcoal etc. not firmly associated with soil minerals.

Flows between the suggested compartments are also indicated in Figure 1. Every flow is mediated by microbial activity under the overall influence of moisture and temperature. Transformation of material generates CO_2, microbial biomass and decomposition products (including residual partly processed material), which are passed on to the receiving pool. Soluble input materials are envisaged to pass to the microbial biomass compartment which in turn provides input mainly to the clay-SOM pool. The decomposition of non-soluble input material is regulated primarily by its lignin content in addition to temperature and moisture. The outcome is again CO_2, microbial biomass and also free LF-OM. The flow between free and intra-aggregate LF-OM depends on soil structural features and forces that interact with aggregation. LF-OM is transformed into clay- and silt-SOM, microbial biomass and CO_2. Microbial biomass, clay- and silt-SOM, and free LF-OM are recycled as decomposition proceeds.

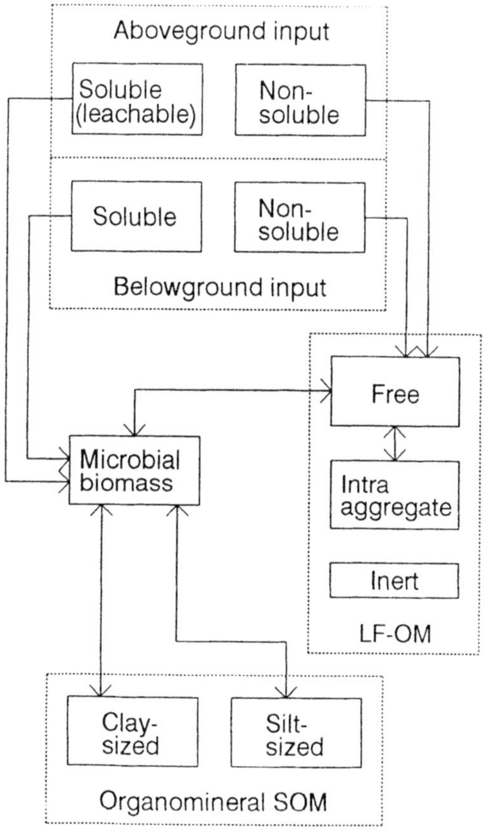

Figure 1 *A first approximation to the compartment structure of a simulation model of carbon turnover.*

To make the proposed model structure operational, rate variables providing turnover times for the various pools have to be worked out. The driving variables are considered to be similar to existing models, whereas rate modifiers must be reconsidered. Rate modifiers will relate to plant input route and to soil structural features, including texture, aggregation and soil disturbance induced by tillage, freeze/thaw and wet/dry cycles, and soil fauna.

The soil fauna needs to be considered as a rate modifier (and perhaps also a state variable) because of faunal activity such as comminution, modification of substrate within the digestive tract, horizontal and vertical transport of substrate and microorganisms in the soil, and grazing on microbial populations. The activity of the soil fauna is intimately linked with soil structure. The interaction of meso- and microfauna with soil structure and SOM dynamics has been discussed by Griffiths (1994) and by Juma (1994) who also offers a conceptual framework to link soil fauna with current SOM turnover models. Rutherford & Juma (1992a, b) present a simulation model of SOM dynamics incorporating protozoa-induced mineralization. In the classical forest floor types of mull and mor, the soil fauna is considered the rate-limiting step in litter transformations.

Concluding remarks

The main emphasis of this contribution is on SOM dynamics over decades to centuries. Clearly the time scale involved is decisive for the structure of SOM simulation models. In any hierarchical system, various levels of organization can be distinguished according to temporal and/or spatial resolution. Feed-back mechanisms operate both within and between levels, and irreversible and discontinuous processes affect material flows on each level. Since predictive models are usually built to address a specified time scale and a specified system, the challenge is to frame the appropriate level of resolution, keeping the balance between structural simplicity, explanatory capability and predictive power. Modellers know that the number of state, rate and driving variables should be kept to a minimum and this has always been gratefully acknowledged by researchers involved in collecting experimental data for modelling.

References

Adams, T.McM. (1980) Macro organic matter content of some Northern Ireland soils. *Record of Agricultural Research* **28**: 1-11.

Ågren, G.I. & E. Bosatta (1987) Theoretical analysis of the long-term dynamics of carbon and nitrogen in soils. *Ecology* **68**: 1181-1189.

Anderson, D.W. (1979) Processes of humus formation and transformation in soils of the Canadian Great Plains. *Journal of Soil Science* **30**: 77-84.

Anderson, J.M. & J.S.I. Ingram (Eds.) (1989) *Tropical Soil Biology and Fertility: A Handbook of Methods*. C.A.B. International, Wallingford, U.K.

Beare, M.H., P.F. Hendrix, & D.C. Coleman (1994a) Water-stable aggregates and organic matter fractions in conventional- and no-tillage soils. *Soil Science Society of America, Journal* **58**: 777-786.

Beare, M.H., M.L. Cabrera, P.F. Hendrix, & D.C. Coleman (1994b) Aggregate-protected and unprotected organic matter pools in conventional- and no-tillage soils. *Soil Science Society of America, Journal* **58**: 787-795.

Berg, B. (1986) Nutrient release from litter and humus in coniferous forest soils - a mini review. *Scandinavian Journal of Forest Research* **1**: 359-369.

Biederbeck, V.O., H.H. Janzen, C.A. Campbell, & R.P. Zentner (1994) Labile soil organic matter as influenced by cropping practices in an arid environment. *Soil Biology and Biochemistry* **26**: 1647-1656.

Boone, R.D. (1994) Light-fraction soil organic matter: origin and contribution to net nitrogen mineralization. *Soil Biology and Biochemistry* **26**: 1459-1468.

Bosatta, E. & G.I. Ågren (1985) Theoretical analysis of decomposition of heterogeneous substrates. *Soil Biology and Biochemistry* **17**: 601-610.

Bosatta, E. & G.I. Ågren (1991) Dynamics of carbon and nitrogen in the organic matter of the soil: a generic theory. *American Naturalist* **138**: 227-245.

Bremer, E., H.H. Janzen, & A.M. Johnston (1994) Sensitivity of total, light fraction and mineralizable organic matter to management practices in a Lethbridge soil. *Canadian Journal of Soil Science* **74**: 131-138.

Buyanovsky, G.A., M. Aslam, & G.H. Wagner (1994) Carbon turnover in soil physical fractions. *Soil Science Society of America, Journal* **58**: 1167-1173.

Cambardella, C.A. & E.T. Elliott (1992) Particulate soil organic-matter changes across a grassland cultivation sequence. *Soil Science Society of America, Journal* **56**: 777-783.

Cambardella, C.A. & E.T. Elliott (1993a) Methods for physical separation and characterization of soil organic matter fractions. *Geoderma* **56**: 449-457.

Cambardella, C.A. & E.T. Elliott (1993b) Carbon and nitrogen distribution in aggregates from cultivated and native grassland soils. *Soil Science Society of America, Journal* **57**: 1071-1076.

Cambardella, C.A. & E.T. Elliott (1994) Carbon and nitrogen dynamics of soil organic matter fractions from cultivated grassland soils. *Soil Science Society of America, Journal* **58**: 123-130.

Christensen, B.T. (1985) Decomposability of barley straw: effect of cold-water extraction on dry weight and nutrient content. *Soil Biology and Biochemistry* **17**: 93-97.

Christensen, B.T. (1992) Physical fractionation of soil and organic matter in primary particle size and density separates. *Advances in Soil Science* **20**: 1-90.

Christensen, B.T. (1995) Carbon in primary and secondary organomineral complexes. *Advances in Soil Science* (in press).

Christensen, B.T. & A.E. Johnston (1995) Soil organic matter and soil quality: Lessons learned from long-term field experiments at Askov and Rothamsted. In: E.G. Gregorich and M.R. Carter (eds.), *Soil Quality for Crop Production.* Elsevier Science Publishers, Amsterdam (in press).

Conti, M.E., R.M. Palma, N. Arrigo, & E. Giardina (1992) Seasonal variations of the light organic fractions in soils under different agricultural management systems. *Communication in Soil Science and Plant Analysis* **23**: 1693-1704.

Elliott, E.T. & C.A. Cambardella (1991) Physical separation of soil organic matter. *Agriculture, Ecosystems and Environment* 34: 407-419.

Gregorich, E.G. & H.H. Janzen (1995) Storage of soil carbon in the light fraction and macroorganic matter. *Advances in Soil Science* (in press).

Griffiths, B.S. (1994) Microbial-feeding nematodes and protozoa in soil: their effects on microbial activity and nitrogen mineralization in decomposition hotspots and the rhizosphere. *Plant and Soil* **164**: 25-33.

Guggenberger, G., B.T. Christensen & W. Zech (1994) Land-use effects on the composition of organic matter in particle-size separates of soil: I. Lignin and carbohydrate signature. *European Journal of Soil Science* **45**: 449-458.

Guggenberger, G., W. Zech, L. Haumaier & B.T. Christensen (1995) Land-use effects on the composition of organic matter in particle-size separates of soil: II. CPMAS and solution ^{13}C NMR analysis. *European Journal of Soil Science* **46**: 147-158.

Hansen, S., H.E. Jensen, N.E. Nielsen, & H. Svendsen (1991) Simulation of nitrogen dynamics and biomass production in winter wheat using the Danish simulation model DAISY. *Fertilizer Research* **27**: 245-259.

Hunt, H.W. (1977) A simulation model for decomposition in grasslands. *Ecology* **58**: 469-484.

IPCC (1995) *Climate Change 1994. Radiative Forcing of Climate Change and An Evaluation of the IPCC IS92 Emission Scenarios.* Intergovernmental Panel on Climate Change. Cambridge University Press, Cambridge, UK

Janssen, B.H. (1984) A simple method for calculating decomposition and accumulation of 'young' soil organic matter. *Plant and Soil* **76**: 297-304.

Jenkinson, D.S. (1990) The turnover of organic carbon and nitrogen in soil. *Philosophical Transactions of the Royal Society of London* B **329**: 361-368.

Jenkinson, D.S. & D.S. Powlson (1976) The effects of biocidal treatments on metabolism in soil - V. A method for measuring soil biomass. *Soil Biology and Biochemistry* **8**: 209-213.

Jenkinson, D.S. & J.H. Rayner (1977) The turnover of soil organic matter in some of the Rothamsted classical experiments. *Soil Science* **123**: 298-305.

Jenkinson, D.S., P.B.S. Hart, J.H. Rayner, & L.C. Parry (1987) Modelling the turnover of organic matter in long-term experiments at Rothamsted. *INTECOL Bulletin* **15**: 1-8.

Juma, N.G. (1994) A conceptual framework to link carbon and nitrogen cycling to soil structure formation. *Agriculture, Ecosystems and Environment* **51**: 257-267.

Meentemeyer, V. (1978) Macroclimate and lignin control of litter decomposition rates. *Ecology* **59**: 465-472.

Melillo, J.M., J.D. Aber, & J.F. Muratore (1982) Nitrogen and lignin control of hardwood leaf litter decomposition dynamics. *Ecology* **63**: 621-626.

Molina, J.A.E., C.E. Clapp, M.J. Shaffer, F.W. Chichester, & W.E. Larson (1983) NCSOIL, a model of nitrogen and carbon transformations in soil: description, calibration, and behavior. *Soil Science Society of America, Journal* **47**: 85-91.

Nicolardot, B., J.A.E. Molina, & M.R. Allard (1994) C and N fluxes between pools of soil organic matter: model calibration with long-term incubation data. *Soil Biology and Biochemistry* **26**: 235-243.

Oades, J.M. (1989) An introduction to organic matter in mineral soil. pp. 89-159. In: J.B. Dixon and S.B. Weed (eds.), *Minerals in Soil Environments, Second Edition.* SSSA Publ. Inc. Madison, WI, USA.

Oades, J.M. (1993) The role of biology in the formation, stabilization and degradation of soil structure. *Geoderma* **56**: 377-400.

Oades, J.M. & A.G. Waters (1991) Aggregate hierarchy in soils. *Australian Journal of Soil Research* **29**: 815-828.

Parton, W.J., D.S. Schimel, C.V. Cole, & D.S. Ojima (1987) Analysis of factors controlling soil organic matter levels in Great Plains grasslands. *Soil Science Society of America, Journal* **51**: 1173-1179.

Parton, W.J., J.W.B. Stewart, & C.V. Cole (1988) Dynamics of C, N, P and S in grassland soils: a model. *Biogeochemistry* **5**: 109-131.

Randall, E.W., N. Mathieu, D.S. Powlson & B.T. Christensen (1995) Fertilization effects on organic matter in physically fractionated soils as studied by ^{13}C NMR: results from long-term experiments. *European Journal of Soil Science* **46** (in press).

Rutherford, P.M. & N.G. Juma (1992a) Simulation of protozoa-induced mineralization of bacterial carbon and nitrogen. *Canadian Journal of Soil Science* **72**: 201-216.

Rutherford, P.M. & N.G. Juma (1992b) Performance of a simulation model describing protozoa-induced mineralization of bacterial C and N in a sandy loam. *Canadian Journal of Soil Science* **72**: 217-228.

Skjemstad, J.O., R.P. Le Feuvre & R.E Prebble (1990) Turnover of soil organic matter under pasture as determined by ^{13}C natural abundance. *Australian Journal of Soil Research* **28**: 267-276.

Spycher, G., P. Sollins, & S. Rose (1983) Carbon and nitrogen in the light fraction of a forest soil: Vertical distribution and seasonal patterns. *Soil Science* 135: 79-87.

Tisdall, J.M & J.M. Oades (1982) Organic matter and water-stable aggregates in soils. *Journal of Soil Science* **33**: 141-163.

van Veen, J.A. & E.A. Paul (1981) Organic carbon dynamics in grassland soils. I. Background information and computer simulation. *Canadian J. Soil Sci.* **61**: 185-201.

van Veen, J.A., J.N. Ladd, & M.J. Frissel (1984) Modelling C and N turnover through the microbial biomass in soil. *Plant and Soil* **76**: 257-274.

Verberne, E.L.J., J. Hassink, P. de Willigen, J.J.R. Groot, & J.A. van Veen (1990) Modelling organic matter dynamics in different soils. *Netherland Journal of Agricultural Science* **38**: 221-238.

Acknowledgements

This work was financially supported by the Danish Environmental Research Programme and the Ministry of Agriculture and Fisheries (BÆR-SP-7, ØKO-SP-1).

[1]Modeling the Measurable or Measuring the Modelable: A Hierarchical Approach to Isolating Meaningful Soil Organic Matter Fractionations

Edward T. Elliott, Keith Paustian and Serita D. Frey
Natural Resource Ecology Laboratory
Colorado State University
Fort Collins, Colorado 80523,
USA

Summary

The approaches of modeling the measurable and measuring the modelable are both valuable for advancing our understanding of soil organic matter (SOM). In the former case, we assume that the measurements we make are the best representation of nature, and that model structure should follow. While this may simplify model testing, it may not yield a particularly useful description of SOM dynamics. There is no question that most of our knowledge about SOM is derived from experimentation, but models are being increasingly used to further our understanding. In the latter case, model structure and parameters are modified in reasonable ways to obtain the best fit of simulation with observation, thus deducing the "correct" structure. In this case, it is conceivable that more than one structure can result in a good fit to the data. Good predictions may be produced, but an incorrect structure may be derived. In either case, there are no clear criteria for obtaining the "truth" except for repeatedly testing our methods/models against new information, preferably obtained from definitive experiments. A combination of approaches is best. We elaborate a particular approach which is mindful of the habitat in which microbes and their substrates reside and relate this theory to methods developed to separately isolate plant and microbially derived SOM, which may be physically protected from microbial attack in the soil.

---

[1] The idea for the title was a product of the Workshop on Management of Carbon in Tropical Soils Under Global Change, Nairobi, Kenya, February, 1994 (Elliott, 1994).

NATO ASI Series, Vol. I 38
Evaluation of Soil Organic Matter Models
Edited by David S. Powlson, Pete Smith and Jo U. Smith

Introduction

Changes in soil organic matter (SOM) and its relationship to soil properties has been an area of intense scientific study over the past decade, and considerable progress has been made in developing the theory of organic matter behavior in soil, especially as it relates to soil fertility (Elliott *et al.*, 1993; Elliott *et al.*, 1994; Swift & Woomer, 1993). Many different approaches and philosophies have emerged and have become intertwined into a complex set of theory, experimental approaches and practice. One important tool that has surfaced in this study is the simulation model, which has been used to integrate various factors (e.g., soil temperature, moisture, aeration, litter chemistry) controlling decomposition processes and SOM dynamics (Bosatta & Agren, 1991; Jenkinson *et al.*, 1987; Molina *et al.*, 1983; Parton *et al.*, 1987, 1988; Paustian, 1994; Paustian *et al.*, 1992; van Veen & Paul, 1981). McGill (this volume) has reviewed a number of these models and Christensen (this volume) has discussed several models in relation to organic matter fractions.

Simulation models of SOM are useful for a number of reasons; they (1) embody multiple facets of our best understanding of SOM dynamics, (2) may be embedded within a whole ecosystem context, (3) may be used to make projections of changes in SOM under different management, climate or with changes in other factors and (4) can be used to investigate the consequences of alterations in the controls on SOM. With regard to this last consideration, there are a number of different approaches that have been used to describe SOM in models. These approaches have utilized a number of SOM pools, each with different properties. Key properties of these pools are their turnover rates, size and the differential effects that changes in the soil environment have on their turnover rates. However, current theory (and models) are weak in two areas of critical importance, (i) the relationship between soil structure and the ability of a soil to stabilize organic matter and (ii) *physical disturbance* effects on organic matter turnover. The focus of this paper will be to present a particular view of the hierarchical nature of soil structure as it relates to the relationship between pools used in simulation models and analytical measurements of SOM pools. Other approaches are presented by Christensen (this volume).

Soil Structure, Models, and Organic Matter Turnover

One of the most often cited deficiencies of current SOM/agroecosystem models is that functional, kinetically-defined pools cannot be directly related to measurable variables.

Consequently, it is difficult to initialize models based on independent measures and therefore site-dependent calibrations based on total SOM levels are necessary. We believe that the link between soil structure and organic matter is crucial for improving the generality of ecosystem models, particularly in the highly disturbed ecosystems created by agriculture.

The interactions between management, soil structure and SOM dynamics are among the most fundamental relationships governing the behavior of agroecosystems; yet these relationships are poorly understood and under represented in our current mathematical models. The accumulated evidence from long-term field experiments, tracer data and soil process studies indicates that as much as half of the total SOM (i) turns over on time scales of 10-50 yr or less, (ii) is the primary determinant of native soil fertility, (iii) is sensitive to a variety of management practices, (iv) is a key determinant of soil structural properties, and (v) is, to a considerable degree, maintained in soil by being physically-protected from rapid decomposition. We believe that recent advances in methodology, as well as theory development, now provide us with the opportunity to improve our understanding by more explicitly representing these organic matter fractions and their controls in our models.

Evidence suggests that a significant proportion of the organic matter in soil is physically-protected from decomposition through occlusion by clay minerals and encapsulation within soil aggregates (Elliott, 1986; Elliott & Cambardella, 1991; Tisdall & Oades, 1982; Strickland *et al.*, 1992). A significant proportion of this organic matter is thought to comprise a pool having an "intermediate" residence time (e.g., 10-50 yr), but which may decompose much faster upon disturbance. For example, the rapid loss of organic matter and natural fertility following plow-out of native grassland, typically as much as 20-40% of the total, should be mainly contained within this pool (Elliott, 1986). More active organic matter, representing microbial biomass and labile organic matter, cycles the most rapidly but makes up only 3-5% of the total organic matter. Its importance in governing the supply rate of minerals to the plant is well recognized but it is of less direct importance to soil physical properties. This "fast" fraction has also received the most attention and is, arguably, the most well characterized as a quantifiable organic matter pool.

At the other extreme is very old material (e.g., the passive fraction in the Century model; Parton *et al.*, this volume) with turnover times on the order of hundreds to thousands of years, based on radiocarbon dating (Campbell *et al*, 1967; Martel & Paul, 1974).

Consequently this material is not subject to rapid change due to management or climate shifts occurring over decadal time scales. Thus, at time scales which are most significant to assessing management and climate change impacts on ecosystem response, we believe the intermediate pools are the most important and least well understood SOM components. By virtue of their large size and relative sensitivity to management, we believe that this intermediate organic matter is a key component which must be better understood to assess agroecosystem sustainability and to predict long-term effects of management and climate change on SOM in agricultural soils.

The theory of soil aggregate hierarchies and its implications for organic matter stabilization through physical protection has emerged as an important new paradigm in soil science (Beare *et al.*, 1994; Elliott, 1986; Elliott & Coleman, 1988; Tisdall & Oades, 1982). However, these concepts have yet to be articulated in quantitative models. While, several models have incorporated the concept of physical protection of organic matter, based on soil texture (Jenkinson *et al.*, 1987; Parton *et al.*, 1987; van Veen *et al.*, 1984), to our knowledge, no existing model relates physical protection and decomposition kinetics explicitly to soil structural properties. This is a significant point, since at a given site, soil texture will presumably not be affected by soil disturbance or other environmental factors (barring changes due to erosion), but soil structure may be highly affected. Thus, a better understanding of SOM dynamics in agroecosystems argues for pool definitions and process formulations which relate to soil structure.

Model-Data Comparison

An interesting dichotomy has arisen regarding the best approach to use for investigating SOM with simulation models. That is, should we "model the measurable or measure the modelable?" Based upon a workshop held to investigate tropical SOM, Elliott (1994) reported on a consensus that one should model the measurable. It is worthwhile to re-examine this dichotomy.

In a strict sense, both analytical measures and model definitions of organic matter pools are based upon hypotheses about what controls the dynamics of SOM. In both cases, we compare the results of our new observations with our existing knowledge (i.e., other observations and concepts). For example, we have come to believe that when a soil developed under natural vegetation is cultivated, there will be a loss of SOM. Therefore, all accurate measures/models will show this decline. Another example is our expectation that SOM will increase as greater amounts of residue are added to the soil. Model results or laboratory measurements that do not meet these underlying assumptions are suspect and, as often as not, there are as many problems with data collection as there are with model development.

As we seek the best approaches for studying SOM, we are faced with some fundamental questions. How many pools should be measured or used in our models? How do we know we are measuring the appropriate pools? How well do the methods/models work for different soils? How do we judge the usefulness of our methods? As approaches are continually tested, we find that our confidence builds or declines depending on the outcome of comparisons. For some analytical measures there is considerable confidence in what we are measuring, e.g., total soil carbon. In this case, several different approaches have been compared with good agreement. Also, there are clear underlying mechanisms for why the approach we are using should give correct results. Confidence is not nearly as high for approaches used to separate SOM into different pools.

The "kinetic" approach for determining of the appropriate number of pools is based upon comparison of SOM loss curves with different model predictions. Discussion regarding the best fit of a model with data on SOM loss have been ongoing for many years (e.g., Jenny, 1941, pp. 256). Nor have contemporary researchers been in full agreement on the best approach. Most agree, however, that there should be a small pool with a rapid turnover rate, but there are differences of opinion regarding the existence of a large and uniform pool with an intermediate turnover-time and a smaller pool with a very slow turnover time (Jenkinson *et. al.*, 1987) vs. intermediate and slow pools of about equal size (Parton *et al.*, 1987). These considerations lead to the general approach of trying to "measure the modelable."

The evaluation of whether SOM model structure and function are appropriate have most often depended upon comparison of model output with measurements of total SOM from treatments within a site (Jenkinson *et al.*, 1987; Parton & Rasmussen, 1994; Paustian *et al.*, 1992; Verberne *et al.*, 1990), and occasionally from data obtained across a series of sites. In general, models have not been explicitly developed to represent experimentally-derived measurements of different pools of SOM. Early on it was found that the long-term dynamics of SOM could not be represented well without employing multiple SOM pools having different turnover times (van Veen *et al.*, 1981). Experimentalists have been and continue to seek the so-called "active fraction" that represents the short-term source of mineralized C but considerable difficulty has been encountered in obtaining a fraction of SOM that can be directly related to the pools represented in models. Motavalli *et all* (1994) compared several methods for estimating soil C pools and used these estimates to initialize model parameters for Century simulations of soil respiration. The simulations were compared to actual respiration patterns observed in a laboratory incubation. Simulations using measured soil C fractions (soluble C, microbial C, and light fraction C) as model input underestimated actual respiration, whereas, use of model pool equilibrium estimates (Century C pools at equilibrium determined by running the model for 3,000 years) resulted in simulations most comparable to the respiration patterns observed in the lab.

In essence, many models require at least two to three pools of SOM with significantly different turnover times to match experimental observation to SOM turnover and experimentalists have found difficulties in isolating such pools. Curve-peeling approaches trying to deconvolute mineralization curves obtained from long-term laboratory incubations has been one approach (Molina *et al.*, 1994), but purely kinetic-based procedures are innately unsatisfying because the "pools" cannot be physically isolated so that other properties, such as chemical composition, can be identified.

Measurement of SOM Pool Turnover Rate

Many experimental approaches have been used to break down the total SOM into fractions. While these approaches have been successful in showing differences in SOM pool sizes for different soils, there has been less work in defining functionally consistent turnover rates of these pools. To make better inter-comparison between models and methods, we must determine the turnover rate of the analytically determined pools.

To determine turnover times under steady-state conditions, one must know the size of the pool plus either the inputs or outputs from the pools. Where the system is not at steady state, then one must know both inputs and outputs, as well as pool size. It is not always easy to determine if SOM pools are at steady-state (i.e., the pool size does not vary significantly over time). One may want to assume that a functional steady-state may be achieved for fast or intermediate pools in ecosystems that have been established for long periods of time. Agroecosystems are less likely than many natural ecosystems to exhibit steady-state behavior, which presupposes a constancy in the plant/soil environment. As new agricultural management systems are adopted, the potential steady-state changes. Because the number of ways these rates can be measured are limited, it is necessary for us to carefully evaluate the various experimental approaches that are available and, most importantly, to critically examine the applicability and interpretation of the result.

Perhaps the most common way for determining SOM pool turnover rate is to add a labeled substrate to the soil and follow the incorporation of the material into the different pools over time. For examination of intermediate turnover pools, the incubations may have to be sampled after a year or more (Christensen & Sorensen, 1985). This approach is limited because it does not actually yield results on the turnover rate of the pools unless it is known that the input to the pools is the same as the output. This equality is unlikely in soil taken from the field and incubated in the laboratory. However, a change in total SOM over the course of the incubation may be measured. Given enough time, the distribution of the label into the different pools may yield valuable information on the differences in dynamics of differentially measured pools (Molina *et al.*, 1994).

Another approach is to measure the incorporation and turnover of ^{14}C derived from above-ground nuclear bomb testing (Hseih, 1993; Jenkinson *et al.*, 1987). It is an intriguing analysis because it can be made within the setting of natural ecosystems. Since above-ground testing occurred more than 40 years ago, the label is deeply incorporated into the SOM fractions and one may discern turnover rates of different pools. This signal is becoming increasingly weak as isotopic dilution continues since cessation of testing. One requirement of this approach is that samples from the sites of interest must be available from before the time of bomb testing so one can determine natural background levels of ^{14}C.

The incorporation of litter derived from plants with a C_3 vs. C_4 photosynthetic pathway can be used as a natural tracer because of differences in the ^{13}C isotopic concentration. Under the best situations, an abrupt shift at a known time from one type of vegetation to the other creates a change in the ^{13}C signature of input material that is sufficient for rates of incorporation into various pools can be discerned. One must also know total SOM levels at the time of sampling. Examples of this are forest to grassland conversions or shortgrass prairie to wheat (Balesdent *et al.*, 1987; Cambardella & Elliott, 1992; Cerri *et al.*, 1985; Tieszen *et al.*, 1979).

Aging of SOM with ^{14}C carbon dating techniques is another approach that has been used (Campbell *et al.*, 1967, Martel & Paul, 1974). However, the resolution of this method may only allow the ages of the most slowly turning over pools to be determined. The more recent pools must be removed prior to analysis so as not to contaminate with bomb ^{14}C. This approach also assumes that it is possible to isolate the "old" C.

Even with this array of methods it may often be difficult to distinguish pools with different turnover times, for a number of reasons. The ^{14}C dating approach gives the information required in the most direct fashion. However, a detectable change in age tends to be restricted to the relatively old pools, thus making it less useful for determining the fast & intermediate turnover rate pools. Measurement of ^{13}C with shifts in vegetation is useful in many cases but misinterpretation may result because of a lack of good information on past management history. Tracing bomb ^{14}C movement requires availability of pre-bomb archived samples, which may have been handled in unknown or inappropriate ways to allow good analysis. Use of synthetically labeled plant material also requires known inputs and outputs from the pool of interest because the system is not likely to be at steady-state. However, it may be possible to determine rates of change in pool size and, combined with knowledge about input rates, one may deduce outputs and thereby obtain an estimate of turnover time.

If knowledge of turnover time is to be valuable, it must be determined on well-defined and meaningful fractions of SOM. Soils have been fractionated in a multitude of ways, some becoming much more widely applied than others, depending upon how well they can be used to investigate phenomena of interest. Clearly, if nature has a structure, and it is related to function, then schemes which are devised for the study of SOM should be related to this structure. In the case of SOM, well defined theory exists that can be related to the structure of the soil.

Physical Protection

While many soil scientists have focussed on the chemical structure of SOM, this approach has generally been less useful in describing functioning of soil than approaches applying physical separations. We believe the reason for the greater success of physical separations is that microorganisms must have physical access to the substrate in order to use it. If they do not have access, the chemical composition of that substrate is inconsequential because it will not be decomposed. In situations where microbial accessibility is not strongly controlled by the structure of the system, as in the litter layers of a forest, for example, then the chemical composition becomes paramount in understanding what controls the dynamics of organic matter. Therefore, our arguments apply mostly to organic matter in mineral soils. For these reasons, we next give a description of the theory of soil structure so that we may later relate it to the methods we propose for obtaining meaningful SOM fractions. Finally, we will relate the theory-derived methods to a model structure; an example of "modeling the measurable."

Soil microorganisms are known to consume most organic substrates falling onto or deposited into the soil, and yet there are tremendous accumulations of material in the soil. Some of this accumulation can be accounted for by the lag that occurs between the time of organic material deposition and the time of consumption. However, most compounds will decompose within a year or less if not in contact with soil. Therefore, the soil must be protecting the incoming organic material in ways that prevent it from being completely consumed. In fact, there are many reasons why substrates may be dislocated from soil microorganisms.

Soil organisms must live between soil particles in the soil pore space which, for an intermediate textured soil, is half filled with water, half with air, on the day following a rain. These soil particles, many of which are composed of coalesced groups of smaller soil particles, range in size over several orders of magnitude. As a result, the size of the pore spaces between the particles exhibit a similar range of sizes. This range in the size of spaces restricts some organisms from entering some spaces depending upon their body size, shape and plasticity. A central concept relating to the physical protection of SOM is that the walls of the pore space are the surfaces of the soil particles, be they primary particles or aggregates (Elliott & Coleman, 1988). Therefore, the protection of SOM from mineralization by soil organisms is determined by the size of the spaces between the soil particles which, to some extent, is determined by the size of the particles themselves. For the smallest particles (clays), physical protection may take the form of adsorption onto surfaces or the covering of organic particles by the clay. The rate of decomposition may be limited if the tortuosity of the route results in a low concentration of enzyme at the site of the substrate.

The pore space/aggregate-particle concept can be used to help elucidate the consequences of various factors on the turnover of SOM, its production of nutrients for plant growth and ways for obtaining meaningful SOM fractions. For example, mixing or stirring of soil during tillage exposes particle surfaces otherwise not accessible to decomposers. From another perspective, losses of SOM from different locations in the soil will reduce the cohesiveness of soil. Positioning of SOM in one place in the soil may result in a fast turnover and nutrient release while its position in another place may result in sequestration for many decades or centuries. In general, when studying SOM, one must be mindful that the soil organisms are the agents for decomposition. Therefore concepts and method developed to increase our understanding of SOM must view the world from their perspective. An elegant conceptual model (Tisdall & Oades, 1982), which continues to be developed, describes the hierarchical structure of soil.

A Conceptual Model for Soil Structure and Organic Matter Relationships

The concept of hierarchical aggregate structure is composed of four distinct structural units, each of which are composed of primary particles and various forms of organic matter. While attempting to avoid use of jargon in our discussion, it is necessary to distinguish among kinds of aggregates and the associated pore space. It appears from recent evidence that this conceptual model only applies to situations where organic matter is the main binding agent. In more highly weathered soils where inorganic constituents are primarily responsible for development of structure (e.g., oxisols), an altered model where the mineral component of the soil is the primary binding agent may better apply (Elliott *et al.*, 1991; Frey *et al.*, in prep.; Oades & Waters, 1991)

The smallest aggregated units in this conceptual model are the silt-sized microaggregates that range in size from 2 to 20 μm. They contain clay- and fine silt-associated organic matter, are bound by persistent SOM (Tisdall & Oades, 1982) and are more aromatic in character than the soil as a whole. Polyvalent metal cations bind the negatively-charged clay with the negatively-charged aromatic organic matter (Edwards & Bremner, 1967). This size aggregate is very stable and is not readily dispersed even by sonication at high energy levels and may constitute a large portion of the silt-sized organomineral complexes separated using the sonication-sedimentation approach to SOM analysis.

A likely explanation for the genesis of these microaggregates is that they are formed at the center of larger aggregates as the result of decomposition of particles of organic matter (Oades, 1984). The microbial community developing around the fragment may result in the formation of a larger aggregate as fungal hyphae and exuded polysaccharides radiate out from the fragment. As decomposition proceeds, an anaerobic microsite may develop which changes the character of the processes determining the stabilization of organic matter. Low oxygen levels will result in a higher solubility of polyvalent metal cations as more reduced forms of iron and manganese come into solution. Products of decomposition are more aromatic in character under anaerobic than aerobic conditions. The pH of the microsite may also be reduced due to the elevated concentration of carbon dioxide. Reduced pH may increase weathering rates and enhance formation of secondary minerals. In combination, these three factors may result in more potential for bridging by polyvalent metal cations between negatively charged aromatic organic matter and newly formed clays, some of which may not yet be crystallized. The end result is formation of a highly stable complex containing amorphous organic matter and perhaps the partially decomposed lignin core of the plant litter fragment.

Sand-sized microaggregates range in size from 20 to 250 μm and largely contain silt-sized microaggregates, course silt and sand. These aggregates are stable with respect to tillage disturbances but are susceptible to dispersion by sonication. The constituent components of these microaggregates are held together by plant and microbially produced polysaccharide that have entered the small pores between the constituent particles (Cambardella & Elliott, 1994; Gregorich *et al.*, 1989). Since the pores in these aggregates are mostly smaller than the size of most decomposers, the inter-constituent organic matter is protected from decomposition. This produces the interesting situation of otherwise labile organic matter potentially having a long turnover-time. It is conceivable that after very long periods of time, as with centuries of cultivation in Europe, some intra-silt sized microaggregate organic matter may be lost with the subsequent effects of loss of soil structure and release of free clays.

Small macroaggregates may be from 250 to 1,000 or 2,000 μm. They are constituted of sand-sized microaggregates and sand plus the agents binding the constituent parts. These binding agents are a combination of plant and microbial derived polysaccharide (comparable to agents in sand-sized microaggregates), bits of particulate organic material, bacterial cells and especially hypha strands. These binding agents are largely accessible to microorganisms and, as a result, have a shorter turnover time than organic matter holding smaller aggregates together. It is this organic matter that makes the largest contribution to soil fertility when previously undisturbed soil is put into cultivation (Elliott, 1986) and may be the first to be formed if tillage is reduced under new management. The mineralized nutrients come from decomposing amorphous (Cambardella & Elliott, 1994) and particulate organic matter (Cambardella & Elliott, 1992). Small soil animals may inhabit aggregates of this size and accelerate mineralization. Larger soil animals, such as acari that feed on hyphae, will not be able to gain access to the interiors of these aggregates. Small macroaggregates are easily disrupted even by gentle sonication and with cultivation. They may remain stable with gentle shaking and even when the soil slakes. They are essentially absent from fields that have undergone long-term cultivation.

The largest aggregates (>2000 μm) are composed of roots, fungi, particulate organic material, sand, small macroaggregates and sand sized microaggregates. These constituents are bound together mostly by roots and mycorrhizal fungi. These large aggregates contain pores large enough to be entered by almost all micro- and mesofauna. Aggregates of this size can demonstrate short term dynamics with significant fluctuation potentially occurring over a growing season (Stefanson, 1971). While the binding agents for these aggregates may demonstrate high rates of turnover, they probably do not contain a very high proportion of the nutrients mineralized over the course of SOM loss after breaking out previously uncultivated soils. However, they may contribute a significant amount of nutrients in the first few years after initial cultivation. These large macroaggregates may be broken down with slaking but retained if soils are vapor or capillary wetted prior to sieving.

We have developed a conceptual model (Figure 1) that is analogous to parts of the Century model (Parton *et al.*, 1987) where plants, litter, and active, slow (our intermediate pool) and passive organic matter pools are the main components. The primary difference in concept is that we differentiate the intermediate pool into material derived directly from residual plant

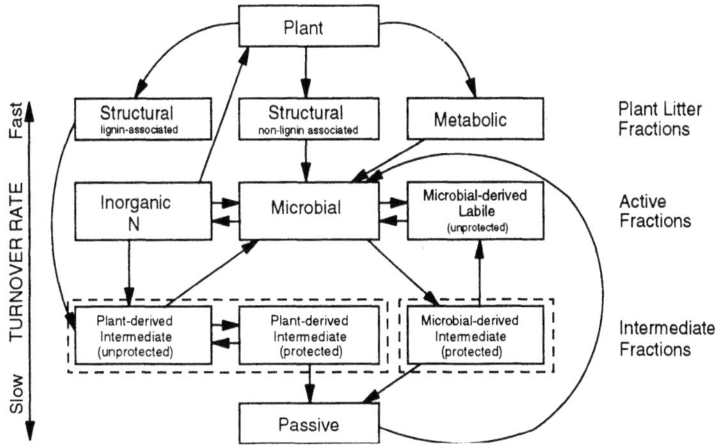

Figure 1 *Conceptual model of SOM dynamics(showing N flows). Pools with intermediate turnover times are divided into plant-derived and microbial-derived pools. Shifts from protected to non-protected states occur with disruption or degradation of soil structure.*

litter vs. material derived from microbial products and we explicitly consider the role of soil aggregation in protecting organic matter from decomposition (currently, these two flows in the Century model enter a single intermediate pool).

Isolating Intermediate Pools of SOM

The plant-derived intermediate fraction (PDI) is comprised of particulate plant matter which is partially decomposed but maintains some structural integrity due to the persistence of a lignified "skeleton" (Cambardella & Elliott, 1992). While recognizing that other compounds conceptualized as microbial gums and mucilages, cell wall constituents and other microbial contribute to litter recalcitrance, lignin is commonly used as a measure of decomposability and a number of studies have shown it to be a key factor in regulating the late stages of litter decomposition (Melillo *et al.*, 1989). Characterization of particulate organic matter (POM) from field experiments at Sidney, NE showed it to consist of approximately 50% lignin (Cambardella & Elliott, 1992). The microbial-derived intermediate fraction (MDI) is debris residing in the spaces surrounding microaggregrates but within macroaggregates thus binding them together (Elliott, 1986; Elliott & Coleman, 1988). This material is analogous to the "temporary" organic matter, which binds microaggregates into macroaggregates, according to the aggregate hierarchy theory of Tisdall & Oades (1982). We believe our understanding, and hence our ability to describe SOM dynamics with simulation models, will be enhanced by keeping these pools conceptually separate because of differences in their origin and differences in the controls on their formation and degradation.

The MDI and PDI pools also correspond to analytically measurable fractions. We believe the MDI pool corresponds to the physically-protected and microbial produced enriched labile fraction (Cambardella & Elliott, 1994) and is isolated through an aggregate size fractionation sequence in which macroaggregates are disrupted into their constituent parts and separated into microaggregate, coarse silt-size, fine silt-size and coarse and fine clay-size fractions. Densiometric separations are then carried out on the resultant size fractions and the C and N concentration of these size-density fractions is determined. After applying a minimal amount of sonic energy to break down the macroaggregates, the resultant 2 to 20 μm size particles with a density of 2.07-2.22 g cm^{-3} were found to be enriched in C (and especially N) and labile when incubated (Cambardella & Elliott, 1994). Other work indicates that this SOM binds microaggregates into macroaggregates (Elliott, 1986), although POM may play a direct role in the stabilization process (Cambardella & Elliott, 1992).

Plant-derived material has often been isolated as "light fraction" using heavy liquids with densities ranging from 1.6 to 2.0 g cm^{-3} (Dalal & Mayer, 1986; Greenland & Ford, 1964; Janzen *et al.*, 1992; Sollins *et al.*, 1984; Spycher *et al.*, 1983). It is thought to be primarily composed of plant material small enough to pass through a 2 mm sieve, but still recognizable as plant constituents, if viewed after magnification with light or scanning electron microscopy (Spycher *et al.*, 1983). Light fraction material is largely composed of plant residues but may also contain fungal hyphae, spores (Spycher *et al.*, 1983) and charcoal. Charcoal may comprise a high proportion of light fraction material in systems that have undergone slash and burn agriculture (Elliott *et al.*, 1991). A further complication with using the light fraction methodology for recovery of the PDI pool is contamination by the heavy liquid used for the separation; this may be an important consideration if subsequent incubations of the light fraction material are planned. An alternative approach, which also yields particulate plant material, involves complete soil dispersion by sonication followed by soil organic C and N analysis of particle size separates (Anderson *et al.*, 1981; Christensen, 1987; Christensen, 1992; Gregorich *et al.*, 1989; Oades *et al.*, 1987; Turchenek & Oades 1979). Most of the SOM isolated in the sand-sized fraction is particulate plant and animal residues (Christensen, this volume). While sonication may minimize chemical transformations associated with the use of chemical dispersants, it introduces problems due to the breakdown of particulate organic material and its redistribution among silt and clay size fractions.

We have used sonication in conjunction with light fraction separations (after Greenland & Ford, 1964) and have found that recovery of light fraction C is highly sensitive to the level of sonication energy (Figure 2). Sonication levels used (20-465 J ml^{-1}) were lower than those commonly used for soil dispersion (~1500 J ml^{-1}). Cambardella & Elliott (1992) developed a different approach for isolating plant-derived SOM. In this method, soil is chemically dispersed using sodium hexametaphosphate and the particulate organic matter (POM) is collected a 53 μm sieve, along with the sand. Carbon and nitrogen analysis is made directly on this fraction. Additionally, the POM can be elutriated or otherwise floated from the sand so that direct analysis of the material can be made. This method avoids the problems associated with sonication and separation via heavy liquid. Also, recent work in our laboratory indicates that POM-C and N may represent substantially more of the total soil C and N than that recovered as light fraction (Figure 2) or the sand-sized SOM obtained from the sonication/primary particle-size separations. Measurement of POM is also much simpler, takes less time and uses more accessible equipment. It can be determined in conjunction with the determination of soil texture. A drawback to the POM separation is that it is not applicable to soils which are not completely dispersed with sodium hexametaphosphate.

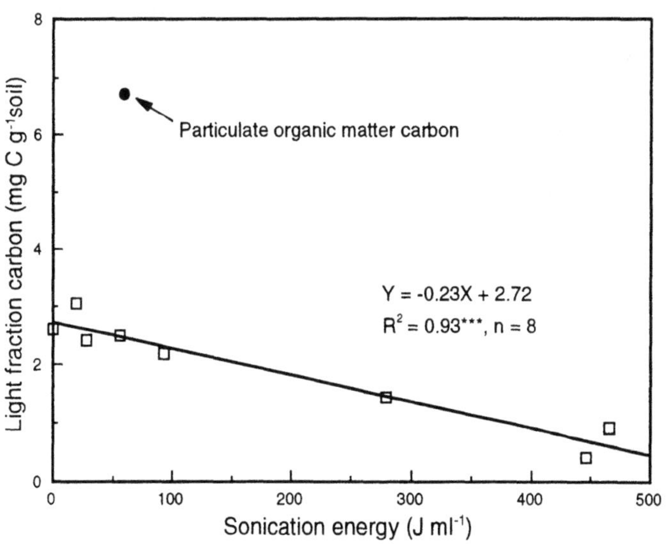

Figure 2 *The effect of increasing sonication energies on recovery of light fraction C from soil collected at a long-term tillage plot at Sidney, NE. Particulate organic matter C is shown for comparison. Plant material >53 μm is recovered with the sonication/primary particle sedimentation method may be extrapolated by moving along the ordinate to ~1500 J ml^{-1}.*

We believe that the presented conceptual model is based on a sound foundation of theory concerning organic matter and aggregate hierarchies, and also on experimental evidence and observation. While it shares several concepts with many kinetic-based SOM models that are currently being used, it more explicitly links theory about the physical/chemical nature of soils to model structure. However, direct experimental knowledge of the factors controlling the size and distribution of these pools and how they respond to management remains minimal.

Conclusions

Simulation models developed with an ecosystem perspective will continue to be increasingly used to develop our knowledge of SOM. They are tools that embody our best understanding about the dynamics of SOM, and as such, can be used to make projections of the future impacts of changes in management, climate and elevated carbon dioxide. For them to be of value, they must be compared with results obtained from the field and laboratory so that we may evaluate our level of confidence in their ability to represent nature. Soil organic matter is described in many of these models as being composed of a number if different pools with a range of turnover rates. Comparison of these pools with measured SOM fractions is a key focus of much ongoing research and requires both new methods and model structures. Working in concert, these two approaches must be iteratively used to derive a better understanding of SOM so that we may apply the knowledge embodied in our models to new situations which may be difficult or impossible to investigate with experimentation alone. We will have the most confidence in the reliability of our predictions/projections if the models we use have been developed and tested with data obtained from methods that closely relate to the structure of the model and from sites that represent regions for which we are interested in making projections. Models will become the powerful tools we hope them to be only if we continue to interplay the development of models and their experimental and methodological counterparts.

References

Anderson, D.W., Saggar, S., Bettany, J.R. & Stewart, J.W.B. (1981) Particle size fractions and their use in studies of soil organic matter: I. The nature and distribution of forms of carbon, nitrogen, and sulfur. *Soil Science Society of America Journal* **45**:767-772.

Balesdent, J., Mariotti, A. & Guillet, B. (1987) Natural ^{13}C abundance as a tracer for studies of soil organic matter dynamics. *Soil Biology and Biochemistry* **19**:25-30.

Beare, M.H., Cabrera, M.L., Hendrix, P.F. & Coleman, D.C. (1994) Aggregate-protected and unprotected organic matter pools in conventional- and no-tillage soils. *Soil Science Society of America Journal* **58**:787-795.

Bosatta, E. & Agren, G.I. (1991) Dynamics of carbon and nitrogen in the organic matter of the soil: a generic theory. *American Naturalist* **138**:227-245.

Cambardella, C.A. & Elliott, E.T. (1992) Particulate soil organic-matter changes across a grassland cultivation sequence. *Soil Science Society of America Journal* **56**:777-783.

Cambardella, C.A. & Elliott, E.T. (1994) Carbon and nitrogen dynamics of soil organic matter fractions from cultivated grassland soils. *Soil Science Society of America Journal* **58**:123-130.

Campbell, C.A., Paul, E.A., Rennie, D.A. & McCallum, K.J. (1967) Applicability of the carbon dating method of analyses to soil humus studies. *Soil Science* **104**:217-224.

Cerri, C., Feller, C., Balesdent, J., Victoria, R. & Plenecasagne, A. (1985) Application du tracage isotopique naturel en ^{13}C l'etude de la dynamique de la matiere organique dans les sols. *C.R. Academy of Science Series 2* **300**:423-428.

Christensen, B.T. and L.H. Sorensen (1985) The distribution of native and labelled carbon between soil particle size fractions isolated from long-term incubation experiments. *Jn. of Soil Sci.* **36**:219-229.

Christensen, B.T. (1987) Decomposability of organic matter in particle size fractions from field soils with straw incorporation. *Soil Biology and Biochemistry* **19**:429-435.

Christensen, B.T. (1992) Physical fractionation of soil and organic matter in primary particle size and density separates. *Advances in Soil Science* **20**:1-90.

Dalal, R.C. & Mayer, R.J. (1986) Long-term trends in fertility of soils under continuous cultivation and cereal cropping in southern Queensland. IV. Loss of organic carbon from different density fractions. *Australian Journal Soil Research* **24**:293-300.

Edwards, A.P. & Bremner, J.M. (1967) Microaggregates in soils. *Journal of Soil Science* **18**:64-73.

Elliott, E.T. (1986) Aggregate structure and carbon, nitrogen, and phosphorus in native and cultivated soils. *Soil Science Society of America Journal* **50**:627-633.

Elliott, E.T. & Coleman, D.C. (1988) Let the soil work for us. *Ecological Bulletin* **39**:23-32.

Elliott, E.T. & Cambardella, C.A. (1991) Physical separation of soil organic matter. *Agriculture, Ecosystems and Environment* **34**:407-419.

Elliott, E.T., Palm, C.A., Reuss, D.E. & Monz, C.A. (1991) Organic matter contained in soil aggregates from a tropical chronosequence: correction for sand and light fraction. *Agriculture, Ecosystems and Environment* **34**:443-451.

Elliott, E. T., Cole, C.V. & Cambardella, C.A. (1993) Modification of ecosystem processes by management and the mediation of soil organic matter dynamics. In: *The Dynamics of Soil Organic Matter in Relation to the Sustainability of Tropical Agriculture.* (ed: R. Merckx & K. Mulongoy), John Wiley and Sons, Chichester: 257-268.

Elliott, E. T., Janzen, H.H., Campbell, C.A., Cole, C.V. & Myers, R.J.K. (1994) Principles of ecosystem analysis and their application to integrated nutrient management and assessment of sustainability. In: *Sustainable Land Management for the 21st Century.* (ed: R.C. Wood & J. Dumanski), University of Lethbridge, Lethbridge, Canada.

Elliott, E. T. (1994) Embodying process information in models evaluated with site network information: Nairobi workshop. In: *Transactions of the 15th World Congress of Soil Science*, Acapulco, Mexico: **9**:163-176.

Greenland, D.J. & Ford, G.W. (1964) Separation of partially humified organic materials from soils by ultrasonic dispersion. In: *Transactions of the 8th International Congress of Soil Science,* Bucharest, Romania: 137-146.

Gregorich, E.G., Kachanoski, R.G. & Voroney, R.P. (1989) Carbon mineralization in soil size fractions after various amounts of aggregate disruption. *Journal of Soil Science* **40**:649-659.

Hsieh, Y. (1993) Radiocarbon signatures of turnover rates in active soil organic carbon pools. *Soil Science Society of America Journal* **57**:1020-1022.

Janzen, H.H., Campbell, C.A., Brandt, S.A., Lafond, G.P. & Townley-Smith, L. (1992) Light-fraction organic matter in soils from long-term crop rotations. *Soil Science Society American Journal* **56**:1799-1806.

Jenkinson, D.S., Hart, P.B.S., Rayner, J.H. & Parry, L.C. (1987) Modelling the turnover of organic matter in long-term experiments at Rothamsted. *INTECOL Bulletin* **15**:1-8.

Jenny, H. (1941) *Factors of Soil Formation*. McGraw-Hill, New York.

Martel, Y.A. & Paul, E.A. (1974) Effects of cultivation on the organic matter of grassland soils as determined by fractionation and radio-carbon dating. *Canadian Journal of Soil Science* **54**:419-426.

Melillo, J.M., Aber, J.D., Likens, A.E., Ricca, A., Fry, B. & Nadelhoffer, K.J. (1989) Carbon and nitrogen dynamics along the decay continuum: Plant litter to soil organic matter. In: *Ecology of Arable Land: Perspectives and Challenges*. (ed: M. Clarholm & L. Bergstrom), Kluwer Academic Publ., Dordrecht, the Netherlands: 53-62.

Molina, J.A.E., Clapp, C.E., Shaffer, M.J., Chichester, F.W. & Larson, W.E. (1983) NCSOIL, a model of nitrogen and carbon transformations in soil: description, calibration and behavior. *Soil Science Society of America Journal* **47**:85-91.

Molina, J.A.E., Nicolardot, B., Houot S., Chaussod, R. & Cheng, H.H. (1994) Biologically active soil organics: a case of double identity. In: *Defining Soil Quality for a Sustainable Environment*. (ed: J.W. Doran, D.C. Coleman, D.F. Bezdicek, & B.A. Stewart), Soil Science Society of America Journal Special Publication Number 35, SSSA, Madison, WI: 169-177.

Motavalli, P.P., Palm, C.A., Parton, W.J., Elliott, E.T. & Frey, S.D. (1994) Comparison of laboratory and modeling simulation methods for estimating soil carbon pools in tropical forest soils. *Soil Biology and Biochemistry* **26**:935-944.

Oades, J.M. (1984) Soil organic matter and structural stability: mechanisms and implications for management. *Plant and Soil* **76**:319-337.

Oades, J.M. & Waters, A.G. (1991) Aggregate hierarchy in soils. *Australian Journal of Soil Research* **29**:815-828.

Oades, J.M., Vassallo, A.M., Waters, A.G. & Wilson, M.A. (1987) Characterization of organic matter in particle size and density fractions from a red-brown earth by solid-state ^{13}C NMR. *Australian Journal of Soil Research* **25**:71-82.

Parton, W.J. & Rasmussen, P.E. (1994) Long-term effects of crop management in wheat-fallow: II. CENTURY model simulations. *Soil Science Society of America Journal* **58**:530-536.

Parton, W.J., Schimel, D.S., Cole, C.V. & Ojima, D.S. (1987) Analysis of factors controlling soil organic matter levels in Great Plains grasslands. *Soil Science Society of America Journal* **51**:1173-1179.

Parton, W.J., Stewart, J.W.B. & Cole, C.V. (1988) Dynamics of C, N, P and S in grassland soils: a model. *Biogeochemistry* **5**:109-131.

Paustian, K., Parton, W.J. & Persson, J. (1992) Modeling soil organic matter in organic-amended and nitrogen-fertilized long-term plots. *Soil Science Society of America Journal* **56**:476-488.

Paustian, K. (1994) Modelling soil biology and biogeochemical processes for sustainable agriculture research. In: *Management of Soil Biota in Sustainable Farming Systems*. (ed: C. Pankhurst, B.M. Doube, V.V.S.R. Gupta & P.R. Grace), CSIRO Publ., Melbourne, Australia.

Sollins, P., Spycher, G. & Glassman, C.A. (1984) Net nitrogen mineralization from light- and heavy-fraction forest soil organic matter. *Soil Biology and Biochemisty* **16**:31-37.

Spycher, G., Sollins, P. & Rose, S. (1983) Carbon and nitrogen in the light fraction of a forest soil: vertical distribution and seasonal patterns. *Soil Science* **135**:79-87.

Stefanson, R. C. (1971) Effect of periodate and pyrophosphate on the seasonal changes in aggregate stabilization. *Australian Journal Soil Research* **9**:33-41.

Strickland, T.C., Sollins, P., Rudd, N. & Schimel, D.S. (1992) Rapid stabilization and mobilization of ^{15}N in forest and range soils. *Soil Biology and Biochemistry* **24**:849-855.

Swift, M.J. & Woomer, P. (1993) Organic matter and the sustainability of agricultural systems: Definition and measurement. In: *Soil Organic Matter Dynamics and Sustainability of Tropical Agriculture*. (ed: K. Mulongoy & R. Merckx), Wiley-Sayce Co., New York: 3-18.

Tieszen, L.L., Hein, D., Quortrup, S.A., Troughten, J.H. & Imbamba, S.K. (1979) Use of ^{13}C values to determine vegetation selectivity in East African herbivores. *Oecologia* **37**:351-359.

Tisdall, J.M. & Oades, J.M. (1982) Organic matter and water-stable aggregates in soils. *Journal of Soil Science* **33**:141-163.

Turchenek, L.W. & Oades, J.M. (1979) Fractionation of organo-mineral complexes by sedimentation and density techniques. *Geoderma* **21**:311-343.

van Veen, J.A. & Paul, E.A. (1981) Organic carbon dynamics in grassland soils. 1. Background information and computer simulation. *Canadian Journal of Soil Science* **61**:185-201.

van Veen, J.A., McGill, W.B., Hunt, H.W., Frissel, M.J. & Cole, C.V. (1981) Simulation models of the terrestrial nitrogen cycle. *Ecological Bulletins (Stockholm)* **33**:25-48.

van Veen, J.A., Ladd, J.H. & Frissel, M.J. (1984) Modelling C and N turnover through the microbial biomass in soil. *Plant and Soil* **76**:257-274.

Verberne, E.L.J., Hassink, J., De Willegen, P., Groot, J.J.R. & van Veen, J.A. (1990) Modelling organic matter dynamics in different soils. *Netherlands Journal of Agricultural Science* **38**:221-238.

Quantitative methods to evaluate and compare Soil Organic Matter (SOM) Models

Jo Smith, Pete Smith & Tom Addiscott

Soil Science Department

IACR-Rothamsted

Harpenden, Herts , AL5 2JQ , U.K.

Summary

Evaluation and comparison of the performance of soil organic matter models is often based upon visual/graphical comparison of the simulated values produced by the model with actual values from field experiments. Such methods provide an immediate qualitative description of the differences, highlighting trends, different types of errors and distribution patterns of simulated and measured values. However, model evaluations or comparisons should ideally incorporate both a qualitative visual/graphical assessment and a quantitative statistical appraisal. Statistical methods have been selected that are suitable for quantitative evaluation and comparison of soil organic matter models. The methods included each provide information on some distinct aspect of the accuracy of the simulation.

Quantifying *association* and *coincidence* provides two different measures of the overall similarity between the simulated and measured values. A high coincidence indicates that the simulated *values* closely correspond to the measured values, whereas a high association indicates that the *shape* of the simulated curve is similar to that measured. The total difference between the simulated values and measurements may be expressed in terms of *consistent* and *inconsistent* errors. Consistent errors quantify the extent of model bias towards either over- or under-prediction of observations. Inconsistent errors correspond to errors which cancel out because the model shows no inclination towards either over- or under-prediction. If the experiments have been replicated, the total

NATO ASI Series, Vol. I 38
Evaluation of Soil Organic Matter Models
Edited by David S. Powlson, Pete Smith and Jo U. Smith

difference between simulated values and measurements can be more usefully expressed in terms of *systematic* and *random* errors. Systematic errors represent the failure of the model to simulate differences between the experiments. Random errors represent experimental error. Finally, the maximum difference between any pair of simulated and measured values can be used to indicate where a small number of very large errors are contributing to a large proportion of the overall error calculated, so that other statistical values may become unreliable. If used in conjunction with visual/graphical methods, the statistical techniques described in this paper provide a rigorous method for the quantitatively evaluating the performance of soil organic matter models, or indeed other models.

Introduction

Evaluation and comparison of the performance of soil organic matter (SOM) models is often based upon the visual/graphical comparison of the simulated values produced by the model with actual measured values from field experiments. Such methods provide a simple, quantitative way of assessing whether or not a model is producing simulated values close to those actually measured. One problem associated with such methods, however, is that the difference between measured and simulated values cannot easily be quantified, and so it is not easy to compare the performance of different SOM models in a quantitative way. Although descriptive model comparisons are very informative (e.g. de Willigen, 1991), it is desirable wherever possible to supplement visual/graphical methods for model evaluation with quantitative measures of model performance (e.g. Otter-Nacke & Kuhlmann, 1991; Vereecken *et al.*, 1991). The use of statistical methods makes it possible to determine common strengths and weaknesses of particular models, and identify datasets which all models fail to simulate adequately.

Statistical measures of model performance do have some limitations (Loague & Green, 1991; Loague *et al.*, 1995). The statistics can be very sensitive to a small number of very large errors, especially in small datasets (Loague & Green, 1991), and it is often difficult to fully interpret the nature of the differences between simulations and measurements using only a single statistic. Visual/graphical methods provide an immediate qualitative description of the differences, highlighting trends, different types of errors and distribution patterns of simulated and measured values (Loague *et al.*, 1995). Despite the limitations of statistical measures, a quantitative evaluation is desirable when comparing the performance of different models using the same datasets. Model evaluation should ideally incorporate both a qualitative visual/graphical assessment and a quantitative appraisal which uses a number of carefully chosen, distinct statistics to describe different aspects of the accuracy of the simulation. The purpose of this paper is to select a range of statistical methods that are appropriate for soil organic matter model evaluation and provide details of how they can be used.

Selection of Statistical Methods

Many methods for the quantitative evaluation of model performance have been suggested and used, for example by Addiscott & Whitmore (1987) , Whitmore (1991) and Loague & Green (1991). In this section, we describe the reasons for selecting particular statistical tests. Each method selected gives information on a distinct aspect of the accuracy of a simulation, and we summarise the information that each statistic provides. The tests are described in greater detail in later sections.

Quantifying the association and coincidence provides two distinct measures of the overall similarity between the simulated and measured values. As illustrated in figure 1, a high coincidence indicates that the simulated *values* closely correspond to the measured values, whereas a high association indicates that the *shape* of the plotted simulation is similar to the measured data.

The association between simulated and measured values is given by the sample correlation coefficient, r (Draper & Smith, 1966). The statistical significance of the association and coincidence may then be obtained using an F-test or Student's t (Snedecor & Cochran, 1989).

The coincidence between measured and simulated values can be assessed by calculating an absolute value for the total difference, expressed as the root mean square error, *RMSE* (Loague & Green, 1991). If sufficient information is available, the statistical significance of the total difference may again be obtained using an F-test. Alternatively, the modelling efficiency and coefficient of determination (*EF* and *CD* respectively, Loague & Green, 1991) provide other comparative measures of the significance of the total difference.

Figure 1 *Schematic Representation of Coincidence and Association; a) High coincidence but low association ; b) Low coincidence but high association. (■ represents measured values; — represents simulated values)*

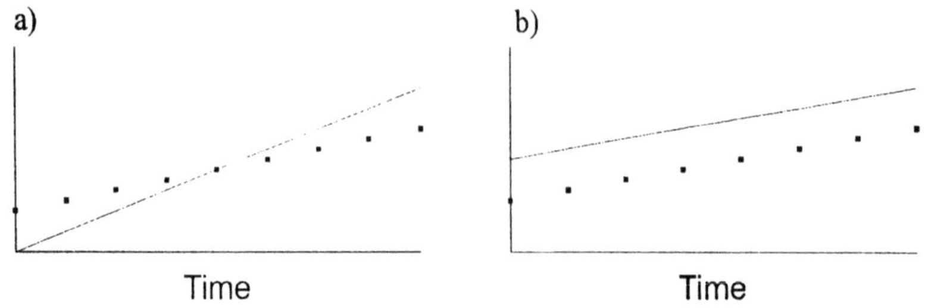

The total difference between simulated and measured values may be further delineated by separating out *consistent errors* (i.e. where the model is biased towards either over- or under-prediction of the observations, see figure 2a) from *inconsistent errors* (i.e. where the model shows no inclination towards either over- or under-prediction of observations, see figure 2b).

Figure 2 *Schematic Representation of Consistent and Inconsistent Errors; a) Consistent errors - model biassed towards either over- or under-prediction of the observations; b) Inconsistent errors - no bias in model errors. (■ represents measured values; — represents simulated values)*

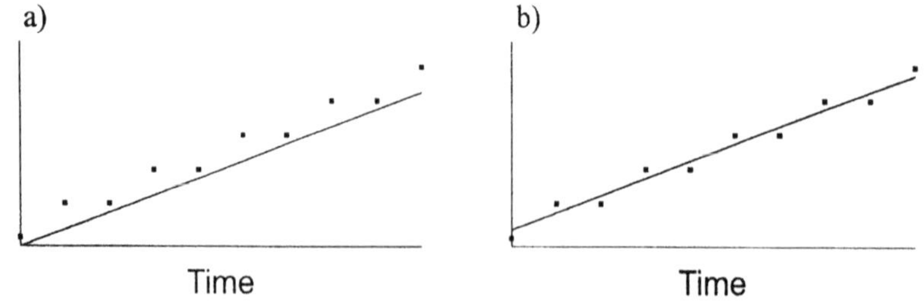

In the calculation of mean difference, *M* (Addiscott & Whitmore, 1987), only *consistent* errors are included since *inconsistent* errors cancel out. The relative error, *E* (Addiscott & Whitmore, 1987) provides a more standardised measure of the consistent errors by expressing the consistent errors as a proportion of the observed values. However, if extreme values are included, *E* is liable to become enlarged in a way that does not truly reflect the accuracy of the simulation, and so *E* should not be considered in isolation of *M*. The coefficient of residual mass, *CRM* (Loague & Green, 1991) provides a different measure of consistent error by determining the consistent error of all simulated values across all measurements with no consideration of the order of the measurements. A high relative error (*E*) but low coefficient of residual mass (*CRM*) implies that the model generates similar values to the measured data but that the model does not place them in the correct chronological order. This information may suggest strategies for future improvements to the model.

If the experiment has been replicated, it is more useful to express the total difference in terms of the *systematic* errors occurring *between* experiments and the *random* errors occurring *within* experiments (see figure 3). The random errors, representing the experimental error, can be obtained by examining the replicated values. The systematic errors, representing the failure of the model to simulate differences between the experiments is then calculated as the lack of fit statistic, *LOFIT* (Whitmore, 1991).

Figure 3 *Schematic Representation of Systematic and Random errors; a) Systematic errors - the failure of the model to simulate differences between experiments b) Random errors - experimental errors. (■ represents measurements from experiment 1; □ represents measurements from experiment 2; — represents simulated values)*

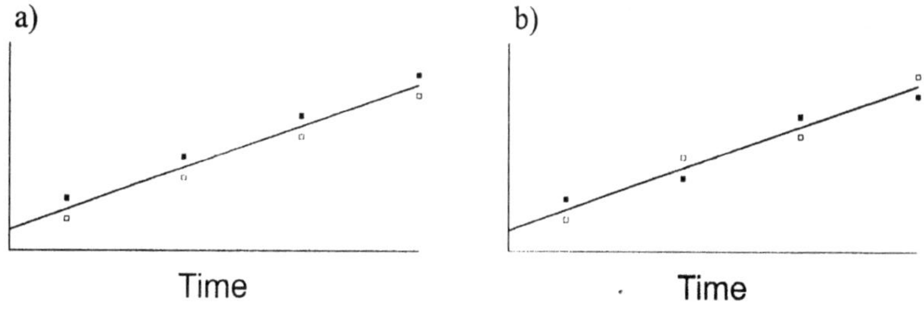

The statistical significance of the systematic errors, expressed as *LOFIT*, may then be calculated using an *F*-test.

Finally, the maximum difference between an individual pair of simulated and measured values may be highlighted by determining the maximum error, *ME* (Loague & Green, 1991). This is of particular value if a maximum allowable error in the simulations can be identified. Alternatively, the maximum error can be used to indicate where a small number of very large errors are contributing to a large proportion of the overall error calculated and so other statistical values may become unreliable.

In all the equations discussed below;

O_i are the observed (measured) values,

P_i are the predicted (simulated) values,

$\bar{O}$ is the mean of the observed (measured) data,

$\bar{P}$ is the mean of the predicted (simulated) data.

n is the number of samples

Association between Simulated and Measured Values

The Sample Correlation Coefficient (r)

$$r = \frac{\sum_{i=1}^{n} (O_i-\bar{O})\,(P_i-\bar{P})}{\left[\sum_{i=1}^{n} (O_i-\bar{O})^2\right]^{1/2} \left[\sum_{i=1}^{n} (P_i-\bar{P})^2\right]^{1/2}} \qquad \textbf{(1)}$$

The sample correlation coefficient (r) provides an estimate of ρ, the correlation coefficient for the whole population (Draper & Smith, 1966). It is a measure of the degree of *association* between simulation and measurement, but not necessarily of *coincidence* (Addiscott & Whitmore, 1987).

Values of r lie between -1 and +1. If $r = +1$, there is perfect positive correlation between simulated and measured values (figure 4a). If $r = -1$, there is a negative correlation (figure 4b). If, however, $r = 0$ there is no correlation between simulations and measurements, and the values are not linearly related (figure 4c).

Figure 4 *Schematic Representation of the plot of measured versus simulated values for a range of correlation coefficients; a) $r = +1$; b) $r = -1$; c) $r \rightarrow 0$. (■ represents the plot of measured versus simulated values; - - - represents the 1:1 correlation line)*

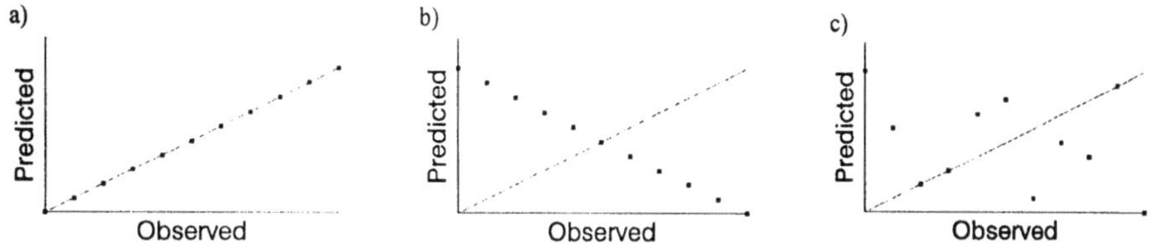

Models are sometimes compared directly with each other by examining values of r (or more commonly r^2). Alternatively, assuming a normal distribution, the statistical significance of r can be assessed using the conventional F-test (as in an analysis of variance from simple linear regression, e.g. Snedecor & Cochran, 1989). The ratio of the *regression mean square* and the *residual mean square* is distributed as F. The residual sum of squares is represented by $(1-r^2)$ with $(n-2)$ degrees of freedom, and the regression sum of squares is represented by r^2 with 1 degree of freedom (Snedecor & Cochran, 1989). Therefore, the significance of r may be tested using

$$F = \frac{(r^2/1)}{(1-r^2)/(n-2)} = \frac{(n-2)r^2}{(1-r^2)} \qquad (2)$$

Tables of r have been calculated using $\sqrt{F}$ (or its equivalent, Student's t) and are published for several levels of probability and a range of degrees of freedom (eg. Fisher & Yates 1964, Snedecor & Cochran, 1989). Alternatively, the probability (P) that the simulated and measured populations are equivalent can be calculated from Student's t or F using published algorithms (e.g. Abramowitz & Stegun, 1965; Cooper, 1968). The test requires a minimum of 5 measured values.

It is most desirable for models to be evaluated against independent data, allowing no input parameters to be adjusted to improve the correlation between simulated and measured values. However, in some cases, a model cannot be run without adjusting input parameters; an example is where no provision has been made within the model to calculate the annual carbon inputs from a particular crop. In this case, the modeller may decide to adjust the carbon inputs to provide the best correlation with measurements. If $k+1$ parameters have been adjusted, $\sqrt{F}$ is no longer equivalent to t, and equation 2 must be expanded to

$$F = \frac{(n-k-1)r^2}{k(1-r^2)} \tag{3}$$

Effectively, one degree of freedom is removed for every parameter that is adjusted to improve the fit, making the test of statistical significance more exacting. This represents an *ad hoc* implementation of the standard F-statistic for a k-parameter multiple regression model (see Snedecor & Cochran, 1989, p.343). It should be noted that when model input parameters are adjusted to improve model fit, simulated and measured values are no longer independent, and so, of the tests presented in this paper, only equation 3 can be reliably used.

Coincidence: Total Difference between Simulations and Measurements

Root mean square error (RMSE)

$$RMSE = \frac{100}{\bar{O}}\sqrt{\sum_{i=1}^{n}(P_i-O_i)^2/n} \tag{4}$$

The root mean square error (*RMSE*) used by Loague & Green (1991) provides a percentage term for the total difference between the predicted and the observed values, proportioned against the mean observed value. The lower limit for *RMSE* is 0, which denotes no difference between measured and simulated values.

RMSE can be used directly to compare the error in the simulations of different models, a lower value of *RMSE* indicating a more accurate simulation. In addition, if the standard deviation, S_d , for the measurements is available, the significance of the root mean square error can be assessed by comparing *RMSE* to the value obtained for different percentage errors $(RMSE_P)$ i.e.

$$RMSE_P = \frac{100}{\bar{O}} \sqrt{\sum_{i=1}^{n} (t_{nP} \times S_d)^2 / n} \tag{5}$$

where

P is the probability that two different populations will have a root mean square error less than $RMSE_P$;

t_{nP} is the t-distribution for n-2 degrees of freedom and probability P.

It is often assumed that if $RMSE$ is less than $RMSE_{0.05}$, the simulated and measured populations are statistically equivalent.

Alternatively, if no estimates for S_d exist, the acceptable root mean square error $(RMSE_{ERR})$ may be obtained by substituting some arbitrary value for acceptable error (O_{ERR}) into the expression for $RMSE$, i.e.

$$RMSE_{ERR} = \frac{100}{\bar{O}} \sqrt{\sum_{i=1}^{n} O_{ERR}^2 / n} \tag{6}$$

The value of O_{ERR} may be derived, for example, from the accuracy possible in experimental measurements or the accuracy necessary in simulations. If $RMSE$ is less than $RMSE_{ERR}$, the error of the simulation is considered to be acceptable.

Modelling efficiency (EF)

$$EF = \frac{\left(\sum_{i=1}^{n} (O_i - \bar{O})^2 - \sum_{i=1}^{n} (P_i - O_i)^2\right)}{\sum_{i=1}^{n} (O_i - \bar{O})^2} \tag{7}$$

Modelling efficiency, EF, provides a standard method for assessing the accuracy of simulations by comparing the variance of predicted from observed values to the variance of observed values from the mean of the observations. This is effectively a comparison of the efficiency of the chosen model to the efficiency of a very simple predictive model: the mean of the observations.

The maximum value for EF is 1 (as illustrated in figure 5a), in which case the predicted values perfectly match the measured values, i.e.

$$\sum_{i=1}^{n} (P_i - O_i)^2 = 0 \tag{8}$$

EF can become negative (as illustrated in figure 5b). If EF is less than 0, the simulated values are worse than simply using the observed mean and the chosen model is not performing adequately (Loague & Green, 1991).

Figure 5 *Schematic Representation of Modelling Efficiency; a) EF = 1; b) EF < 0. (■ represents measured values; — represents predicted values; - - - represents the mean of the measured values)*

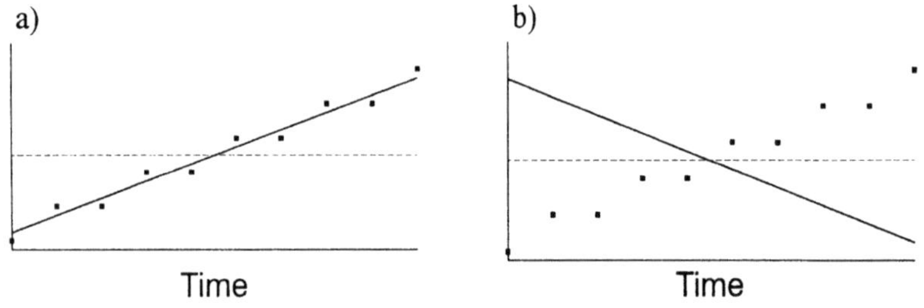

Coefficient of determination (CD)

$$CD = \sum_{i=1}^{n} (O_i - \bar{O})^2 \;/\; \sum_{i=1}^{n} (P_i - \bar{O})^2 \tag{9}$$

The coefficient of determination, CD (Loague & Green, 1991), compares the variance of the observations from the mean observed value to the variance of the predictions from the mean observed value. CD is effectively a measure of the proportion of the total variance in the observed data that is explained by the predicted data.

The lower limit for CD is 0, in which case the observed values do not deviate from the mean. The statistic can then provide no information about the variance of the predictions and should be ignored.

A *CD* value of 1 denotes no difference between the variances in the measured and predicted values. However, it should be noted that this does not necessarily indicate equivalence between the measured and predicted values since the statistic does not use matched data pairs (see figure 6).

Figure 6 *Schematic Representation of Coefficient of Determination Value 1; a) Predicted and measured values not equivalent; b) Measured and predicted values equivalent.* *(■ represents measured values; — represents simulated values)*

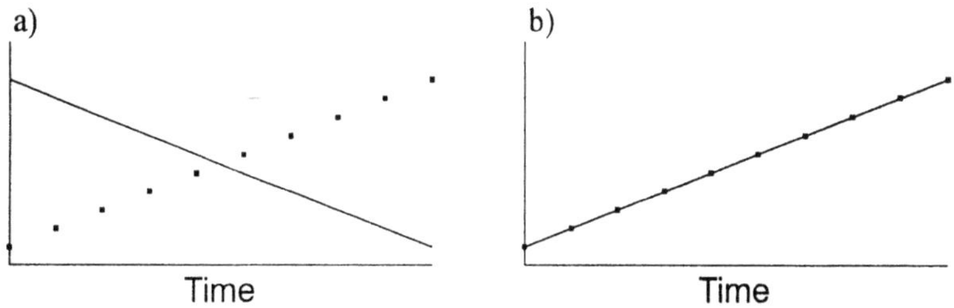

If *CD* is between 0 and 1, the deviation of the predictions from the mean observation is greater than observed in the measurements. This could be due to a bias or systematic error in the model, or alternatively might indicate that descriptions of processes in the model of the correct type are "over-active" and should be using different parameters. Visual/graphical methods can be used to fully interpret the meaning of this test result.

If *CD* is greater than 1, the deviation of the predictions from the mean observation is less than observed in the measurements. This may indicate that the model does not adequately describe extreme values or could merely reflect the lower variation in the modelled values than in the measurements due to greater errors in experimental measurements. Again, visual/graphical methods can be used to further investigate the nature of any experimental errors and so fully interpret this test result.

Consistent Errors in the Model

Mean difference between measurement and simulation (M)

$$M = \sum_{i=1}^{n} (O_i - P_i) / n \tag{10}$$

The mean difference between measured and simulated values (Addiscott & Whitmore, 1987) gives an indication of the *bias* or *consistent* errors. Because *M* does not include a square term, simulated values above and below the measurements cancel out and so any *inconsistent* errors are ignored.

M can be positive or negative. If simulated and measured values are identical, *M* will be zero. A non-significant, small standard error in *M* shows a satisfactory degree of coincidence but not necessarily of association.

If no model parameters have been adjusted to minimise the mean difference, the coincidence between the simulated values and the population of measured values can again be tested using Student's *t*. Assuming a normal distribution for the whole population and using the null hypothesis that the mean difference between the simulated and measured populations is zero, Student's *t* can be calculated as shown below:

$$t = \frac{M}{s_d / \sqrt{n}} = \frac{\sum_{i=1}^{n} (O_i - P_i) / n}{s_d / \sqrt{n}} \tag{11}$$

where

$$s_d^2 = \frac{\sum_{i=1}^{n} (d_i - M)^2}{n-1} = \frac{\sum_{i=1}^{n} ((O_i - P_i) - (\sum_{i=1}^{n} (O_i - P_i) / n))^2}{n-1} \tag{12}$$

d_i is the difference between measured and simulated values

M is the mean difference between measured and simulated values

The *t* statistic (here after Chatfield, 1983) can then be related to the probability that the simulated and measured populations are statistically equivalent using standard statistical tables (e.g. Fisher & Yates, 1964) or published algorithms (e.g. Abramowitz & Stegun,

1965; Cooper, 1968). The number of degrees of freedom for the two-tailed *t*-test is n-2. The test requires a minimum of 3 measured values.

Relative Error (E)

$$E = \frac{100}{n} \sum_{i=1}^{n} (O_i - P_i)/O_i \tag{13}$$

The relative error, used by Addiscott & Whitmore (1987) gives the error in the simulation as a proportion of the measurement, and expresses this as the mean percentage over all measurements. However, because E is expressed as a proportion of the measurements, extreme values can enlarge E in a way that does not truly reflect the accuracy of the simulation, and so should be excluded when computing this statistic.

As for the root mean square error (*RMSE*), E can be used directly to compare the consistent errors in the simulations of different models, or its significance can be assessed by comparing E to the value obtained for different percentage errors (E_P) i.e.

$$E_P = \frac{100}{n} \sum_{i=1}^{n} (t_{nP} \times S_d)/O_i \tag{14}$$

where

P is the probability that two different populations will have a relative error less than E_P ;

t_{nP} is the t-distribution for n-2 degrees of freedom and probability P.

Again, it is often assumed that if E is less than $E_{0.05}$, the simulated and measured values are statistically equivalent.

Alternatively, the acceptable relative error (E_{ERR}) can be obtained by substituting some arbitrary value for acceptable error (O_{ERR}) into the expression for E, i.e.

$$E_{ERR} = \frac{100}{n} \sum_{i=1}^{n} O_{ERR}/O_i \tag{15}$$

If E is less than E_{ERR}, the consistent error of the simulation is considered to be acceptable.

Coefficient of residual mass (CRM)

$$CRM = \left(\sum_{i=1}^{n} O_i - \sum_{i=1}^{n} P_i\right) / \sum_{i=1}^{n} O_i \qquad (16)$$

The coefficient of residual mass, *CRM* (Loague & Green, 1991), gives an indication of the consistent errors in the distribution of all simulated values across all measurements with no consideration of the order of the measurements.

CRM can be positive or negative: a negative value indicating that the majority of predicted values are greater than the measured values; a positive value indicating that the majority of predicted values are less than the measured values. A CRM value of 0 denotes no bias in the distribution of predicted values with respect to measured values. However, as for *CD*, it should be noted that since this is not a paired test, *CRM = 0* does not necessarily indicate that the actual values obtained by prediction and measurement are equivalent.

Systematic Errors in the Model

Sum of squares attributable to lack of fit (LOFIT)

$$LOFIT = \sum_{j=1}^{N} m_j \bar{d}_j^{\,2} = \sum_{j=1}^{N} m_j (\bar{O}_j - P_j)^2 \qquad (17)$$

where:

N is the number of experiments,

m_j is the number of replicates within each experiment,

P_j is the simulation for the *j*th experiment,

$\bar{O}_j$ is the mean of the measurements in the *j*th experiment

$\bar{d}_j$ is the mean deviation $= \bar{O}_j - P_j$.

LOFIT, representing the systematic errors in the model (Whitmore, 1991), differs from the statistics discussed earlier in that the replicates of each set of measurements have been used to eliminate the inherent variability in the measured data (the random error, *SSE*) from the total error sum of squares (*RSS*) i.e.

$$LOFIT = RSS - SSE \tag{18}$$

where

$$RSS = \sum_{j=1}^{N} \sum_{i=1}^{m_j} (O_{ij} - P_j)^2 \tag{19}$$

$$SSE = \sum_{j=1}^{N} \sum_{i=1}^{m_j} \left((O_{ij.} - P_j) - (\overline{O}_j - P_j) \right)^2 \tag{20}$$

and O_{ij} is the *i*th measurement in the *j*th experiment.

The statistical significance of *LOFIT* can be obtained by comparing the mean square due to lack of fit (*MSLOFIT*) to the mean square due to the random error (*MSE*) to derive the variance ratio (*F*).

$$F = \frac{MSLOFIT}{MSE} = \frac{\sum_{j=1}^{N} (m_j - 1) \sum_{j=1}^{N} m_j (\overline{O}_j - P_j)^2}{N \sum_{j=1}^{} N \sum_{i=1}^{n} \left((O_{ij} - P_j) - (\overline{O}_j - P_j) \right)^2} \tag{21}$$

This F statistic can then be related to the probability that the simulated and measured populations are statistically equivalent using standard statistical tables (e.g. Fisher & Yates, 1964) or published algorithms (e.g. Abramowitz & Stegun, 1965; Cooper, 1968).

Maximum Difference between Simulations and Measurements

Maximum error (ME)

$$ME = \max |P_i - O_i| \Big|_{i=1}^{n} \tag{22}$$

The maximum error observed between any of the pairs of predicted and observed values, *ME,* is different from the statistics considered previously in that it highlights differences between *individual* pairs of predicted and observed values. *ME* can be used in a number of ways. The lower limit for *ME* is 0, which denotes no difference between any of the measured and simulated values (Loague & Green, 1991). A value of *ME* that is much greater than $RMSE \times \bar{O}/100$ could indicate an erroneous observation or may point to a particular aspect of the model that requires further development. Given reliable experimental data, the model may be inadequate if *ME* is much greater than some chosen value for acceptable error.

Conclusions

In the preceding sections, statistical methods that can be used to compare and evaluate SOM models have been discussed. The successful application of these methods depends on the information available and the nature of the simulation.

The sample correlation coefficient, *r*, is the first statistic that should be calculated when evaluating the accuracy of the simulations. The value of *r* will indicate if the model has the appropriate structure. The statistical significance of the *r* value can be calculated using an *F*-test. If any model input parameters have been adjusted to improve the fit, the expression for *F* should be expanded to allow for the loss of one degree of freedom for each parameter that has been adjusted. Note that none of the other tests discussed can be used to assess statistical significance if model parameters have been adjusted to improve the fit of the simulation since measured and simulated values are no longer independent.

If the experiments have been replicated, the lack of fit statistic, *LOFIT*, should always be calculated. This statistic allows the experimental errors to be distinguished from the failure of the model. The significance of the *LOFIT* should be calculated using an *F*-test. If the test indicates a probability of greater than 95% that the modelled and the simulated populations are different, the model could be improved by further refinement.

However, long-term experiments, started many years ago, are often not replicated. In this case, the statistical analysis becomes more cumbersome and is often less informative. The total difference between the simulated and the measured values should first be calculated as the root mean square error, *RMSE*. If the standard deviations of the measurement are known, the statistical significance of *RMSE* should be assessed by comparing to the value obtained assuming 95% error. If no estimates for the standard deviations exist, the value of *RMSE* should be compared to the value obtained for some arbitrary value of acceptable error, and by calculating the modelling efficiency, *EF*, and the coefficient of determination, *CD*. Taken together, these tests allow *RMSE* to be interpreted, despite not knowing the standard deviations of measurements.

The total difference should be analyzed further by calculating the relative error, *E*. If the standard deviations of the measurements are known, the significance of *E* should again be determined by comparing to the value obtained assuming 95% error, or, if the standard deviations of the measurements are unknown, using some arbitrary value for acceptable error. If the relative error is significant, calculate the coefficient of residual mass, *CRM*. A *CRM* value close to zero could suggest that although the processes are adequately described in the model, the relative timing of modelled events is incorrect. Reference back to a graphical comparison of simulated and measured values should help to interpret the nature of these errors.

Finally, the maximum error, *ME*, should always be calculated. A value of *ME* that is much greater than $RMSE \times \bar{O}/100$ could indicate that a small number of very large errors are contributing to a large proportion of the overall error. This may indicate measurement errors, suggesting that other statistical tests are unreliable.

All of the statistical methods and qualitative visual/graphical techniques discussed above have been brought together in a computer program, MODEVAL. The MODEVAL program provides a simple framework that allows the user to, a) read-in simulated values from SOM models and actual measured data, b) compare simulated and measured values visually/graphically using screen-plotting routines, and c) derive a range of statistics that give a quantitative indication, for the purposes of model comparison, of how well SOM models simulate measured data. If used with care, the range of statistics provided by MODEVAL can be used for the rational and quantitative evaluation and comparison of SOM model performance. Details of MODEVAL and a DOS or Windows version of the program can be obtained from the authors.

References

Abramowitz, M. & Stegun, I.A. (eds.) (1965) *Handbook of mathematical functions*. Dover Publications, New York, 1046pp.

Addiscott, T.M. & Whitmore, A.P. (1987) Computer simulation of changes in soil mineral nitrogen and crop nitrogen during autumn, winter and spring. *Journal of Agricultural Science, Cambridge* **109**: 141-157.

Chatfield, C. (1983) *Statistics for technology. Third Edition*. Chapman & Hall, London. 381pp

Cooper, B.E. (1968) The integral of Student's *t*-distribution. Algorithm AS3. *Applied Statistics* **17**: 189-190.

de Willigen, P. (1991) Nitrogen turnover in the soil-crop system; comparison of fourteen simulation models. *Fertilizer Research* **27**: 141-149.

Draper, N.R. & Smith, H. (1966) *Applied regression analysis*. Wiley & Sons, New York. 407 pp.

Fisher, R.A. & Yates, F. (1964) *Statistical tables for biological, agricultural, and medical research* (6th ed.), Hafner Publishing, New York.

Loague, K. & Green, R.E. (1991) Statistical and graphical methods for evaluating solute transport models: Overview and application. *Journal of Contaminant Hydrology* **7**: 51-73.

Loague, K., Bernknopf, R.L., Giambelluca, T.W. & Green, R.E. (1995) The impact of data uncertainty upon regional scale leaching assessments on non-point source pollutants. In: *Applications of GIS to the modeling of non-point source pollutants in the vadose zone.*, ASA-CSSA-SSSA Bouyoucos Conference, Mission Inn, Riverside, California, May 1-3, 1995: 494-535.

Otter-Nacke, S. & Kuhlmann, H. (1991) A comparison of the performance of N simulation models in the prediction of N min on farmers' fields in the spring. *Fertilizer Research* **27**: 341-347.

Snedecor, G.W. & Cochran, W.G. (1989) *Statistical Methods, Eighth Edition*. Iowa State University Press. pp.491

Vereecken, H., Jansen, E.J., Hack-ten Broeke, M.J.D., Swerts, M., Engelke, R., Faberwitz, S. & Hansen, S. (1991) Comparison of simulation results of five nitrogen models using different datasets. In: *Soil Groundwater Research Report II. Nitrate in soils*. Commission of the European Communities: 321-338.

Webster,R., & Oliver,M.A. (1990) *Statistical Methods in Soil and Land Resource Survey*. Oxford University Press, New York. pp.316.

Whitmore, A.P. (1991) A method for assessing the goodness of computer simulations of soil processes. *Journal of Soil Science* **42**: 289-299.

Acknowledgements

IACR receives grant-aided support from the Biotechnology and Biological Sciences Research Council of the United Kingdom. We acknowledge funding for P.S. from the Terrestrial Initiative in Global Environmental Research (TIGER) programme of the U.K. Natural Environment Research Council (NERC), and NATO for funding the Advanced Research Workshop for which this paper was prepared. We are grateful to Jon Arah (ITE, Edinburgh) for suggestions regarding the relationship between r and F, and to Mick Whelan (Exeter University) for suggestions on an earlier draft of the manuscript. We would also like to thank Moira Mugglestone and Gavin Ross of IACR-Rothamsted Statistics Department for invaluable statistical advice and access to computer code for calculating p values from t and F.

Soil Organic Matter Models and Global Estimates of Soil Organic Carbon

W. M. Post, A. W. King, and S. D. Wullschleger
Environmental Sciences Division
Oak Ridge National Laboratory
P. O. Box 2008, Bldg. 1000
Oak Ridge, Tennessee 37831-6335
USA

Summary

The large size and potentially long residence time of the soil organic matter pool make it an important component of the global carbon cycle (Schlesinger, 1977; Post *et al.*, 1982, 1985, 1990). The input rates and decomposition rates for different terrestrial ecosystems vary over several orders of magnitude resulting in widely different amounts and turnover rates of soil organic matter. The amounts of carbon stored in soils and the rates of exchange of soil carbon with the atmosphere depend on many factors related to the chemistry, biology, and physics of soil and soil organic matter.

We review the major factors that influence soil organic matter dynamics that need to be explicitly considered in development of global estimates of carbon turnover in the world's soils. We discuss current decomposition models that have been used to represent global soil organic matter dynamics. We use a global scale implementation of the Rothamsted model explore the potential transient response of soil organic matter under climate change where the climate conditions change substantially faster than the length of time required for soil organic matter dynamics to equilibrate. These transient simulations indicate that after 100 years of a continuously changing climate to a projected 2xCO_2 climate, global soil organic matter change is only about one-third of the way toward the eventual equilibrium under a 2xCO_2 climate. Using the UKMO predicted climate change, there is a small net sequestration of carbon in soil early in the climate transition period as the result of tropical ecosystem soil carbon increases. This is followed by large carbon releases from arctic and boreal soils later in the 100 year climate transition period. The largest net release of carbon from soil is at the end of the 100 year climate transition after which these net releases gradually decrease as the soil carbon pools approach equilibrium under the 2xCO_2 climate regime.

NATO ASI Series, Vol. I 38
Evaluation of Soil Organic Matter Models
Edited by David S. Powlson, Pete Smith and Jo U. Smith

Factors influencing soil organic matter dynamics

The amount of carbon stored in soil is determined by the balance of two biotic processes: productivity of terrestrial vegetation and decomposition of organic matter. Both of these processes have strong physical and biological controlling factors. These include climate; soil chemical, physical, and biological properties; and vegetation composition. Interactions among these controlling factors are of particular importance.

Organic Matter Inputs

Quantity. The amount of carbon stored in soils and the amount of trace gases produced through decomposition is to a great extent determined by the rate of organic matter input through litterfall, root exudates and root turnover. The main factors that influence vegetation production are suitable temperatures for photosynthesis, available soil moisture

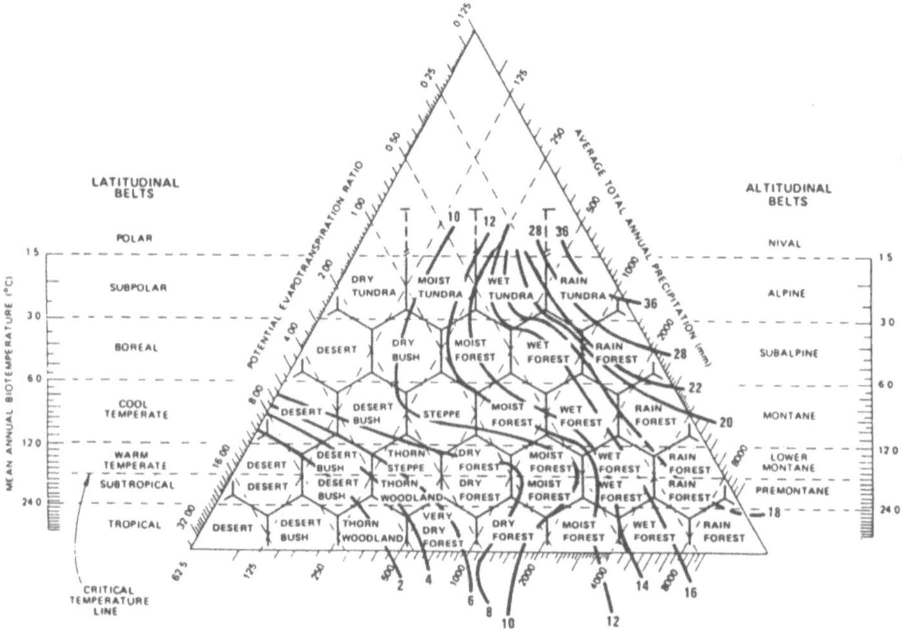

Figure 1. *Soil carbon concentration (kg m⁻³) as a function of Holdridge Life Zone (from Post et al., 1985 and reprinted by permission from Nature vol. 317 p. 614. Copyright 1985 MacMillan Magazines Ltd.).*

and rates at which the plant can exchange CO_2 and H_2O. Dry and / or cold climates support low vegetation production rates and soils under such climates have low organic matter contents. Exceptions occur where soils are so wet or frozen that decomposition is limited and even small rates of vegetation production accumulate over long periods as dead organic matter. Where climates are warm and moist, vegetation production is high and soil organic matter contents are correspondingly high (Figure 1).

Vegetation production depends not only on climate but also nutrient supply. Walker & Adams (1958) hypothesized that the level of available phosphorus during the course of soil development is the primary determinant of terrestrial net primary production. Numerous workers have examined this hypothesis. Tiessen *et al.*, (1984) and Roberts *et al.*, (1985) found that available phosphorus explained about 25% of the variance in soil organic matter in many different soil orders. The relationship between phosphorus and carbon is strongest during the aggrading stage of vegetation-soil system development (Anderson, 1988). Initially, the production of acidic products by pioneer vegetation promotes the release of phosphorus by weathering of parent material. Organic matter builds up in the soil, increasing the storage of phosphorus in decomposing organic compounds. Nitrogen-fixing bacteria populations, which depend on a supply of organic carbon and available phosphorus, can grow to meet ecosystem demands for nitrogen. Plant growth is enhanced by this increasing nitrogen and phosphorus cycling, resulting in increased rates of weathering. This process continues until the vegetation is constrained by other factors affecting phosphorus availability (such as leaching losses becoming larger than the weathering inputs (Jenny, 1980), or an increasing fraction of the phosphorus becoming unavailable by adsorption or precipitation with secondary minerals (Walker & Syers, 1976)) or by factors affecting nitrogen availability (such as denitrification or leaching reaching or exceeding nitrogen inputs and fixation (Schlesinger, 1991). In mature soils, net primary production is more likely to be limited by nitrogen. Availability of other nutrients that are largely derived from parent materials, such as most base cations, may also influence soil organic matter accumulation during early soil development. Soils derived from base cation rich volcanic parent materials (basic extrusive) have much higher carbon contents on average than soils from other parent materials (Zinke *et al.*, 1984, Table 1).

Species Composition. Biotic factors, in particular, plant species composition also affect soil organic matter dynamics. Decomposition is to some degree controlled by species composition.

Each terrestrial plant species produces and sheds a characteristic blend of carbon compounds, each of varying decomposability. This range of decomposability may be summarized by the lignin and nitrogen content of the organic material (Aber & Melillo, 1982, Meentenmeyer, 1978). Nitrogen is made available to plants during the decomposition process. Nitrogen is a limiting element for productivity in most terrestrial ecosystems so the rate at which it is released during decomposition is an important factor in ecosystem production. Thus the interactions between processes regulating plant populations and their productivity, and microbial processes regulating nitrogen availability result in some of the observed variation in soil carbon and nitrogen storage and so also in the rate of carbon exchange between the soil and atmosphere.

Table 1. *Soil carbon concentration in relationship to parent material (modified from Zinke et al., 1984).*

Parent Material	*Mean Carbon Density (kg m^{-1})*
Acid intrusive	10.9
Basic intrusive	12.2
Ultrabasic	10.6
Acid extrusive	15.1
Basic extrusive	21.8
Sedimentary (consolidated)	12.6
Sedimentary (weakly consolidated)	14.6
Metamorphic	11.2

Placement. The deeper that fresh detritus is placed in the soil, the slower it decomposes. This is a result of declining decomposer activity and increased protection from oxidation with depth in the soil. Prairies have a somewhat lower productivity than forests, and produce no slowly decomposing woody material. Nevertheless, prairies have a very high soil organic matter content because prairie grasses allocate twice as much production to below-ground roots and tillers than to above-ground leaves (Sims & Coupland, 1979). The result is high soil organic matter contents with a uniform distribution in the upper 1 meter of soil (Zinke *et al.*, 1985). In contrast, a spruce-fir forest contains 50 % of its soil organic matter in the top 10 cm. There are interesting exceptions to the rule that above-/below-ground plant allocation determines soil organic matter distribution patterns in soil. Tropical moist forest soils show a uniform depth distribution similar to the depth distribution of temperate grasslands, however, in tropical forests this is largely due to a long-term accumulation of recalcitrant organic materials at lower depths in the soil rather than increased allocation to roots. Alpine tundra soils support a largely herbaceous flora but show a similar depth distribution as forest soils because of inhibition of surface litter decomposition by low temperatures and high water saturation.

Decomposition

Climate. Organic matter decay rates can be related to environmental parameters such as temperature and soil moisture. There are a number of detailed models of microbial processes (Bunnell & Dowding, 1974; Hunt, 1977; Bunnell *et al.*, 1977; Juma & Paul, 1981; Smith, 1982) that attempt to capture the details of the relationship between these environmental factors and decomposition. These models characteristically work on a short time step (one day or less), predict decomposition by modeling microbial populations, and have high data requirements. A simpler approach is predicting decay rates from more easily measured climatic and chemical indices rather than modeling the detailed microbial dynamics. Climatic indices that correlate well with decay rates include soil moisture and temperature indices (Fogel & Cromack, 1977), linear combinations of temperature and rainfall (Pandey & Singh, 1982), and actual evapotranspiration (AET; Meentemeyer, 1978). These climatic indices are estimates of potential rather than actual microbial perception of soil water availability and temperature.

Organic Matter Quality. On global scale, climate may be the most important factor controlling decay rates, but within a given region, substrate chemistry is the more important factor (Meentemeyer, 1978; Flanagan & VanCleve, 1983; McClaugherty *et al.*, 1984). Decay rate is often negatively related to substrate C:N ratio. Litter C:N is initially much greater than microbial C:N but approaches microbial C:N as the microbes release the carbon as CO_2 while taking up nitrogen (nitrogen immobilization). The further the initial litter C:N is from microbial C:N, the slower the decay rate. Lignin content or lignin:nitrogen ratios are better predictors of decay rates because lignin itself is difficult to decompose, and shields nitrogen and other more easily degraded chemical fractions from microbes. Concise and simple models of decay rate can be based on a combination of chemical and climatic indices (Pastor & Post, 1985).

Soil Physical Structure. As organic matter is fragmented into smaller components and converted into humic molecules, the physical structure of soil particles becomes an important factor in the rate of decomposition. The formation of complexes between organic compounds and clay particles plays a central role in stabilizing organic matter in soil (Anderson, 1979). Newly formed humic acids have a relatively high aliphatic component, much of it as long side chains. This structure and high molecular weight results in a tendency for humic acids to be adsorbed by fine mineral colloids (Greenland, 1965). Soil clay fraction includes a diversity of minerals with different surface areas and properties, and different arrangements of individual clay particles. The formation of soil particle aggregates through polycation bridging, organism glues and mucilages, or organic-inorganic bonds, results in micropore structure that can entrap organic matter and protect it from organisms and their extracellular enzymes.

Using tracer techniques, McGill & Paul (1976) found that much recently formed humic material was associated with mineral colloids <0.04µm in size. In an eight year long decomposition experiment the loss of carbon due to decomposition was strongly related to clay content (Ladd *et al.*, 1985; Amato & Ladd, 1992). In four experimental plots kept free of vegetation, the content of soil organic matter was followed. The soils were all calcareous and influenced by the same climate. Soils with 5%, 12%, 20%, and 42% clay lost 40%, 25%, 15% and 0% organic carbon respectively. In most models of soil organic matter usually around 50% of the soil humus is assumed to be a physically stabilized fraction (Paul & van Veen, 1978; Jenkinson & Rayner, 1977; Parton *et al.*, 1987).

Kilburtus (1980) reports that bacteria within soil aggregates only exist in pores at least three times their own diameter. The greater the clay content, the greater the proportion of small pore spaces, and therefore less pore space is accessible to bacteria. As much as 90% of the pore space in clayey soils may harbour organic matter from which bacteria and other organisms are physically excluded (Oades, 1988).

Interactions between organic matter and inorganic minerals, involving various sized mineral particles and organic colloids with the formation of aggregates, can result in the physical protection of organic matter and increase its retention in soil (Oades, 1988, 1995). Cultivation breaks these natural aggregates through mechanical stress. The decrease in physical protection of organic matter by such disturbances then can lead to an increase in decomposition rates.

Soil organic matter models

Soil organic matter models that infer carbon fluxes and amounts from current compartment sizes and biotic and environmental conditions are often refered to as process models. In actuality all soil organic matter models contain some degree of parameterization based on phenomenological observations so it is difficult to draw sharp distinctions between models that are based on correlations and those that are based on descriptions of biological and physical processes. We will, however, eliminate from the discussion here, strictly correlative models where future soil organic matter is not determined by rates of organic matter input and losses to one or more compartments or pools. While such models have been useful in examining the potential changes in soil carbon resulting from paleoclimates (Prentice & Fung, 1990, Freidlingstein *et al.*, 1992) and potential future climates (Emanuel *et al.*, 1985; Smith & Shugart, 1993), this aproach assumes soil organic matter content is in some sort of equilibrium with respect to inputs and losses. It is becoming increasingly important that we have models that are capable of evaluating the role of soil as sources or sinks of carbon over periods of years to decades during which environmental and biological conditions are changing and equilibrium conditions are not applicable.

Table 2 presents a summary of soil organic matter models that have been applied at the global scale in order to explore the potential role of soils in carbon uptake or release during the period of time over which the burning of fossil fuel (the last 150 years and into the next several centuries) is involved in altering global environmental conditions. The earliest was that of Goudriaan & Ketner (1984) who were interested in the release of carbon from soil as the result of land-use at the global scale. This coarse spatial scale model did not incorporate the effect of many environmental factors directly. Rather, different environmental conditions are represented by the assignment of parameter values expressing flux rates between the different compartments. Future climate projections were not included in Goudrian & Ketner (1984), however, a vegetation/soil model derived from their work forms the terrestrial component of the IMAGE 2 model (Goldewijlk *et al.*, 1994). IMAGE 2 improves the parameterization of temperature and moisture representations which make climate change projections possible.

Table 2. *Model derived global estimates of soil organic carbon and responses to global change.*

Reference [Model Used]	*T*	H_2O	*NPP*	*Nutrients/ Parent Material*	*Nutrient Cycling*	*Litter Quality*	*Placement*	*Physical Protection*	*Spatial Resolution*	*Current SOC (Total Pg C)*	$2xCO_2$ *(Total Pg C)*	*dC/dT (Pg/°C)*	*After 100 yr. Climate Trans.*
Goudriaan & Ketner, 1984	i	i	+				+	+	6 ecosystems	1377			
Esser *et al.*, 1990 [OBM]	+	+	+						2.5° grid	1536	1415[1]	-32.5[1]	
Jenkinson *et al.*, 1991 [Roth.]	+	+	+				+	+	18 ecosystems	1394[2]		-33.7[7]	1294 to 1353[7]
Polglase & Wang [Roth.]	+	+	+				+	+	10 ecosystems	1393			
Potter *et al.*, 1993 [CENTURY]	+	+	+		+	+	+	+	1° grid	455 (20 cm)			
Goto *et al.*, 1993 [Nakane]	+	+	+						5° grid	701[3]			
Goldewijk *et al.*, 1994 [IMAGE]	+	+	+	i					0.5° grid	1499			2058[4]
Schimel *et al.*, 1994 [CENTURY]	+	+	+		+	+	+	+	37 sites interp. to 0.5° grid			-11.1 (20cm)	
Kaduk & Heimann, 1994 [OBM]	+	+	+						2.5° grid	1364			
Melillo *et al.*, 1995 [TEM]	+	+	+		+	+			0.5° grid	1301	1090 - 1121[5] 1308 - 1330[6]	-49 to -64[5] -2 to +76	
this study [Rothamsted]	+	+	+			+		+	0.5° grid	1288[3]	948[7] 1189[5] 1388[6]	-64.0[7] -18.6[5] +18.9[6]	1187[7] 1273[5] 1369[6]

i. indirectly with parameterized representation
1. includes land use effects
2. inputs adjusted to achieve this amount
3. active fractions only
4. all cummulative NEP appears to be SOC for 1975 to 2075
5. climate change only
6. climate change and CO_2 fertilization
7. litter inputs at $1xCO_2$ rate

The next several models presenting global results show two different approaches to handling the representation of soil organic carbon. The Osnabruck Biosphere Model (Esser *et al.*, 1990) used a fairly simple model but at relatively fine spatial resolution. The work of Jenkinson *et al.*, (1991) and Polglase & Wang (1992), on the other hand, employed the Rothamsted turnover model, a model that was developed largely for application to specific locations and contains explicit consideration of many soil processes. These publications applied their models globally with relatively coarse spatial resolution compared to the global spatial variation in climate, vegetation, and key soil properties.

The development of the representation of soil organic carbon dynamics at the global scale after these initial landmark publications has been to increase the spatial resolution and utilization of increasingly detailed and site validated soil models. This presents several difficulties that must be overcome. The first is a technical difficulty. Considerable computational resourses and increased levels of data management of inputs and model results are required. These resources are increasingly available due to advances in computer hardware and software particularly for workstations and parallel processing computers.

A more difficult constraint is that considerable quantities of information concerning biotic, environmental and soil conditions must be supplied at the required spatial resolution. The problem of supplying organic matter inputs at fine spatial resolution and of the appropriate qualities and quantities is usually solved by implementing a plant production model either as a component of the overall terrestrial ecosystem model or in parallel with the decomposition model. Inputs of climate conditions likewise can be provided with the use of global general circulation models. The coupling of additional models has drawbacks in that these models are an additional source of uncertainty. Care must be taken in subsequent interpretation of results. Additional global datasets on vegetation and soil properties at considerably finer spatial scales are currently being developed, improved, and utilized in global models.

The CASA model (Potter *et al.*, 1993) takes an interesting approach in providing many of these inputs from direct observation. Plant production is developed monthly for input into a version of the Century model by a sophisticated inference from global NDVI sattelite images. While this approach has the drawback that it is limited to periods of time for which we have observations, it has great potential for evaluating actual ongoing changes in soil carbon stocks and fluxes. As a result, however, climate change projections are not possible.

Rather than go through each of the models in Table 2, we will, instead, provide an example of the application of the Rothamsted model at a relatively fine spatial resolution for global change studies. This example will provide insight into current trends in understanding soil organic carbon dynamics in the context of global change, in particular for analyses involving the global carbon cycle.

We begin by partitioning the Earth's surface with a square lattice of 0.5 degree latitude by 0.5 degree longitude resolution. Each half degree cell is labeled either land or water. We describe the carbon dynamics of each half-degree terrestrial cell with a compartment model of carbon in vegetation and soil (Figure 2). Each terrestrial grid cell is assigned to one of 31 ecosystem types according to the vegetation map of Matthews (1983) and correspond to the Formation Groups of the Unesco (1973) classification of world vegetation. Each terrestrial grid cell was also assigned a soil type according to the Zobler (1986) digitization of the FAO Soil Map of the World.

The model for Net Primary Production (NPP) is based on the Miami model (Leith, 1975). We assumed a carbon:dry matter ratio of 0.45 to convert the dry matter of the Miami model to carbon. A modified Miami model is often used in carbon cycle studies (Esser, 1991; Esser & Lautenschlager, 1994; Kaduk & Heimann, 1994; Friedlingstein *et al.*, 1992; Dai & Fung, 1993). Furthermore, Foley (1994) showed that the Miami model's estimates of NPP are in good agreement with those from a process-based model of NPP (DEMETER).

Polglase & Wang (1992) developed a formulation representing the effect of CO_2 fertilization by deriving a CO_2 response coefficient from the biochemical model of C-3 photosynthesis described by Farquhar & von Cammerer (1982). This coefficient is an increasing function of CO_2 and an increasing function of temperature consistent with theory and observations of leaf photosynthetic response and applicable to annual ecosystem NPP at half-degree spatial scales (King *et al.*, submitted). We use this coefficient to scale the NPP of the Miami model in response to atmospheric CO_2 concentration. We use the NPP as input to a carbon model of vegetation adaptated from Goudriaan & Ketner (1984) which partitions NPP among four plant parts: leaf, branch, stem and root. We use partitioning coefficients and life span parameters estimated from Goudriaan & Ketner (1984) and Polglase & Wang (1992) for each of our model's vegetation types. Litter production for each plant part is derived from life span parameters and steady state biomass of each plant part.

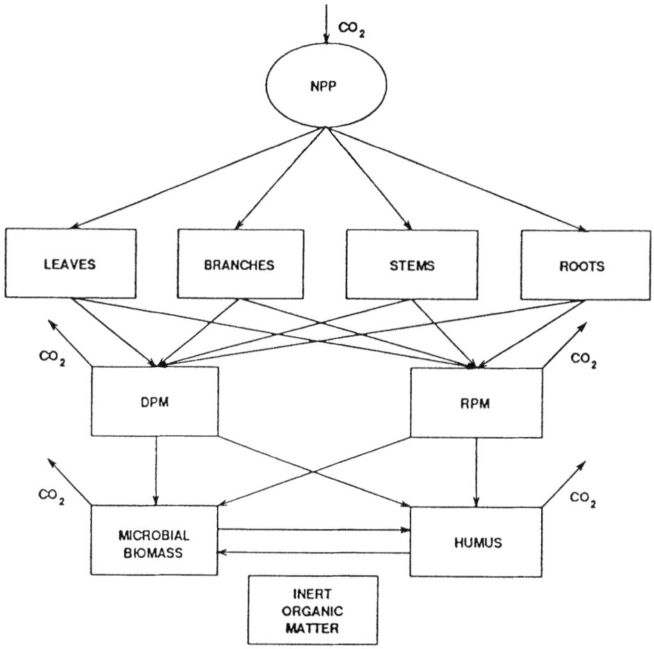

Figure 2. *Model structure for each spatial grid cell to provide organic matter inputs to the Rothamsted turnover model.*

We describe the dynamics of litter and soil carbon with an implementation of the Rothamsted soil carbon turnover model. This model is described in detail elsewhere (Jenkinson, 1990). Each compartment decomposes at it's own characteristic rate modified by environmental factors which include temperature, soil moisture deficit, and a plant protection factor. We needed to modify the temperature function slighty. For monthly temperatures below -15.0 degrees C we set this function to 0.0 since below -18.3 degrees C the temperature function does not behave well. The soil moisture function varies from 0.0 to 1.0. It is 1.0 for pressures ranging from field capacity (mm water held at -33 kPa pressure) to a pressure of -100 kPa. At this point the soil moisture function decreases linearly from 1.0 to 0.2 at a soil moisture corresponding to a pressure of -1500 kPa. We estimate actual soil water pressure for each grid cell using the climatic water budget model of Thornthwaite & Mather (1957). The soil hydraulic characteristics for the Rothamsted soil moisture function and the Thornthwaite & Mather (1957) water balance calculations are derived from soil depth and texture information for each FAO soil type (Webb *et al.*, 1991), and relationships between soil texture and water content at the critical pressures (Rawls *et al.*, 1982).

We generated baseline maps of monthly temperature and precipitation using data from 8034 globally distributed meteorological stations (Emanuel *et al.*, 1985). Monthly values were interpolated at the centroid of each 0.5 degree grid cell using a triangle based, piecewise linear interpolation designed for application on the surface of a sphere (Renka, 1982; Renka & Cline, 1984). We also generated digital maps of monthly temperature and precipitation from the 2x CO_2 equilibrium climates of the U.K Meterological Office (UKMO) atmospheric general circulation model (Jenne, 1992). The monthly temperature and precipitation from the 1xCO_2 and 2xCO_2 climates of the UKMO model were interpolated to the centroid of each 0.5 degree grid. Change factors were calculated for each cell and applied to the baseline temperature and precipitation data interpolated from the meteorological stations. This results in a 5.3 degree C increase in global mean annual temperature and a 14% increase in annual global precipitation for the 2xCO_2 UKMO climate.

The model was run for each terrestrial grid cell using current climate conditions for 20,000 simulated years in order to reach equilibrium. The inert organic matter (IOM) soil pool was set to 0.0 so the results are for the active soil carbon pool. The predicted equilibrium under current conditions results in an NPP of 56.2 Pg yr^{-1} and total active soil carbon of 1288 Pg. The model was then run for an additional 20,000 simulated years under climate conditions projected by the UKMO atmospheric general circulation model with an atmospheric CO_2 concentration of 625 $\mu mol\ mol^{-1}$ CO_2 (current CO_2 levels are 358 $\mu mol\ mol^{-1}$ and pre-industrial levels were 280 $\mu mol\ mol^{-1}$). The global NPP predicted by our model is 63.8 Pg yr^{-1} and active soil carbon pools total 1189 Pg. The ecosystem distribution of active soilcarbon simulated under current climate is shown in Table 3. The change in active soil carbon between equilibrium at a 1xCO2 climate and a 2xCO2 climate (Table 3) indicates that the decrease observed in the global total is a spatially complex summation of increases in some regions and decreases in others. The sensitivity of the overall model to a climate change is low compared to the other model results (Table 3). The UKMO GCM projects a 5.3 degree C temperature increase and a 14% precipitation increase. The soil carbon sensitivity to

Table 3. *Model predicted amount of soil organic carbon by ecosystem type in equilibrium with current climate and in equilibrium with a 2xCO_2 climate.*

Vegetation Type	*Current Climate (Pg C)*	*2xCO_2 Climate (Pg C)*	*Change in Soil C (Pg C)*
tropical evergreen rainforest	89.64	113.10	23.46
tropical/subtrop. evergreen seasonal broadleaved forest	31.00	34.18	3.18
subtropical evergreen rainforest	2.50	2.56	0.05
temperate/subpolar evergreen rainforest	3.73	2.84	-0.89
temperate evergreen seasonal broadleaved forest	8.51	10.12	1.61
evergreen broadleaved sclerphyllous forest	5.74	4.70	-1.03
tropical/subtropical evergreen needleleaved forest	4.20	3.63	-0.58
temperate/subpolar evergreen needleleaved forest	120.97	102.70	-18.26
tropical/subtropical drought deciduous forest	29.58	36.76	7.18
cold deciduous forest, with evergreens	85.08	73.56	-11.52
cold deciduous forest, without evergreens	60.51	59.01	-1.50
xeromorphic forest/woodland	25.64	26.58	0.94
evergreen broadleaved sclerophyllous woodland	18.02	17.59	-0.43
evergreen needleleaved woodland	36.21	29.76	-6.45
tropical/subtropical drought deciduous woodland	32.60	45.05	12.44
cold deciduous woodland	42.49	43.80	1.32
evergreen broadleaved shrubland/thicket	8.97	6.98	-1.99
evergreen needleleaved/sclerophyllous shrubland/thicket	12.23	8.77	-3.46
drought deciduous shrubland/thicket	8.27	7.16	-1.10
cold deciduous subalpine/subpolar shrubland	8.29	7.61	-0.69
xeromorhic shrubland/dwarf shrubland	64.26	51.57	-12.69
arctic/alpine tundra, mossy bog	188.73	152.41	-36.32
tall/medium/short grassland with 10-40% woody cover	79.21	84.45	5.24
tall/medium/short grassland with < 10% woody cover	39.35	39.04	-0.30
tall/medium/short grassland with shrub cover	93.94	73.30	-20.64
tall grassland, no woody cover	13.47	14.41	0.94
medium grassland, no woody cover	18.82	15.26	-3.57
meadow, short grassland, no woody cover	90.48	63.73	-26.75
forb formation	5.16	3.23	-1.92
desert	60.24	54.75	-5.48
Totals	1287.83	1188.59	-99. 23

temperature change is -18.6 Pg °C^{-1}. Some of this reduced sensitivity is the result of increasing NPP with climate change. If NPP is held at it's 1xCO_2 level, as was done in Jenkinson *et al.* (1991) the 2xCO_2 climate change results in a large decline in soil organic carbon of 340 Pg or -64 Pg °C^{-1}. If NPP is stimulated through the mechanism of CO_2 fertilization, then there is a projected increase of 200 Pg in soil organic carbon. Thus while the Rothamsted model shows some sensitivity to potential CO_2 induced temperature changes, it is also quite sensitive to changes in the amount of organic matter inputs.

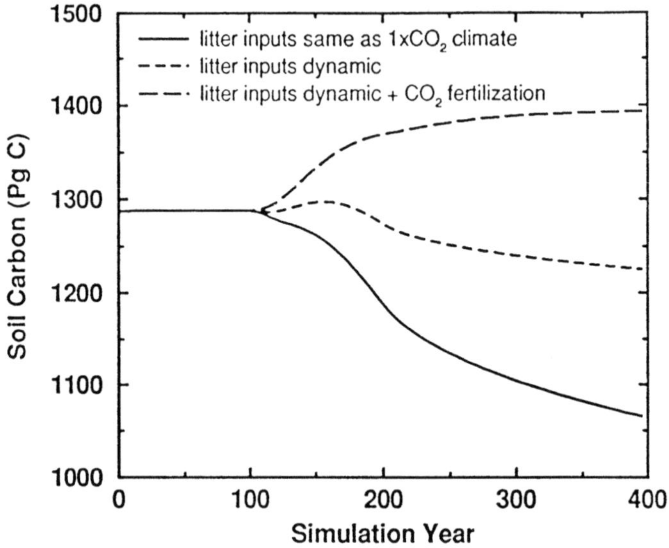

Figure 3. *Changes in global soil organic carbon storage as climate parameters are linearly changed from 1xCO2 conditions in simulation year 100 to 2xCO2 conditions (as projected by the UKMO GCM) in simulation year 200.*

Long term equilibrium responses may be quantitatively different than shorter term transient responses. As Schimel *et al.* (1994) point out, the sensitivity of a multicompartment model to a perturbation or forcing is considerably smaller than a corresponding one pool approximation. We performed simulations with our model starting at the 1xCO_2 equilibrium and at simulation year 100 linearly modifying monthly temperatures, soil moisture deficits, and NPP over the next 100 simulation years until the projected 2xCO_2 climate was obtained in year 300. The model was run for an additional 200 simulation years. The global change in soil organic carbon for these transient simulations is plotted in Figure 3. Holding the organic matter inputs at their 1xCO_2 level shows a loss of 101 Pg carbon over 100 years, considerably less than the eventual equilibrium loss of 340 Pg (Table 4). Spreading the transient climate change period over 200 years instead of 100 years slows the annual rate of soil carbon change but the overall patterns remain the same. At the end of the 200 year climate transient the soil organic carbon pool is still far from climate equilibrium (Table 4). This model shows a temperature sensitivity of -18.6 Pg $°C^{-1}$, nearly half the temperature sensitivity of -33.7 Pg $°C^{-1}$ for a 60 year transient simulation reported by Jenkinson *et al.*, (1991). This difference can be attributed to soil moisture deficit changes being included in our simulations.

Table 4. *Simulated global net primary production (NPP) and soil carbon content for current climate and climate projected with a doubling of atmospheric CO_2 by the UKMO model.*

	Current Climate	*100 Year Transient*	*200 Year Transient*	*2xCO_2 Climate Equilibrium*
Climate Change Only				
NPP (Pg yr^{-1})	56.2	56.2	56.2	56.2
SOC (Pg)	1288	1187	1154	948
Climate Change and Comittant NPP Change				
NPP (Pg yr^{-1})	56.2	63.8	63.8	63.8
SOC (Pg)	1288	1273	1268	1189
Climate Change, NPP a Function of Climate+CO_2				
NPP (Pg yr^{-1})	56.2	75.9	75.9	75.9
SOC (Pg)	1288	1369	1392	1388

The transient simulations show an interesting pattern of an early small increase followed by larger declines as the simulations approach the 2xCO_2 climate. Analysis of the results by ecosystem show that this is due to tropical ecosystems, in particular tropical evergreen rainforest, show rapid increases with climate change while boreal and arctic ecosystems show declines that become greater as the 2xCO_2 climate is achieved (Figure 4). In a study of climate change impacts on grassland vegetation, Parton *et al.*, (1995) found similar responses. When climate change alone was considered they found a decrease in soil carbon globally with small increases in tropical and humid regions and large decreases in cold desert steppe. They attribute the soil carbon increases for the warm humid grasslands as a result of increased N mineralization and its effect on production. Our model does not curently consider nitrogen availability. Simulated climate change impacts alone, and its effect on soil water availablity is sufficient to result in increases in soil carbon in some tropical ecosystems with our model without any consideration of changes in nitrogen availability.

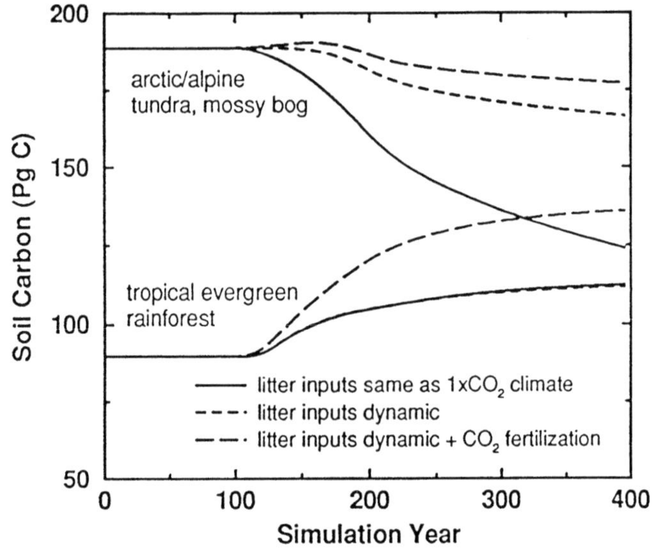

Figure 4. *Changes in soil organic carbon storage for the ecosystem types that have the greatest inpact on the changes in global totals as the result of climate changes.*

When these transient simulations are plotted as net changes in soil carbon per year (Figure 5) climate change can be seen to result in a relatively rapid increase in net soil carbon storage reaching a maximum of 0.5 Pg yr^{-1} followed by fairly substantial losses of carbon to the atmosphere approaching -1.0 Pg yr^{-1} before relaxing as equilibrium is slowly approached. Results of the most responsive ecosystems are plotted individually in Figure 6. This plot further emphasizes the inital increase in tropical/subtropical soil organic carbon storage followed by ever larger decreases in boreal and arctic ecosystems. It should be noted that these result represent a summation of responses from many individual grid cells that may individually show considerably larger or smaller responses to projected climate change.

Conclusions

The balance between carbon assimilation by photosynthesis, its partioning among several terrestrial pools of varying turnover times, and releases of C from dead material through decompostion dictates the magnitude of the net exchange of C between the atmosphere and the world's terrestrial ecosystems. The global soil organic matter pool consists of an array of materials with widely differing turnover times which collectively, are very large and have a potentially important effect on the net storage and release of C at several time scales. To quantitatively evaluate the role of SOM in the global carbon cycle many physical, chemical and

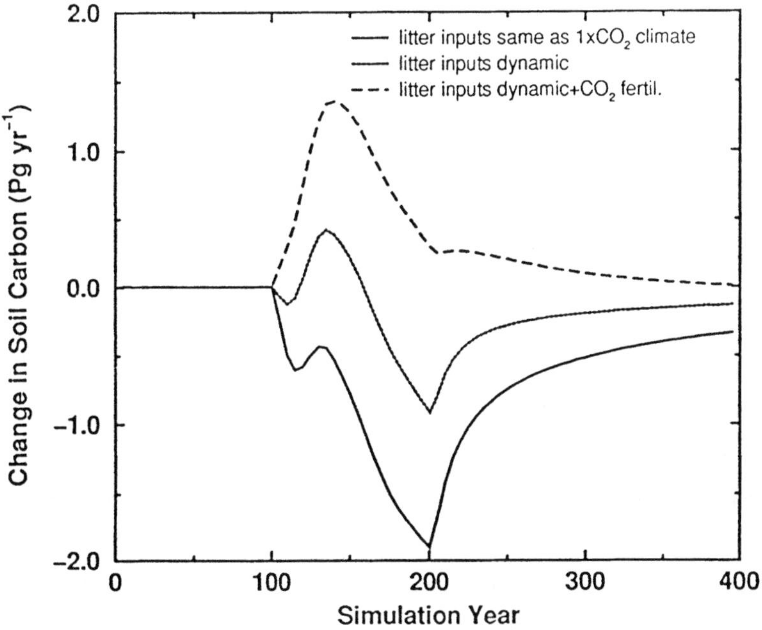

Figure 5. *Net annual changes in soil organic carbon. Positive values indicate a net sequestration of carbon in soil. Negative values indicate a loss of carbon as CO_2 to the atmosphere.*

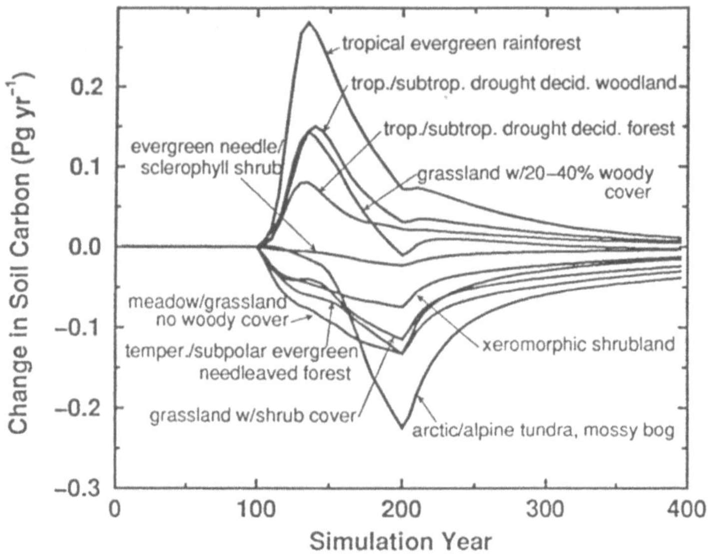

Figure 6. *Net annual changes in soil orgnic carbon storage reported for the ecosystems with the largest contributions to changes in the global total as a result of climate change and associated changes in NPP.*

biotic factors that vary widely from place to place over the earth's surface must be considered. The most important factors include (1) the magnitude of organic matter inputs which depend largely on climate conditions, especially soil water status, nutrient availability, and growth/allocation patterns of species; and (2) the rate of decomposition which depends mainly on climate and its variation, soil chemical composition, and soil physical structure. Simulation models have recently been developed that are quite general, consider most of the important processes regulating organic matter turnover, and may be suitable for extrapolation to most terrestrial systems. The first challenge in applying these simulation models globally is assembling the necessary information, parameters, and environmental inputs to make global calculations. A spatial scale of 0.5 degrees or 50 km is large enough to make this task feasible. It is not clear, however, whether this is an appropriate spatial scale for capturing the natural variation in soil properties. Many soil and vegetation characteristics vary at a much finer scale of spatial resolution. The second and more difficult challenge for global carbon cycle research results will be to capture, in some way, the important aspects of such fine scale variation, especially in regions where it is important. Such regions include those with large topographic variation, dispersed wetlands, a variety of contrasting parent materials, and patchworks of different human utilizations.

References

Aber, J.D. & Melillo, J.M. (1982) Nitrogen immobilization in decaying hardwood leaf litter as a function of initial nitrogen and lignin content. *Canadian Journal of Botany* **58**:416--421.

Amato, M. & Ladd, J.N. (1992) Decomposition of ^{14}C-labeled glucose and legume material in soils: Properties influencing the accumulation of organic residue C and microbial biomass C. *Soil Biology and Biochemistry* **24**:255-464.

Anderson, D.W. (1979) Processes of humus formation and transformation in soils of the Canadian Great Plains. *Journal of Soil Science* **30**:77--84.

Anderson, D.W. (1988) The effect of parent material and soil development on nutrient cycling in temperate ecosystems. *Biogeochemistry* **5**:71--97.

Bunnell, F.L. & Dowding, P. (1974) ABISKO - a generalized model for comparisons between tundra sites.. In: *Soil Organisms and Decomposition in Tundra.* (ed: A.J. Holding, O.W. Heal, S.J.F. MacLean Jr. & P.W. Flanagan) Tundra Biome Steering Committee, Stockholm, Sweden: 227--247

Bunnell F.L., Tact, D.E.N., Flanagan, P.W., & VanCleve, K. (1977) Microbial respiration and substrate weight loss. I. A general model of the influences of abiotic variables. *Soil Biology and Biochemistry* **9**:33-40.

Dai, A. & Fung, I.Y. (1993) Can climate variability contribute to the "missing" CO_2 sink? *Global Biogeochemical Cycles* **7**:599-609.

Esser, G. (1990) Modeling global terrestrial sources and sinks of CO_2 with special reference to soil organic matter. In: *Soils and the Greenhouse Effect.* (ed: A.F. Bouwman) John Wiley and Sons, New York: 247-261.

Esser, G. (1991) Osnabruck biosphere model: structure, construction, results. In: *Modern Ecology - Basic and Applied Aspects*, (G. Esser and D. Overdieck, eds.), Elsevier, Amsterdam: 679-709.

Esser, G. & Lautenschlager, M. (1994) Estimating the change of carbon in the terrestrial biosphere from 18,000 BP to present using a carbon cycle model. Environmental Pollution 83:45-53.

Farquhar, G.D., von Cammerer, S. & Berry, J.A. (1980) A biochemical model of photosyntetic CO_2 assimilation in leaves of C_3 species. *Planta* **149**:78-90.

Flanagan, P.W. & VanCleve, K. (1983) Nutrient cycling in relation to decomposition and organic matter quality in tiaga ecosystems. *Canadian Journal of Forest Research* **13**:795--817.

Fogel, R. & Cromack, K. (1977) Effect of habitat and substrate quality on Douglas fir litter decomposition in western Oregon. *Canadian Journal of Botany* **55**:1632--1640.

Foley, J.A. (1994) Net primary productivity in the terrestrial biosphere: the application of a global model. *Journal of Geophysical Research* **99**:20,773-20,783.

Freidlingstein, P., Delire, C., & Muller, J.F. (1992) The climate induced variation of the continental biosphere: A model simulation of the last glacial maximum. *Geophys. Res. Lett.* **19**:897-900.
Goldewijk, K.K., van Minnen, J.G., Kreileman, G.J.J., Vloedbeld, M. & Leemans, R. (1994) Simulating the carbon flux between the terrestrial environment and the atmosphere. *Water, Air and Soil Pollution* **76**:199-230.
Goto, N., Sakoda, A., & Suzuki, M. (1993) Modelling of soil carbon dynamics as a part of carbon cycle in terrestrial ecosystems. *Ecological Modelling* **74**:183-204.
Goudriaan, J. & Ketner, P. (1984) A simulation study for the global carbon cycle, including man's impact on the biosphere. *Climatic Change* **6**:167-192.
Greenland, D.J. (1965) Interactions between clays and organic compounds in soils. *Soils Fertil.* **28**:415-25 and 521-32.
Hunt, H.W. (1977) A simulation model for decomposition in grasslands. *Ecology* **58**:469--484.
Jenkinson, D.S. (1990) The turnover of organic carbon and nitrogen in soil. *Phil. Trans. R. Soc. Lond.* B **329**:361--369.
Jenkinson, D.S. & Rayner, J.H. (1977) The turnover of soil organic matter in some of the Rothamsted classical experiments. *Soil Science* **123**:298--305.
Jenkinson, D.S., Adams, D.E. & Wild, A. (1991) Global warming and soil organic matter. *Nature* **351**:304--306.
Jenny, H. (1980) *The Soil Resource*. Springer, Berlin Heidelberg New York.
Juma, N.G., & Paul, E.A. (1981) Uses of tracers and computer simulation techniques to asses mineralization and immobilization of soil nitrogen. In: *Simulation of Nitrogen Behavior of Soil-Plant Systems* (ed: J.M. Fissel & J.A. VanVeen), Centre for Agricultural Publishing and Documentation, Wageningen, The Netherlands: 145-154.
Kilburtus, G. (1980) Etudes des microhabitats contenus dans les agregats du sol, leur relation avec la biomasse bacterienne et la taille des procaryotes presents. *Rev. Ecol. Biol. Sol* **17**:545--557.
King, A.W., Post, W.M., & Wullschleger, S. (submitted) The potential response of terrestrial carbon storage to changes in climate and atmospheric CO_2. Climatic Change.
Ladd, J.N., Amato, M., & Oades, J.M. (1985) Decomposition of plant material in Australian soils. III. Residual organic matter decomposing under field conditions. *Australian Journal of Soil Research* **23**:603-611.
Lieth, H. (1975) Modelling the primary productivity of the world. In: *Primary Productivity of the Biosphere* (H. Leith & R.H. Whittaker, eds.) Springer-Verlag, Berlin: 237-263.
McClaugherty, C.A., Pastor, J., Aber, J.D., & Melillo, J.M. (1984) Forest litter decomposition in relation to soil nitrogen dynamics and litter quality. *Ecology* **66**:266--275.
McGill, W.B. & Paul, E.A. (1976) Fractionation of soil and ^{15}N nitrogen to separate the organic and clay interactions of immobilized N. *Canadian Journal of Science* **56**:203--212.
Meentemeyer, V. (1978) Macroclimate and lignin control of litter decomposition rates. *Ecology* **59**:465--472.
Melillo, J.M., Kicklighter, D.W., McGuire, A.D., Peterjohn, W.T., & Newkirk, K.M. (1995) Global change and its effects on soil organic carbon stocks. In: *Role of Nonliving Organic Matter in the Earth's Carbon Cycle* (ed: R.G. Zepp & C.H. Sonntag), John Wiley & Sons, New York: 175-189.
Oades, J.M. (1988) The retention of organic matter in soils. *Biogeochemistry* **5**:35--70.

Oades, J.M. (1995) An overview of processes affecting the cycling of organic carbon in soils. In: *Role of Nonliving Organic Matter in the Earth's Carbon Cycle* (ed: R.G. Zepp & C.H. Sonntag), John Wiley & Sons, New York: 293-303.

Pandey, V. & Singh, J.S. (1982) Leaf-litter decomposition in an oak-conifer forest in Himalaya: the effects of climate and chemical composition. *Forestry* **55**:47--59.

Parton, W.J., Scurlock, J.M.O., Ojima, D.S., Schimel, D.S., Hall, D.O., & Scopegram Group Members. (1995) Impact of climate change on grassland production and soil carbon worldwide. *Global Change Biology* **1**:13-22.

Parton, W.J., Schimel, D.S., Cole, C.V. & Ojima, D.S. (1987) Analysis of factors controlling soil organic matter levels in Great Plains grasslands. *Soil Science Society of America Journal* **51**:1173--1179.

Pastor, J. & Post, W.M. (1985) *Development of a Linked Forest Productivity-Soil Process Model.* ORNL/TM-9519, Oak Ridge, TN.

Pastor, J. & Post, W.M. (1986) Influence of climate, soil moisture, and succession on forest carbon and nitrogen cycles. *Biogeochemistry* **2**:3--27.

Paul, E.A. & vanVeen, H. (1978) The use of tracers to determine the dynamic nature of organic matter. *Transaction of the 11th Congress of the International Society of Soil Science, Edmonton* **3**:61-102.

Polglase, P.J. & Wang, Y.P. (1992) Potential CO_2-induced carbon storage by the terrestrial biosphere. *Austrialian Journal of Botany* **40**:641-656.

Post, W.M., Emanuel, W.R., Zinke, P.J. & Stangenberger, A.G. (1982) Soil carbon pools and world life zones. *Nature* **298**:156--159.

Post, W.M., Pastor, J., Zinke, P.J. & Stangenberger, A.G. (1985) Global patterns of soil nitrogen storage. *Nature* **317**:613--616.

Post, W.M., Peng, T.-H., Emanuel, W.R., King, A.W., Dale, V.H., DeAngelis, D.L. (1990) The Global Carbon Cycle. *American Scientist* **78**:310--326.

Potter, C.S., Randerson, J.T., Field, C.B., Matson, P.A., Vitousek, P.M., Mooney, H.A. & Klooster, S.A. (1993) Terrestrial ecosystem production: a process model based on satellite and surface data. *Global Biogeochemical Cycles* **7**:811-841.

Prentice, K.C. & Fung, I.Y. (1990) The sensitivity of terrestrial carbon storage to climate change. *Nature* **346**:48-51.

Rawls, W.J., Brakensiek, D.L. & Saxton, K.E. (1982) Estimation of soil water properties. *Transaction of the ASAE* **25**:1316-1328.

Renka, R.J. (1982) *Interpolation of Data on the Surface of a Sphere*. Technical Report ORNL/CSD-108, Oak Ridge National Laboratory, Oak Ridge, Tennessee.

Renka, R.J. & Cline, A..K. (1984) A triangle-based C1 interpolation method. *Rocky Mountain Journal of Mathematics* **14**:223-237.

Roberts, T.L., Stewart, J.W.B., Bettany, J.R. (1985) The influence of topography on the distribution of organic and inorganic soil phosphorus across a narrow environmental gradient. *Canadian Journal of Soil Science* **65**:651--665.

Schimel, D.S., Braswell Jr., B.H., Holland, E.A., McKeown, R., Ojima, D.S., Painter, T.H., Parton, W.J. & Townsend, J.R. (1994) Climatic, edaphic, and biotic controls over storage and turnover of carbon in soils. *Global Biogeochemical Cycles* **8**:279-293.

Schlesinger, W.H. (1977) Carbon balance in terrestrial detritus. *Annual Reviews of Ecology and Systematics* **8**:51--81.

Schlesinger, W.H. (1991) *Biogeochemistry: An Analysis of Global Change*. Academic Press, New York.

Sims, P.L., Coupland, R.T. (1979) Producers. In: *Grassland Ecosystems of the World: Analysis of Grasslands and Their Uses* (R.T.Coupland ed.), Cambridge University Press, Cambridge.

Smith, O.L. (1982) *Soil Microbiology: A Model of Decomposition and Nutrient Cycling*. CRC Press, Boca Raton, Florida.

Smith, T.M. & Shugart, H.H. (1993) The transient response of terrestrial carbon storage to a perturbed climate. *Nature* **361**:523-526.

Thornthwaite, C.W. & Mather, J.R. (1957) Instructions and Tables for Computing Potential Evapotanspiration and the Water Balance. *Publications in Climatology* **10**: 183-311.

Tiessen, H.J., Stewart, W.B. & Cole, C.V. (1984) Pathways of phosphorus transformations in soils of differing pedogenesis. *Soil Science Society of America Journal* **48**:853--858.

Townsend, A.R.,Vitousek, P.M. & Holland, E.A. (1992) Tropical soils could dominate the short-term carbon cycle feedbacks to increased global temperatures. *Climatic Change* **22**:293-303.

UNESCO (1973) *International sification and Mapping of Vegetation*. UNESCO, Paris.

Walker, T.W., Adams, A.F.R. (1958) Studies on soil organic matter: I. Influence of phosphorus content of parent materials on accumulations of carbon, nitrogen, sulfur, and organic phosphorus in grassland soils. *Soil Science* **85**:307--318.

Walker, T.W. & Syers, J.K. (1976) The fate of phosphorus during pedogenesis. *Geoderma* **15**:1-19.

Webb, R.S. Rosenzweig, C.E. & Levine, E.R. (1991) *A global data set of particle size properties. NASA Technical Memorandum, 4286*. NASA, Goddard Space Flight Center, Institute for Space Studies, New York.

Zinke, P.J., Stangenberger, A.G., Post, W.M., Emanuel, W.R. & Olson, J.S. (1984) *Worldwide Organic Soil Carbon and Nitrogen Data*. Oak Ridge National Laboratory, Oak Ridge, Tennessee (ORNL/TM-8857).

Zobler, L. (1986) *A World Soil File for Global Climate Modeling. NASA Technical Memorandum, 87802*. NASA, Goddard Space Flight Center, Institute for Space Studies, New York.

Acknowledgements

The authors thank Kevin Coleman and Dr. David Jenkinson of IACR-Rothamsted for careful and detailed explanations of the Rothamsted turnover model calculations that made this work possible. Research sponsored by the Department of Energy, Carbon Dioxide Research Program, Atmospheric and Climate Research Division, Office of Health and Environmental Research, under contract DE-AC05-84OR21400 with Lockheed Martin Energy Systems, Inc. Publication No. 4461 , Environmental Science Division, Oak Ridge National Laboratory.

SOIL ORGANIC MATTER MODELS

The Soil Submodel of the ITE (Edinburgh) Forest and Hurley Pasture Models

J.R.M. Arah
Institute of Terrestrial Ecology (Edinburgh)
Bush Estate, Penicuik
Midlothian EH26 0QB, UK

Introduction

The ITE (Edinburgh) Forest (EF) and Hurley Pasture (HP) models combine physiologically-based treatments of carbon, nitrogen and water dynamics in plants with a functional-compartment treatment of soil organic matter turnover. They are intended to be investigative tools, capable of answering questions about changing dynamics in changing environments, and to that end they were originally constructed to capture as much as possible of the plant physiology which dominates nutrient cycling in unamended ecosystems (Johnson & Thornley, 1985; Thornley, 1972, 1981, 1991; Thornley & Verberne, 1989). Soil processes were treated very simply at first, the soil submodel becoming more complex as the focus shifted to include systems receiving organic and mineral inputs (Thornley & Cannell, 1992).

The models presume an areally homogeneous soil-plant system. Trees are divided into leaves, branches, stems, coarse roots and fine roots (including mycorrhizae); grass consists of four age-categories of shoots and roots. Structural growth of the various compartments is dictated by local nutrient availability, water potential and temperature. Movement of substrates between adjacent compartments occurs in response to concentration gradients, transfer coefficients being functions of size. Since plants respond rapidly to short-term changes in their environment the basic timestep must be short (around fifteen minutes to ensure stability); where environmental data are not available at such frequencies (the usual case) they are interpolated or generated from longer-term averages and diurnal or seasonal sine waves.

The short calculation interval, coupled with the large number of variables required for even a minimally adequate representation of the complexities of the soil-plant system, makes for much computation. Both models, programmed in ACSL (1981), take around two minutes to simulate one year on a fast PC 486. Both are still under development.

NATO ASI Series, Vol. I 38
Evaluation of Soil Organic Matter Models
Edited by David S. Powlson, Pete Smith and Jo U. Smith

Soil submodel

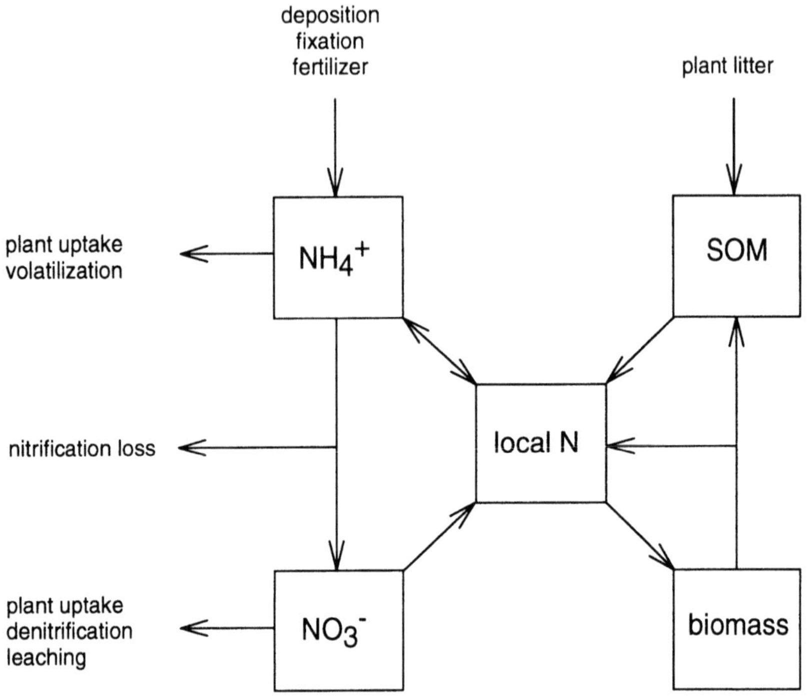

Figure 1 *SOM pools and transformations*

Both models assume a single homogeneous soil layer, taken by default to be 1 m deep. Their below-ground components are very similar, differing primarily in the number and nature of the litter pools. Ignoring these, Figure 1 illustrates the simple nitrogen (N) cycle common to both models. The carbon (C) cycle is even simpler; litter inputs to soil organic matter (SOM) are utilised for soil biomass growth (with yield coefficient 0.4) and returned on biomass death (with a fraction 0.8 lost to respiration). Turnover rates depend on temperature, moisture and substrate pool concentrations as discussed below.

A feature peculiar to the EF and HP models is the division of mineral N into pools accessible either to plants or to the biomass, but not to both. This is an attempt to reflect the spatial heterogeneity of the soil-plant system; slow exchange between these pools denotes significant separation (the default exchange constant is 0.01 d^{-1} for forests and 0.05 d^{-1} for pastures). The pool labelled "local N" in Fig.1. represents the fraction of the soil mineral N to which the biomass has privileged access; it is fed by diffusion-mediated exchange with the bulk solution and by mineralisation on biomass death, and depleted by biomass growth. Ammonium (NH_4^+) in the bulk soil is available for volatilization, nitrification (with a fixed fraction, taken to be 0.1 by default, of the nitrification flux lost as gaseous oxides of nitrogen) and plant uptake; atmospheric deposition, nitrogen fixation and fertilizer application add to the pool. Atmospheric deposition and fertilizer application also contribute to the nitrate (NO_3^-) pool, as does nitrification; plant uptake, denitrification and leaching deplete it.

Biomass dynamics

The size of the biomass pool m (kg m^{-2}) is governed by the following equations:

$$\frac{\partial m}{\partial t} = f_T f_\psi \left(v_g - v_d\right) \qquad [1]$$

$$v_g = k_g m^{2/3} \left(\frac{C_{SOM}}{K_{Cg} + C_{SOM}} \frac{N_{loc}}{K_{Ng} + N_{loc}} \right) \left(1 - \frac{m}{\eta_m C_{SOM}} \right) \qquad [2]$$

$$v_d = k_d m \left(1 - \frac{m_{\min}}{m} \right) \qquad [3]$$

where f_T and f_ψ are temperature and moisture potential modifiers of the growth rate v_g and the death rate v_d, C_{SOM} is the concentration of (dead) organic matter C (kg C m^{-2}) and N_{loc} is the concentration of accessible mineral N (kg N m^{-2}). The 2/3 exponent in Eq.[2] guarantees less-than-exponential biomass growth (with intrinsic growth rate constant k_g taken to be 0.025 $kg^{1/3}$ $m^{-2/3}$ d^{-1}), the first term in parentheses arises from dual-substrate Michaelis-Menten kinetics (Michaelis constants K_{Cg} and K_{Ng} are 10 kg C m^{-2} and 0.002 kg N m^{-2} respectively), and the second term sets an upper limit η_m on the ratio between living biomass C and dead SOM C (0.04). The intrinsic death rate constant k_d in Eq.[3] is taken to be 0.002 d^{-1}; the term in parentheses ensures that the biomass never falls below a minimum m_{min} (0.001 kg m^{-2}). The rate modifiers f_T and f_ψ are discussed below.

Nitrogen transformations

Leaching (v_{leach}: kg N m^{-2}) occurs whenever the single soil layer is above field capacity:

$$v_{leach} = k_{drain} N_{nit} \quad [4]$$

where k_{drain} is the drainage exchange rate (d^{-1}) and N_{nit} is the nitrate concentration (kg N m^{-2}) in the bulk soil. Ammonia volatilization is an inorganic process, independent of biomass:

$$v_{vol} = k_{vol} f_T f_\psi N_{amm} \quad [5]$$

where k_{vol} is an intrinsic volatilization rate constant (taken to be 0.005 d^{-1} for forests and 0.02 d^{-1} for pastures, which are likely to less acidic), N_{amm} is the concentration (kg N m^{-2}) of the bulk ammonium pool, and f_T and f_ψ are the temperature and moisture tension modifiers encountered above and discussed below. Nitrification and denitrification are microbial transformations, so their rates depend on the biomass m (kg m^{-2}):

$$v_{nit} = k_{nit} f_T f_\psi m N_{amm} \quad [6]$$

$$v_{denit} = k_{denit} f_T f_\psi S^{20} m N_{nit} \quad [7]$$

where the intrinsic rate constants k_{nit} for nitrification and k_{denit} for denitrification are taken to be 0.1 m^2 kg^{-1} d^{-1} and 0.05 m^2 kg^{-1} d^{-1} respectively, S is the ratio between moisture content and field capacity ($0 \leq S \leq 1$), and the exponent of 20 in Eq.[7] ensures that denitrification occurs only in near-saturated soil.

Plant uptake

Low temperatures favour ammonium uptake. An effective mineral N concentration N_{eff} is calculated from:

$$N_{eff} = N_{amm} + \left(\frac{T}{20}\right) N_{nit} \quad [8]$$

where T is the soil temperature (Celsius); where T exceeds 20°C N_{eff} is just the sum of N_{amm} and N_{nit}. Uptake rate depends on N_{eff}, root mass, and substrate levels in the roots:

$$v_{up} = k_{up} f_T f_{\psi r} \left(\frac{N_{eff}}{K_{eff} + N_{eff}} \right) \left(\frac{M_r}{\phi_r} \right) \qquad [9]$$

where the rate modifier $f_{\psi r}$ depends on root rather than soil tension (see below), M_r is the mass of root (kg m^{-2}), the intrinsic nitrogen uptake rate constant k_{up} is taken to be 0.1 kg N kg^{-1} d^{-1} for forests and 0.25 kg N kg^{-1} d^{-1} for pastures, the half-saturation mineral N concentration K_{eff} is 0.005 kg N m^{-2}, and the root state factor ϕ_r is given by:

$$\phi_r = 1 + \left(\frac{K_{Cr}}{C_r} \right) \left(1 + \frac{N_r}{K_{Nr}} \right) \qquad [10]$$

where C_r is the root C concentration (kg C kg^{-1}), N_r the concentration of root N (kg N kg^{-1}), and the constants K_{Cr} and K_{Nr} are 0.05 kg C kg^{-1} and 0.005 kg N kg^{-1} respectively. The first term in parentheses in Eq.[10] serves to reduce root uptake when root C is low (the process requires energy); the second term reflects inhibition of N uptake when root N is high.

Temperature and moisture potential

Temperature affects nutrient cycling via the rate modifier f_T:

$$f_T = 0.075T - 0.00125T^2 \qquad [11]$$

which is zero when the temperature T (Celsius) is zero, unity when T equals 20°C, and at a maximum value of 1.125 when T is 30°C; above that temperature f_T declines once more.

The moisture potential rate modifier f_ψ takes the form:

$$f_\psi = \left[\exp\left(\frac{\mu_w \psi}{RT_{abs}} \right) \right]^{10} \qquad [12]$$

where R is the universal gas constant (8.314 J K^{-1} mol^{-1}), T_{abs} is the absolute temperature, μ_w is the molar mass of water (18 x 10^{-3} kg mol^{-1}) and ψ is the moisture potential (J kg^{-1} ≡ kPa). Since ψ is always less than or equal to zero, f_ψ is always less than or equal to unity, falling off sharply as the moisture potential drops (*ie* in dry soils). Soil moisture characteristic (θ–ψ) curves are approximated by a one-parameter power law:

$$\psi = -0.375\left(\frac{0.557}{\theta}\right)^q \qquad [13]$$

where θ is the volumetric moisture fraction ($m^3\ m^{-3}$) of the soil layer and the parameter q varies from around 3 to 10 depending on texture (more clay means higher q).

Conclusion

The EF and HP models share a common soil submodel. There is one homogeneous soil layer with seven compartments (the five shown in Fig.1 plus SOM and biomass C). Other compartments might be added (perhaps a "resistant" or an "immobilised" OM fraction) and additional soil layers introduced; the representation of water dynamics might be improved. However, the complexity of the above-ground parts of the models (and their consequent lack of speed) makes it difficult to assess the direct impact of any such modification. For the same reason, parameter optimisation using measured data is not really an option. The models are synthetic rather than analytic tools.

References

ACSL (1981) *Advanced continuous simulation language. User guide / reference manual, 3rd edn.* Mitchell and Gauthier Associates, Concord, MA.

Johnson, I.R. & Thornley, J.H.M. (1985) Dynamic model of the response of a vegetative grass crop to light, temperature and nitrogen. *Plant, Cell and Environment* **8**: 485-499.

Thornley, J.H.M. (1972). A balanced quantitative model for root-shoot ratios in vegetative plants. *Annals of Botany* **36**: 431-441.

Thornley, J.H.M. (1981). Organogenesis. In: *Mathematics and Plant Physiology* (eds: D.A. Rose & D.A. Charles-Edwards), Academic Press, London: 49-65.

Thornley, J.H.M. (1991). A transport-resistance model of forest growth and partitioning. *Annals of Botany* **68**: 211-226.

Thornley, J.H.M. & Cannell, M.G.R. (1992) Nitrogen relations in a forest plantation-soil organic matter ecosystem model. *Annals of Botany* **70**: 137-151.

Thornley, J.H.M. & Verberne, E.L.J. (1989) A model of nitrogen flows in grassland. *Plant, Cell and Environment* **12**: 863-886.

Acknowledgements

Thanks to Pete and Jo Smith for their time, and to John Thornley for encouragement.

SOMM - a model of soil organic matter and nitrogen dynamics in terrestrial ecosystems

O.G. Chertov & A.S. Komarov [1]

St.-Petersburg Forest Technical Academy,

Institutsky 5,

194018, St.-Petersburg,

Russia,

Summary

SOMM - a simulation model of soil organic matter mineralisation, humification and nitrogen release has been further developed. The model takes into account the rate of processes that are dependent upon the nitrogen and ash content of litter fall, temperature and moisture. The activity of the main groups of soil decomposers is reflected in the model. The model represents a system of linear differential equations with variable coefficients.

Key words: Soil litter and humus; Soil micro and meso fauna; Mineralization; Humification; Nutrients; Simulation model

Model description

SOMM represents a quantification of the humus types concept (Mor, Moder and Mull) existing in forest pedology since last century (Müller, 1887; Wilde, 1958; Duchafour, 1961). It has been created as a tool of theoretical analysis and natural ecosystems' simulation firstly.

---

[1] *Institute of Soil Science and Photosynthesis of Russian Academy of Sciences, 142292, Pushchino, Moscow Region, Russia.*

NATO ASI Series, Vol. I 38
Evaluation of Soil Organic Matter Models
Edited by David S. Powlson, Pete Smith and Jo U. Smith

The experimental basis for the SOMM model is a set of classical laboratory works on the rate of organic litter decomposition in controlled conditions in dependence of temperature, moisture and chemical composition of the material (Kostychev, 1889; Waxman & Tenney, 1927; Waxman & Gerretsen, 1931; Kononova, 1951; Mikola, 1954; Alexandrova, 1970). The experiments allowed quantification of the activity of the complex of micro-organisms (*Fungi*, *Actinomycetes* and *Bacteria*) with micro fauna. The procedure of obtaining the kinetic parameters from the data set has been described in detail previously (Chertov, 1985; 1990). Data from the laboratory experiments of Chernova (1978) for soil arthropods and Perel & Sokolov (1964) and Striganova et al. (1987) for earthworm activity have been used to quantify the role of decomposers in organic debris humification.

The idea of a complex of humic substances with undecomposed plant debris (CHS) was proposed for use in the model. This complex is responsible for the slow decomposition phase of plant debris, which still contains many easily mineralisable substances. This occurs because CHS, being impregnated with humic acids, is transforming at the rate of "true humus". It allows a new explanation for the "soil protective effect" (Van Veen & Paul, 1981), which has previously been attributed to the increasing lignin content in decomposing matter. CHS is believed to be the main inhibiting or modifiying factor of decomposition and is the "forgotten" compartment in the studies of SOM biochemistry. The CHS consists of sub-horizons F and H in forest floor with a superficial input of litter fall. It corresponds to a fraction of "labile (active) humus" in soils with root litter input.

The main specific features of the SOMM model are as follows:

1. The biomass of soil organisms represents a negligible part of all decomposed matter and has a high rate of decomposition (Chernova, 1978; Crossley *et al.*, 1991), and we do not take it into consideration as a separate pool.

2. The number and species composition of decomposing organisms is dependent on the biochemical properties of organic debris and on hydrological and thermal conditions. We postulate that there are no barriers for a rapid invasion of new organisms. Thus, it is possible to calculate the decomposition coefficients for the communities as a function of the biochemical properties of litter, temperature and moisture.

3. The rate of nutrient release due to mineralization corresponds to the rate of organic matter decomposition with the only exception being nitrogen kinetics in organic layers (forest floor and peat). It is well documented that the rate of nitrogen mineralization is sufficiently lower than that of carbon when the decomposing material is poor in nitrogen. The mechanism for this phenomenon is nitrogen re-utilization by soil microorganisms. Detailed discussion, basic data and numerical dependencies reporting this effect have been described (Chertov, 1985, 1990). In mineral horizons the rate of nitrogen release corresponds to the rate of carbon mineralization only when the C/N ratio is less than 8 (Alexandrova 1970).

4. On the basis of previous studies (Chertov, 1985) we postulated that the community of *Bacteria* and *Arthropoda* produces humus with a C/N ratio of 15, and the community of *Lumbricidae* forms humus with a C/N ratio of 8. These complexes of organisms produce 20% of CHS nitrogen from their biomass, and 80% of nitrogen goes into humus formation. These figures have been used for the calculation of the humification rate. The nutrients in matter consumed by these orgnaisms are fully mineralized and are available for plants for one year because of the short ontogenesis of *Arthropoda* and *Lumbricidae*.

5. Nitrogen and elements being released through mineralisation are fully consumed by plants and they are not immobilised in the soil. They do not, therefore, affect the rate of soil organic matter transformation directly. They influence the transformation being returned with litter fall only.

6. Leaching and diffusion of soil organic matter, input of nitrogen from the atmosphere, nitrogen fixation, and denitrification are not taken into account in the model. Leaching and atmospheric nitrogen input have been incorporated when using the model for ecosystem simulation (Chertov, 1990).

7. The rate variables (coefficients of the equations) are dependent on litter fall nitrogen and ash content, and the temperature and moisture content of decomposing material and soil; every rate variable has specific dependencies on these factors.

8. The role of the chemical composition of mineral horizons (humus and clay content, acidity etc.) in decomposition is not taken into account directly.

9. The specific functions of *Isoptera* in subtropical and tropical soils are not taken into account.

In correspondence with these postulates, the kinetics of organic litter and humus transformation in the soil can be expressed by the following system of linear first-order differential equations with variable coefficients:

$$dL/dt = L_0 - (k_1 + k_3)L, \quad (1)$$

$$dF/dt = k_3L - (k_2 + k_4 + k_5)F, \quad (2)$$

$$dH/dt = (k_4 + k_5)FM_f - k_6H, \quad (3)$$

where L is undecomposed litter content; L_0 is the input of L on/in soil; F is CHS content, which corresponds approximately to the F-layer of forest floors; H is the fully humified material (humus) content on/in a soil; k_1 is the rate of litter mineralization by a first complex of soil destructors; k_2 is the rate of CHS mineralization; k_3 is a rate of litter transformation to CHS; k_4 is the rate of CHS consumption by a second community of *Bacteria* and *Arthropoda*; k_5 is the rate of CHS consumption by earthworms; k_6 is the rate of true humus mineralization; M_f is a coefficient describing a part of CHS transforming into true humus. The units of L, F, H are kg m^2 and the unit of k_i is day^{-1}. The value of the k_i coefficients change according to both substrate properties (ash and nitrogen content) and temperature and moisture. A similar system of equations has been developed also for nitrogen and other elements bonded in soil organic matter. The model structure is expressed in Figure 1.

The SOMM model has been programmed in TurboPascal 5.5 for MS-DOS on an IBM compatible PC. The iteration step in the model is a day, but earlier in model development we used monthly averages as input values and presented the outputs as monthly or yearly data. The model inputs required are: number of years of simulation, nitrogen and ash contents of litter fall (%), initial amounts of soil organic matter and nitrogen in litter (*L*), CHS (*F*) and humus in the whole solum (*H*), litter input (amount and time), temperature of topsoil (monthly means), moisture (mass %) of organic layers (*L* and *F*) and mineral topsoil (monthly means). The model outputs include simulated annual amounts of *L*, *F*, *H* (organic matter and nitrogen), the C/N ratio of humus, brutto production of carbon dioxide and available nitrogen due to organic matter mineralization.

Figure 1 *Flow chart of SOMM model. Symbols on the chart correspond to equations 1 to 3 describing the model*

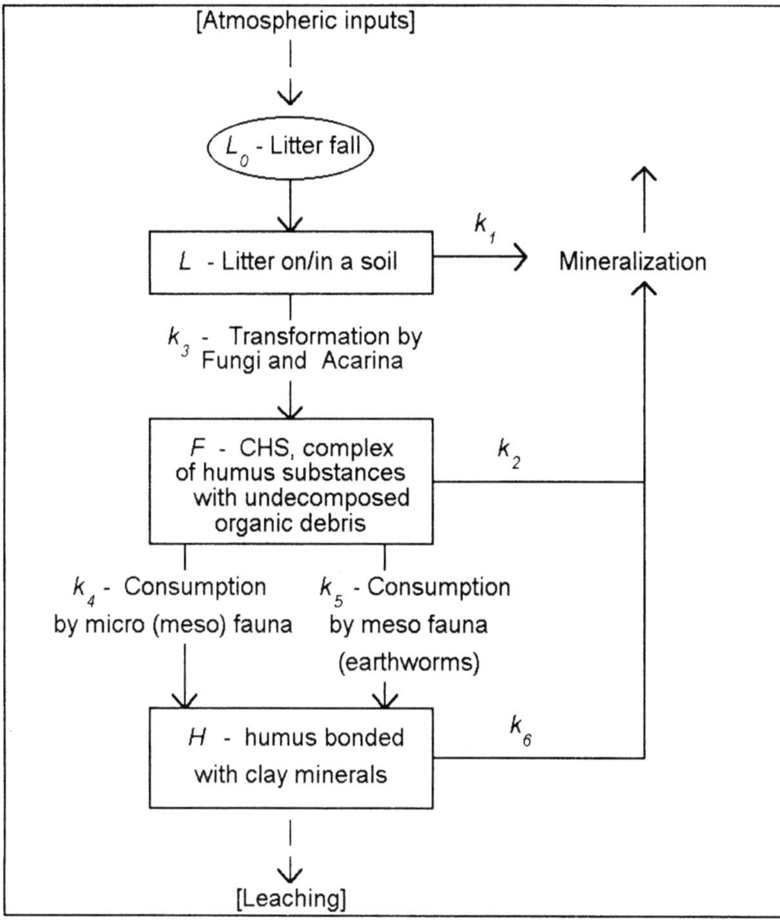

The SOMM model has undergone prelimenary verification using a set of independent laboratory experiments. The first version of the model (Mor model) and the present form have been described in detail elsewhere (Chertov, 1985; 1990; Chertov & Komarov, 1995a; 1995b).

References

Alexandrova, L.N. (1970) Istochniki gumusovyh veshchestv v pochve. *Transactions of Leningrad Agr. Institute* **142**: 5-25 (in Russian).

Chernova N.M. (1978) *Ekologicheskie suktsessii pri razlodzenii rastitelnyh ostatkov*. Nauka, Moscow (in Russian).

Chertov, O.G. (1985) Imitatsionnaya model mineralizatsii i gumifikatsii lesnogo opada i podstilki. *Zhurn. Obshchey Biologii (Moscow)* **46**: 794-804 (in Russian with English summary).

Chertov, O.G. (1990) SPECOM - a single tree model of pine stand/raw humus soil ecosystem. *Ecological Modelling* **50**: 107-132.

Chertov, O.G. & Komarov, A.S. (1995a) On mathematical theory of soil forming processes. I. Theoretical bachground. II. SOMM - a model of soil organic matter dynamics. III. Basic ideas of a mineral phase modelling. Pushchino Res. Center of Russian Academy of Sciences, Pushchino. Preprint. 41 pp.

Chertov, O.G. & Komarov, A.S. (1995b) SOMM - a model of soil organic matter dynamics. *Ecological Modelling* (in press).

Crossley, Jr. D.A., Coleman, D.S., Hendrix, P.F., Cheng, W., Wright D.H., Beare, M.H., Edwards, C.A. (Eds.) (1991) *Modern Techniques in Soil Ecology*. Elsevier, Amsterdam.

Duchaufour, Ph. (1961) *Precis de pedologie*. Paris. Russian translation Moscow, Mir, 1970.

Kononova, M.M. (1951) *Problema Pochvennogo Gumusa*. Moscow: Nauka. English translation: 1966, Pergamon Press, Oxford.

Kostychev, P.A. (1889) Obrazovanie i svoistva peregnoya. St.- Peterburg (Cit.: 1951. Selected Works. Nauka, Leningrad: 251-296, in Russian).

Mikola, P. (1954) Kokeellisia tutkimuksia metsäkarikkeiden hajaantumisnopeudesta. *Comm. Inst. For. Fenn.* **43** 50 pp (in Finnish with English summary).

Müller, P.E. (1887) *Studien uber die naturlichen Humusformen und deren Entwicklung auf Vegetation und Boden*. Berlin.

Perel, T.S., Sokolov, D.F. (1964) Kolichestvennaya otsenka uchastia dozhdevyh chervey v pererabotke opada. *Zool. J. (Leningrad)* **43**: 1618-1625 (in Russian with English summary).

Striganova, B.R., Kudriashova, I.V., Tiunov, A.V. (1987) Pishchevaya aktivnost dozhdevych chervey *Eisenia hordenskioldi* (Eisen) (*Oligochaeta, Lumbricidaea*). *Pochvovedenie (Moscow)*. **1**: 72-77 (in Russian with English summary).

VanVeen, J.A., Paul, E.A. (1981) Organic carbon dynamics in grassland soils. I. Background information and computer simulation. *Canad. J. Soil Sci.* **61**: 185-201.

Waxman, S.A., Gerretsen, F.S. (1931) Influence of temperature and moisture upon the nature and extent of decomposition of plant residues by microorganisms. Ecology. 12(1): 33-60.

Waxman, S.A.,Tenney, F.G. (1927) The composition of natural organic materials and their decomposition in the soil. *Soil Sci.* **24**: 317-324.

Wilde, S. A. (1958) *Forest Soils*. John Wiley and Sons, New York.

RothC-26.3 - A Model for the turnover of carbon in soil

K Coleman and DS Jenkinson
IACR - Rothamsted,
Harpenden,
Herts. AL5 2JQ. U.K.

Introduction

Rothc-26.3 is a model of the turnover of organic carbon in non-waterlogged soils that allows for the effects of soil type, temperature, moisture content and plant cover on the turnover process. It uses a monthly time step to calculate total organic carbon (t C ha^{-1}), microbial biomass carbon (t C ha^{-1}) and $\Delta^{14}C$ (from which the radiocarbon age of the soil can be calculated) on a years-to-centuries timescale (Jenkinson, 1990; Jenkinson and Coleman, 1994; Jenkinson *et al.*, 1987; Jenkinson *et al.*, 1991; Jenkinson *et al.*, 1992). It needs few inputs and those it needs are easily obtainable. It is an extension of the earlier model described by Jenkinson and Rayner (1977), and by Hart (1984). Needless to say, it has many ideas in common with other contemporary turnover models, notably CENTURY (Parton *et al.*, 1988) and Van Veen and Paul's model (Van Veen and Paul, 1981).

Data requirements

The only data required to run the model are:-

1) Monthly rainfall (mm).
2) Monthly open pan evaporation (mm).
3) Average monthly air temperature (°C).
4) Clay content of the top soil (as a percentage).
5) An estimate of the decomposability of the incoming plant material - the DPM/RPM ratio.

NATO ASI Series, Vol. I 38
Evaluation of Soil Organic Matter Models
Edited by David S. Powlson, Pete Smith and Jo U. Smith

6) Soil cover - Is the soil bare or vegetated in a particular month ?.

7) Monthly input of plant residues (t C ha^{-1}).
As the below-ground part of this input is rarely known, the model is most often run 'in reverse' generating monthly (or more often yearly) inputs from known soil, site and weather data.

8) Monthly input of farmyard manure (FYM, in t C ha^{-1}), if any.

Model Structure

Soil organic carbon is split into four active compartments and a small amount of inert organic matter (IOM). The four active compartments are Decomposable Plant Material (DPM), Resistant Plant Material (RPM), Microbial Biomass (BIO) and Humified Organic Matter (HUM). Each compartment decomposes by a first-order process with its own characteristic rate. The priming action is taken to be zero and the proportion of an input of organic matter that is decomposed in a given time is taken to be independent of the amount added. The IOM compartment is resistant to decomposition. The structure of the model is shown in Figure 1.

Figure 1 - Structure of the Rothamsted Carbon Model

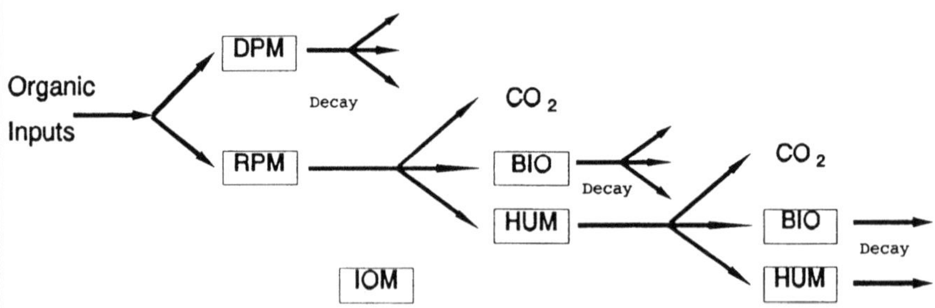

RPM : Resistant Plant Material

DPM : Decomposable Plant Material

BIO : Microbial Biomass

HUM : Humified OM

IOM : Inert Organic Matter

Incoming plant carbon is split between DPM and RPM, depending on the DPM/RPM ratio of the particular incoming plant material. For example, for most agricultural and improved grassland, we use a DPM/RPM ratio of 1.44, i.e. 59% of the plant material is DPM and 41% is RPM. For a deciduous or tropical woodland a DPM/RPM ratio of 0.25 is used, so 20% is DPM and 80% is RPM. All incoming plant material passes through these two compartments once, but only once.

Both DPM and RPM decompose to form CO_2 (lost from the system), BIO and HUM. The proportion that goes to CO_2 and to BIO + HUM is determined by the clay content of the soil - see below. The BIO + HUM is then split into 46% BIO and 54% HUM, a partition established by tuning the model to data from some of the Rothamsted long-term field experiments (Jenkinson, 1990). BIO and HUM both decompose to form more CO_2, BIO and HUM.

FYM is assumed to be more decomposed than normal plant material. It is split in the following way : DPM 49%, RPM 49% and HUM 2%.

Decomposition of an active compartment

If an active compartment contains Y t C ha^{-1}, this declines to Y e^{-abckt} t C ha^{-1} at the end of the month.

where a is the rate modifying factor for temperature

b is the rate modifying factor for moisture

c is the plant retainment rate modifying factor

k is the decomposition rate constant for that compartment

t is 1 / 12, since k is based on a yearly decomposition rate.

So Y (1 - e^{-abckt}) is the amount of the material in a compartment that decomposes in a particular month.

Decomposition rate constants

The decomposition rate constants (k), in $years^{-1}$, for each compartment are set at: -

DPM, 10.0; RPM, 0.3; BIO, 0.66; HUM, 0.02. These values were originally set by tuning the model to data from some of the long-term field experiments at Rothamsted (Jenkinson *et al.*, 1987; Jenkinson *et al.*, 1992) : they are not normally altered when using the model.

Calculation of the rate modifying factors

Temperature : the rate modifying factor (*a*) for temperature is given by: -

$$a = \frac{47.9}{1 + e^{\left(\frac{106}{tm + 18.3}\right)}}$$

where *tm* is the average monthly air temperature (°C). Air temperature is used rather than soil temperature because it is more often obtainable.

Moisture : Rainfall and open pan evaporation are used to obtain soil moisture deficit (SMD), as it is easier to obtain rainfall and pan evaporation data, from which SMD can be calculated, than monthly measurements of the actual topsoil water deficit. The rate modifying factor (*b*) for SMD is calculated in the following way:-

The maximum SMD for the 0-23 cm layer of a particular soil is first calculated from

$$\text{Maximum SMD} = -(20.0 + 1.3\ (\%\text{clay}) - 0.01\ (\%\text{clay})^2)$$

So for Rothamsted (%clay = 23.4), the maximum SMD = - 44.94

The maximum SMD thus obtained is that under actively growing vegetation : if the soil is bare during a particular month, this maximum is divided by 1.8, to allow for the reduced evaporation from a bare soil.

Next, the accumulated SMD is calculated from the first month when 0.75 (evaporation) exceeds rainfall until it reaches the max. SMD, where it stays until the rainfall starts to exceed 0.75 (evaporation) and the soil wets up again. The factor 0.75 is conventional for converting open pan evaporation to evapotranspiration from a growing crop.

Finally, the rate modifying factor (*b*) used each month is calculated from :-

if acc. SMD < 0.444 max. *SMD*,

$$b = 1.0$$

otherwise,

$$b = 0.2 + (1.0 - 0.2) * \frac{(\text{max. } SMD - acc.\ SMD)}{(\text{max. } SMD - 0.444 \text{ max. } SMD)}$$

Plant retainment factor : The plant retainment factor (c) slows decomposition if growing plants are present, since decomposition is faster in fallow soil than in cropped soil, even when the cropped soil is not allowed to dry out (Jenkinson *et al.*, 1987; Sommers *et al.*, 1981; Sparling *et al.*, 1982).

If soil is vegetated $c=0.6$

If soil is bare $c=1.0$

Partitioning of carbon between that lost from the soil and that remaining : the CO_2 / (BIO+HUM) ratio

The model adjusts for soil texture by altering the partitioning between CO_2 evolved and (BIO+HUM) formed during decomposition, rather than by using a rate modifying factor, such as that used for temperature. The ratio CO_2 / (BIO + HUM) is calculated from the clay content of the soil using the following equation: -

$$x = 1.67\ (1.85 + 1.60 \exp(-0.0786\ \%\text{clay}))$$

where x is the ratio CO_2 / (BIO+HUM)

Then $x / (x + 1)$ is evolved as CO_2

and $1 / (x + 1)$ is formed as BIO + HUM

The scaling factor 1.67 is used to set the CO_2 / (BIO+HUM) ratio in Rothamsted soils (23.4% clay) to 3.51 : the same scaling factor is used for all soils.

Calculation of the radiocarbon age

Radiocarbon measurements are commonly expressed in one of two ways, as % modern, i.e. 100 (specific activity of the sample) / (specific activity of the standard) or as the $\Delta^{14}C$ value, i.e. 1000 (specific activity of the sample - specific activity of the standard) / (specific activity of the standard).

Radiocarbon age is related to $\Delta^{14}C$ in the model by the following equation

$\Delta^{14}C = 1000 \exp(-\text{radiocarbon age} / 8035) - 1000$

using the conventional half-life for ^{14}C (5568 years)

Before 1860, the model assumes that the radiocarbon age of the plant material entering the soil each year is zero, ie its $\Delta^{14}C$ value is zero and it is 100 % modern. After 1860 the radiocarbon content of the incoming plant carbon (expressed as % modern) in a particular year is set from an internal table, constructed from data on the radiocarbon content of atmospheric CO_2 : for the 1860 - 1949 period from Baxter & Walton (1971), for 1950 - 1984 from Harkness *et al.*, (1986) and for 1987 - 1993 from Levin *et al.*, (1994). The radiocarbon content of each year's input of plant carbon is taken to be the same as that of atmospheric CO_2 for the same year. The 'radiocarbon activity scaling factor' in the model print-out is the radiocarbon activity of the input for a particular year, expressed as either (%modern) / 100 or $(\Delta^{14}C + 1000) / 1000$, i.e. taking the value for 1859 as 1.

The age of the IOM fraction is set by default to 50,000 years, implying that it contains virtually no ^{14}C ($\Delta^{14}C = -998.0$) and that it is of geological age rather than pedological age.

An example of the use of the model

The use of the model will be illustrated using data from one of the Rothamsted long-term field experiments, on the continuous cultivation of spring barley. This experiment was started on Hoosfield in 1852 and was designed to study the effects of fertilizers and FYM on the yield of barley. A detailed account can be found in the booklet *Guide to the classical field experiments* (Rothamsted Experimental Station, 1991). None of the data from this experiment were used in setting the model parameters, so the fit obtained between model and data is an objective test of the model.

In modelling the Hoosfield data, it is first necessary to run the model to produce a starting soil organic C content that is the same as that originally present in the soil (33.8 t C ha^{-1} in 1852, which includes 3.8 t C ha^{-1} in IOM). This is done by assuming that the soil organic C had reached equilibrium in 1852 and running the model iteratively with different annual inputs of plant C from a spring barley crop (DPM/RPM ratio 1.44) until the value of 33.8 t C ha^{-1} is reached (within $\pm$ 0.5 t C ha^{-1}).

It should be noted that the annual input finally chosen (1.63 t C ha^{-1} $year^{-1}$) is distributed between April, May, June and July, with no inputs in other months. This distribution is no more than an educated guess for a particular crop, spring barley, sown in February or March and harvested in early August. It makes little difference to the calculated equilibrium value for total organic C or to radiocarbon age how the annual input is distributed, or even if it is all added in a single pulse. Only if the model is being used to predict annual changes in fractions with short turnover times (notably biomass and DPM) will the input distribution appreciably affect the results.

The IOM content of the Hoosfield soil was set at 3.8 t C ha^{-1} throughout the run to establish the equilibrium state. The same value for IOM was used in all the subsequent runs with this soil. This value gives a radiocarbon age of 1055 years for the whole soil organic C in 1852, a little less than the mean (1290 years) for all the pre-1900 Rothamsted topsoils (0 - 23 cm) analysed for radiocarbon (Jenkinson *et al.*, 1992). No measurements of radiocarbon are available from the Hoosfield experiment : had they been, the IOM content of the soil *and* the annual inputs of plant C would have been iteratively adjusted to give both the correct organic C content and the correct radiocarbon age for a particular sampling date.

Once the starting C content has been established, the model was used to predict changes in the organic carbon of three of the plots in the Hoosfield experiment : plot 7-2 (FYM annually), plot 7-1 (FYM annually 1852-1871, nothing thereafter) and a mean of plots 6-1 and 6-2 (both unmanured).

For the unmanured treatment, 1.76 t C ha^{-1} of plant residues was used (distributed, again arbitrarily, with 0.16 t C ha^{-1} in April, 0.16 in May, 0.32 in June and 1.12 in July). A soil cover factor of one was used in April, May, June and July, zero in the other months. These input figures were used from 1852 to 2000, except in the years which were fallow (1912, 1933, 1943 and 1967). For the fallow years the plant input was set at zero (bare cultivated fallow) and a soil cover factor of zero was used in all twelve months.

For the treatment receiving FYM annually (plot 7-2), the annual input of plant residues from the barley was taken to be 2.6 t C ha^{-1} (0.21 t C ha^{-1} in April, 0.37 in May, 0.53 in June and 1.49 in July) Again a soil cover factor of one was used in April, May, June and July, zero in the other months. As with the unmanured treatment, this input was used from 1852 to 2000, except in the four fallow years (1912, 1933, 1943 and 1967). The FYM (containing 3.0 t C ha^{-1}) was applied in February each year from 1852-1911 and from 1913-01930. In 1931 FYM containing 6.0 t C ha^{-1} was applied (3.0 in Feb and 3.0 in Nov). From 1932 to 2000, FYM containing 3.0 t C ha^{-1} was applied in November each year.
The third treatment received FYM (3 t C ha^{-1}) every February from 1852 to 1871 and nothing thereafter (plot 7-1). From 1852 to 1876, plant residues were set at 2.6 t C ha^{-1} (split in the same way as plot 7-2), with the same soil cover factor of one in April, May, June and July, zero in the other months. From 1877 to 2000, plant residues were set at 1.76 t C ha^{-1} (split in the same way as the unmanured plot), with the same soil cover factor, except in the four fallow years of 1912, 1933, 1943 and 1967. Figure 2 shows the results.

Figure 2 – Carbon Changes in the Hoosfield Barley Experiment
Data modelled by RothC-26.3 (Solid lines)

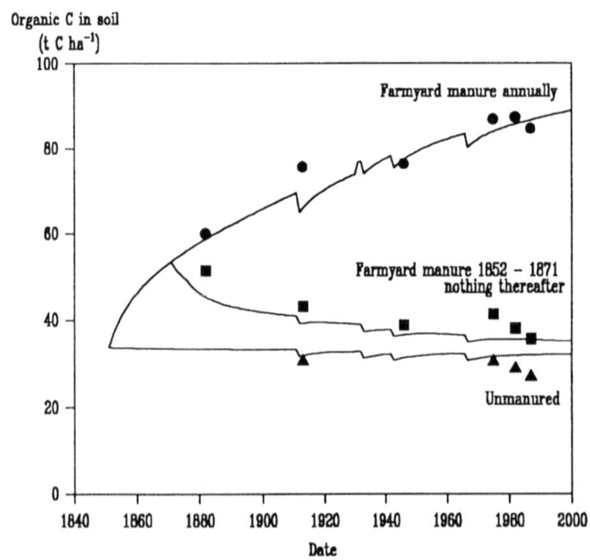

Advantages and Disadvantages of RothC-26.3

RothC-26.3 is solely concerned with soil processes; it does not contain a sub-model for plant production, unlike models such as CENTURY. Its main advantage is that it runs on data that are readily available. As with all such models, it is at its best when operating in situations similar to those for which its was originally parameterised : for RothC-26.3, arable soils in the temperate zone. It should be used with the greatest caution in situations where there are as yet no good long-term measurements of decomposition rates - tundra and taiga soils for example. It deals exclusively with processes in the topsoil, which includes litter, if present.

References

Baxter MS and Walton A (1971) Fluctuation of atmospheric carbon-14 concentrations during the past century. *Proceedings of the Royal Society of London, A.* **321**: 105-127.

Harkness DD, Harrison AF and Bacon PJ (1986) The temporal distribution of 'bomb' ^{14}C in a forest soil. *Radiocarbon* **28**: 328-337.

Hart PBS (1984) Effects of soil type and past cropping on the nitrogen supplying ability of arable soils. PhD thesis, University of Reading, UK.

Jenkinson DS (1990) The turnover of organic carbon and nitrogen in soil. *Philosophical transactions of the Royal Society, B.* **329**: 361-368.

Jenkinson DS, Adams DE and Wild A (1991) Model estimates of CO2 emissions from soil in response to global warming. *Nature*, **351**: 304-306.

Jenkinson DS and Coleman K (1994) Calculating the annual input of organic matter to soil from measurements of total organic carbon and radiocarbon. *European Journal of Soil Science*, **45**: 167-174.

Jenkinson DS, Harkness DD, Vance ED, Adams DE and Harrison AF (1992) Calculating net primary production and annual input of organic matter to soil from the amount and radiocarbon content of soil organic matter. *Soil Biology & Biochemistry* **24**: 295-308.

Jenkinson DS, Hart PBS, Rayner JH and Parry LC (1987) Modelling the turnover of organic matter in long-term experiments at Rothamsted. INTECOL Bulletin **15**: 1-8.

Jenkinson DS and Rayner JH (1977) The turnover of soil organic matter in some of the Rothamsted classical experiments. *Soil Science* **123**: 298-305.

Levin I, Kromer B, Schoch-Fischer H, Bruns M, Münnich M, Berdau D, Vogel JC and Münnich KO (1994) $\Delta^{14}CO_2$ records from sites in Central Europe. 203-222. In Boden TA, Kaiser DP, Sepanski RJ and Stoss FW (eds.), *Trends '93: A compendium of Data on Global Change.* ORNL/CDIAC-65. Carbon Dioxide Information Analysis Center, Oak Ridge National Laboratory, Oak Ridge, Tenn., USA.

Parton WJ, Stewart JWB and Cole CV (1988) Dynamics of C, N, P and S in grassland soils: a model. *Biogeochemistry* **5**: 109-131.

Rothamsted Experimental Station (1991) Guide to the classical field experiments. Lawes Agricultural Trust, Harpenden.

Rothamsted Experimental Station (1991) Guide to the classical field experiments. Lawes Agricultural Trust, Harpenden.

Sommers LE, Gilmour CM, Wilding RE and Beck SM (1981). Effect of water potential and decomposition processes in soils. *Water Potential Relations In Soil Microbiology*, p, 97-117. Special publication of the Soil Science Society of America No. 9, Madison.

Sparling GP, Cheshire MV and Mundie CM (1982). Effect of barley plants on the decomposition of ^{14}C-labelled soil organic matter. *Journal of Soil Science* **33**: 89-100.

Van Veen JA and Paul EA (1981) Organic carbon dynamics in grassland soils. I Background information and computer simulation. *Canadian Journal of Soil Science* **61**: 185-201.

Modelling approaches of soil organic matter turnover within the CANDY system

Uwe Franko
Centre for Environmental Research Ltd.,
Hallesche Straße 44,
D-06246 Bad Lauchstädt
Germany

Introduction

The development of the CANDY system (**CA**rbon and **N**itrogen **DY**namics) has been based on experience of organic matter turnover and nitrogen dynamics gained over a long period, a major part of the scientific work in Bad Lauchstädt. The main objective in developing the model was to give farmers a tool for calculating short term dynamics of nitrogen transformations and long term changes in the carbon content of the soil. For this reason the system consists of a database interface and several model components. The main component is the CANDY module, written in TURBO - PASCAL which calculates daily changes of water, temperature, carbon and nitrogen for a 2 m deep soil profile. Alongside this daily-timestep nitrogen simulation, there is a long term carbon module that calculates the amount of decomposable carbon in steady state for a given crop rotation. The calculation is specific to the site and management practices and is based on averaged turnover conditions, yields and inputs of organic material.

Other modules of the CANDY system are designed to

1. analyse the specific turnover conditions of a given combination of soil, climate and management,
2. calculate an environmentally compatible nitrogen fertilizer recommendation using the simulated amount of mineral nitrogen in soil at a given time and the predicted nitrogen mineralization from organic pools and
3. display graphics of simulation results.

NATO ASI Series, Vol. I 38
Evaluation of Soil Organic Matter Models
Edited by David S. Powlson, Pete Smith and Jo U. Smith

Model description

The CANDY system is an integration of different simulation modules with a database system. The database provides model parameters, measured values, initial values, weather data and management data. A simulation can be characterized by the set of initial values, the specification of site specific information including soil parameters and weather conditions and a list of management activities to be considered during the simulation.

The soil parameters required are bulk density, particle density, water capacity, wilting point and a texture parameter (TEX) describing the fraction of particles with a size ≤6 µm (clay+fine silt) per unit volume of soil. These parameters have to be specified for each horizon. For some soil types standard parameter records are available.

The meteorological module of the CANDY system requires daily values for precipitation, air temperature and global radiation or sunshine duration. These inputs may be read from a data-base or can be generated with the help of a special weather generator (Oelschlägel, 1992). Outputs are daily values for corrected precipitation, global radiation, air temperature and potential evapotranspiration.

The soil water module uses the corrected precipitation and potential evapotranspiration as input values. State variables are soil water content and the water equivalent of the snow cover. This submodel is based on a capacity model (Koitzsch & Günther 1990) and calculates snow accumulation and melting, crop interception, evapotranspiration and drainage.

The next step in the chain of submodels is the soil temperature module, where the heat flow equation is solved using an approach of Suckow (1986). At the lower boundary it is assumed that no heat flow occurs. The soil surface temperature, calculated by an empirical approach using air temperature of the last three days and crop coverage, is used as the upper boundary condition.

Another module of the system calculates the changes of the soil state variables as a result of tillage, addition of nitrogen fertilizer and manure or other fresh organic matter, irrigation, plant emergence or harvest.

Using an empirical approach special crop state variables like height, coverage, root depth and nitrogen demand are calculated. The nitrogen demand can be calculated by means of an "s-shaped" function depending on the expected yield or on the water consumption of the crop.

The crop module can be switched off in order to simulate the carbon dynamics only. In this case the yield is used as an input value to calculate the amount of roots and plant residues.

The module of organic matter turnover and nitrogen dynamics is the main part of CANDY. The model includes three kinds of organic material: fresh organic matter (FOM, material added to the soil), activated soil organic matter (the most important fraction for all turnover processes) and stabilized soil organic matter. Stabilized soil organic matter and the activated soil organic matter together make up the pool of decomposable soil organic matter (Fig. 1).

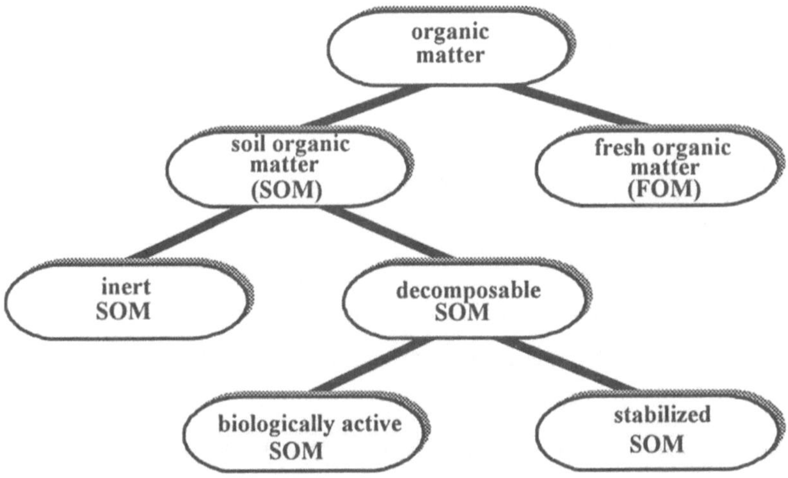

Figure 1 *Pools of organic matter in soil*

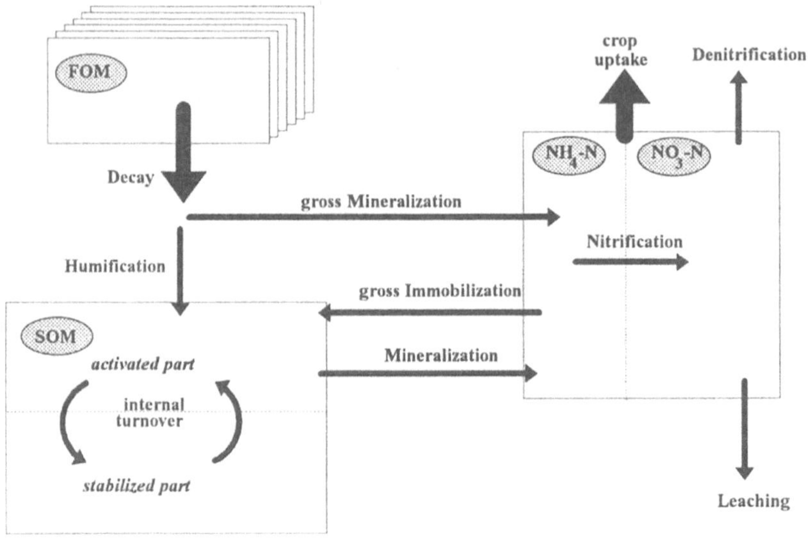

Figure 2 *Scheme of Carbon and Nitrogen turnover modeling in CANDY*

The inert organic matter pool is not included in the turnover processes. The amount of inert carbon C_i is calculated with respect to the soil texture parameter TEX:

$$C_i = const * TEX$$

The constant factor *const* in most cases is observed within the range of 0.4 to 0.6. The difference between total organic carbon and inert organic carbon is considered to be decomposable. The source of all decomposable carbon is fresh organic matter that enters the soil via manuring or plant debris. It is assumed, that an input of plant residues occurs only after harvest. The amount of this fresh organic matter is calculated depending on the crop yield.

The model is able to handle up to six different pools of fresh organic matter. The fresh organic matter undergoes the decay process shown in Fig.2. As a result of this decay new soil organic matter is produced (humification). The amount of newly synthesized soil organic carbon (C_{REP}) depends on the efficiency factor η_{FOM} of the FOM under consideration. Nitrogen mineralization and immobilization are dependent on the C to N ratio of the fresh organic matter.

The decay of the activated soil organic matter also results in mineralization. Exchange between the activated (C_A) and the stabilized (C_S) pools of soil organic matter has an important effect on long term dynamics. This exchange does not influence the mineral nitrogen pool but makes the decomposable organic matter available for mineralization over a long period of time.

All turnover processes are modelled using first order kinetics.

$$\frac{dC_{FOM}(t)}{dt} = -k_{FOM}\, C_{FOM}(t)$$

$$\frac{dC_{REP}(t)}{dt} = \eta_{FOM}\, k_{FOM}\, C_{FOM}(t)$$

$$\frac{dC_{A}(t)}{dt} = \frac{dC_{REP}}{dt} - (k_m + k_S)\, C_A(t) + k_A\, C_S(t)$$

$$\frac{dC_{S}(t)}{dt} = k_S\, C_A(t) - k_A\, C_S(t)$$

where k_{INDEX} are the rate constants referring to

FOM: fresh organic matter

REP : reproduction of soil organic matter

A : activated soil organic matter

S : stabilized soil organic matter

m : decay (mineralization) of activated soil organic matter

The rates of reaction processes are adjusted with respect to environmental conditions such as soil moisture, soil temperature and soil aeration. The effect of these environmental conditions is described by means of reduction functions (r_T, r_W and r_{AIR}) with an output range between 0 and 1:

temperature:

$$r_w(\theta) = \begin{cases} 4\dfrac{\theta}{PV}\left(1 - \dfrac{\theta}{PV}\right) & if\ \dfrac{\theta}{PV} \leq 0.5 \\ 1 & if\ \dfrac{\theta}{PV} > 0.5 \end{cases}$$

moisture:

$$r_w(\theta) = \begin{cases} 4\frac{\theta}{PV}\left(1-\frac{\theta}{PV}\right) & if\ \frac{\theta}{PV} \leq 0.5 \\ 1 & if\ \frac{\theta}{PV} \rangle\ 0.5 \end{cases}$$

aeration:

$$r_{AIR}(d, TEX, T, \theta) = \exp\left(-d\sqrt{\frac{S(TEX)\ r_T\ (T)\ r_w\ (\theta)}{\varepsilon_{AIR}\left(\varepsilon_{AIR}-\varepsilon_p\right)}}\right)$$

where T: soil temperature; θ: soil moisture; PV: pore volume; ε_{AIR}: fraction of air filled pore space; d: depth of soil layer; $S(TEX)$: texture depending parameter; ε_P, Q_{10}: constants

In contrast to other models, these reduction functions are applied to the time step as opposed to reaction coefficients - thus calculating the biological active time:

$t := t^* R(x, t^*)$

where t: biological active time,
t^*: standard time

$$R(x, t) = R(d, TEX, T, \theta) = r_T\ (T)\ r_w\ (\theta)\ r_{AIR}\ (d, TEX, T, \theta)$$

The sum of these daily values of biological active time for one year can be used to characterize turnover activities with respect to different land uses or soil conditions. The long term average over each year (Σt_{bat}) can be used for rough calculations of carbon accumulation. In this case the flux of synthesized carbon is replaced by the amount of carbon reproduced (C_{REP}) per year (in terms of biological active time):

$$\frac{d\,C_{REP}}{dt} = \frac{C_{REP}}{\Sigma t_{bat}} = \sum_{FOM} \frac{C_{FOM} * \eta_{FOM}}{\Sigma t_{bat}}$$

With this simplification the time course of the accumulation of decomposable carbon (C_{DEC}) can be described as follows:

$$C_{DEC} = C^0_{DEC} * \left(A_1 * e^{\lambda_1 * t} + A_2 * e^{\lambda_2 * t}\right) + \sum_{FOM} \frac{C_{FOM} * \eta_{FOM}}{\Sigma t_{bat}} * \left(B_1 * e^{\lambda_1 * t} + B_2 * e^{\lambda_2 * t} + \frac{1+\xi}{k_m}\right)$$

where $\xi = \frac{k_S}{k_A}$

$$\lambda_{1/2} = -\frac{k_m + k_A + k_S}{2} \pm \sqrt{\frac{(k_m + k_A + k_S)^2}{4} - k_m k_A}$$

$$A_1 = -\frac{k_A * k_m + \lambda_2 * (k_A + k_S)}{(\lambda_1 - \lambda_2) * (k_A + k_S)} \qquad A_2 = \frac{k_A * k_m + \lambda_1 * (k_A - k_S)}{(\lambda_1 - \lambda_2) * (k_A + k_S)}$$

$$B_1 = \frac{1 + \lambda_2 * \frac{k_A + k_S}{k_m * k_A}}{\lambda_1 - \lambda_2} \qquad B_2 = -\frac{1 + \lambda_1 * \frac{k_A + k_S}{k_m * k_A}}{\lambda_1 - \lambda_2}$$

C^{O}_{DEC} : initial amount of decomposable carbon
k_{INDEX} are the rate constants referring to
FOM: fresh organic matter
REP : reproduction of soil organic matter
A : activated soil organic matter
S : stabilized soil organic matter
m : decay (mineralization) of activated soil organic matter

The calculation of the steady state level of decomposable carbon can easily be applied to crop rotations in order to estimate their impact on soil organic mater:

$$C_{DEC} = \frac{1 + \xi}{k_m} * \frac{C_{REP}}{\Sigma t_{bat}}$$

These simplifications were used to create two very simple models for the accumulation of decomposable carbon, the first working with an annual time step and the second calculating only the steady state level. In this case it is possible to estimate the biological active time from soil texture (particles < 6 µm), the long term average of precipitation and air temperature.

The daily timestep model also describes nitrification and denitrification with respect to nitrate concentration and the amount of activated soil organic matter and nitrate leaching by convective water movement. This assumes that only the dissolved nitrogen in the water content above the wilting point is available for transport (Franko & Oelschlägel, 1993). The performance of the model concerning temperature, water and nitrogen dynamics has been reported by Franko *et al.* (1995a).

All functions for data input, running simulations and analysis of results are integrated in a user-interface, which allows non-modellers an easy access to the system. When the model is applied to a large area, the heterogenity of the soil must be taken into account. For this purpose the model has been integrated within a Geographic Information System, supplying soil and management data for the model inputs and creating thematic maps from the simulation results (Franko *et al.* 1995b).

Refererences

Franko, U. & Oelschlägel, B. (1993) CANDY. -In: *Simulationsmodelle zur Stickstoffdynamik*. Schriftenreihe Agrarinformatik **25:** 99-110 .

Franko, U.; Oelschlägel, B.; Schenk, S. (1995a) Simulation of Temperature-, Water- and Nitrogen dynamics using the Model CANDY. *Ecological Modelling* **81**: 213-222.

Franko,U.,Oelschlägel,B.,Schenk,S. (1995b) Modellierung von Bodenprozessen in Agrarlandschaften zur Untersuchung der Auswirkungen möglicher Klimaveränderungen. *UFZ-Bericht 3/1995*. Leipzig 1995.

Koitzsch, R.& Günther, R. (1990) Modell zur ganzjährigen Simulation der Verdunstung und der Bodenfeuchte landwirtschaftlicher Nutzflächen mit und ohne Bewuchs. *Arch. Acker-Pflanzenbau Bodenkd.* **34**: 803-810.

Oelschlägel, B. (1992) Zur Anwendung eines Wettergenerators bei der Modellierung von Bodenprozessen. In: *Tag.Ber. zum Symposium Dauerfeldversuche und Nährstoffdynamik*, Leipzig, 131-140.

Suckow, F. (1986) Ein Modell zur Berechnung der Bodentemperatur unter Brache und unter Pflanzenbestand. *Dissertation A*, AdL der DDR, Berlin.

Organic matter dynamics simulated with the 'Verberne'-model

Huub Klein Gunnewiek
Research Institute for Agrobiology and Soil Fertility,
AB-DLO Haren,
PO-Box 129
9750 A.C. Haren
The Netherlands

Introduction

It is widely accepted that direct empirical studies can not give all the answers to the complex and highly interacting systems involved in organic matter dynamics. Mechanistic models offer the opportunity to predict the behavior of the system under any given combination of environmental conditions.

The objective of the present work is to give a mechanistic description for the various processes of grass growth, nitrogen uptake, soil nitrogen dynamics and soil moisture dynamics on the basis of underlying morphological and physiological processes in relation to the environment. Nutrients other than nitrogen are assumed to be non-limiting, and soil pH and aeration are not included in the model.

The structure of the model is outlined in Figure1: it comprises four submodels. The first is a crop growth model with responses to light, temperature and nitrogen based on Johnsson & Thornley (1985), the second is a soil moisture submodel (Van Keulen & Seligman, 1987), the third describes soil organic matter dynamics (Verberne *et al.*, 1990) and the fourth submodel is related to inorganic nitrogen.

NATO ASI Series, Vol. I 38
Evaluation of Soil Organic Matter Models
Edited by David S. Powlson, Pete Smith and Jo U. Smith

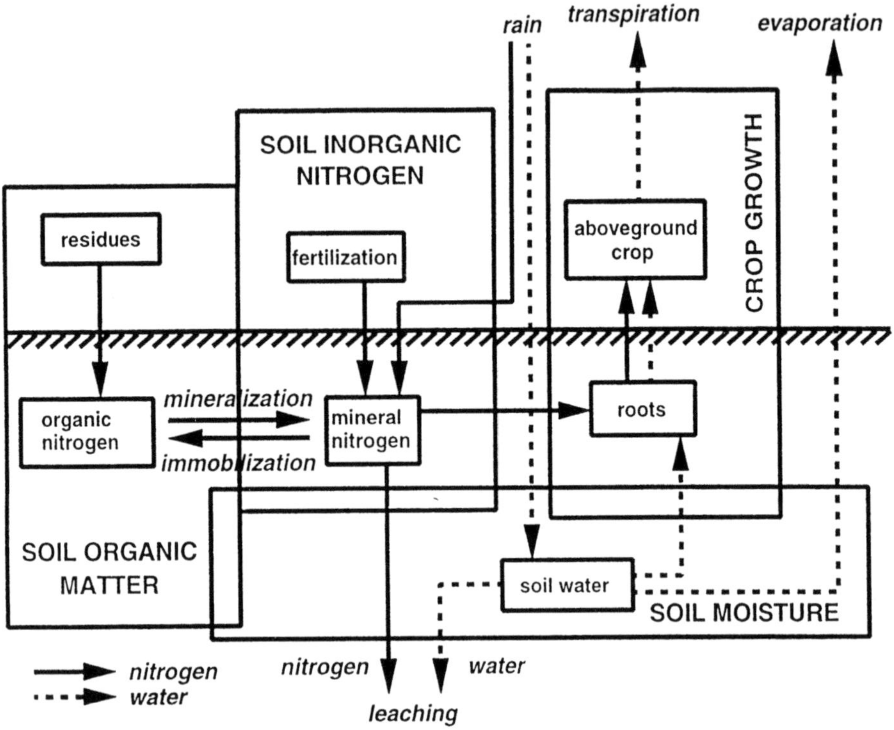

Figure 1 *Schematic representation of the model structure, which comprises four submodels simulating crop growth and the dynamics of soil moisture, soil organic matter and soil inorganic nitrogen.*

Crop submodel

A scheme for the crop submodel is given in Johnsson *et al.* (1983, 1985) and Thornley *et al.* (1989). Some modifications were made concerning light interception, photosynthesis, distribution of reserves and nitrogen demand (Spitter *et al.*,1989; McGrath, 1988; Van Keulen *et al.*,1987). In the model, the below-ground crop and the above-ground crop, four age categories are distinguished which each have unique features. The model is calibrated for perennial ryegrass (Verberne, 1992). Residues from the plant (aboveground as well as below-ground) are the major input for the soil organic matter submodel.

Soil moisture / soil inorganic nitrogen submodel

The soil is treated as a series of horizontal layers of differing thicknesses and each layer is assumed to be homogeneous. The change in the amount of soil moisture of a specific layer is calculated from the combined effect of daily water infiltration, soil surface evaporation rate, water uptake rate by the roots and daily downward movement of water through the layers. Each soil layer can be filled with water until at field capacity; this is the maximum water storage capacity of a soil under field conditions. Excess water drains to the next soil layer.

The soil inorganic nitrogen submodel is integrated in the soil water submodel. No distinction is made between ammonium and nitrate nitrogen. The amount of inorganic nitrogen in each soil layer is calculated from fertilizer application, atmospheric deposition, decomposition rate of crop residues and native soil organic matter, nitrogen uptake rate by the roots, and downward transport of nitrogen by leaching.

Soil Organic Matter Submodel

Soil organic matter dynamics models vary in complexity. Most models distinguish different pools, each representing material with a different stability with regard to decomposition (Jenkinson, 1977; Frissel & Van Veen, 1981). Although adequate experimental methods to establish the partitioning of soil organic matter among the model pools have not been developed, this approach seems the most promising to describe soil organic matter dynamics on a field scale (Jenkinson, 1977; Verberne *et al.*, 1990). Figure 2 shows the flow chart of the model, given in greater detail by Verberne *et al.* (1990). The main partition of organic matter is between recently added organic materials such as crop residues and manure, and native soil organic matter. Both are further subdivided into different components.

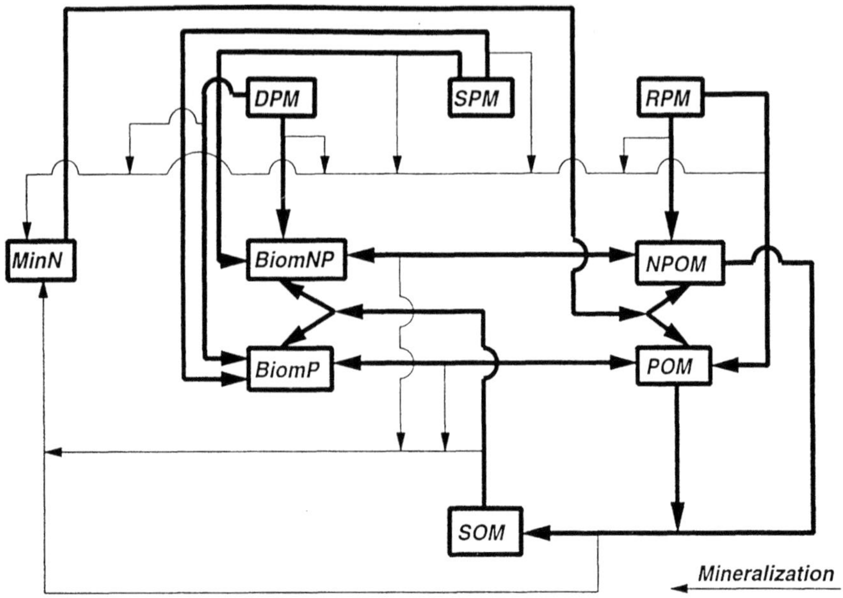

Figure 2 *Overview of the soil organic matter submodel.*

Each type of residue (leaves, sheath plus stems, and roots) is considered to consist of three fractions each with a unique resistance to biological decomposition: (1) decomposable material (DPM), i.e. carbohydrates and proteins; (2) structural material (SPM), i.e. cellulose and hemicellulose; (3) resistant material (RPM), i.e. lignified structural material. Each fraction is assumed to have a fixed C/N ratio. The residues are split into the three fractions (DPM, SPM and RPM) taking the C/N ratio of each fraction into account as well as the C/N ratio of the entire residue. The RPM of the residues is calculated from the C/N ratio and a maximum value of RPM. When the RPM is calculated, the decomposable and structural fractions are then determined by splitting the remaining residues into two parts according to their C/N ratios and the C/N ratios of the decomposable and structural fractions.

The rates of change in the amounts in carbon for the three fractions result from residue inputs minus the decay of residual material already present. The nitrogen fluxes are assumed to be proportional to the carbon fluxes and depend on the C/N ratio of the fractions.

Exchange of soil organic matter and microbial biomass between soil layers, mainly caused by the soil fauna, is not described explicitly. It is modelled by assuming that dead leaves and sheath plus stem material are distributed over the soil layers and is equivalent to the total root dry weight distribution throughout the soil layers.

Decomposition of the different fractions are considered to be first-order reactions, with specific first-order decomposition constants. A yield efficiency factor defines the fraction of the decomposing carbon incorporated into microbial biomass; the remainder (1-yield coefficient) leaves the system as CO_2. The decomposition constants and efficiency factors are assumed to be independent of the type of residue. Decomposable and structural material are considered to be decomposed by the microbial biomass, while the lignin fraction (RPM) directly enters the soil organic matter pool (Van Veen & Paul, 1981). The decomposition constants for structural and resistant material are calculated as a function of the lignin fraction. Both decomposition constants decrease as the lignin component increases. This is based on the assumption that as the lignin component increases, the accessibility, and thus the decomposability of structural materials such as hemicellulose and cellulose, decreases rapidly (Parton *et al.*, 1987).

Protection

Net mineralization and immobilization are more rapid in sandy soils than in clay soils (Hassink *et al.*, 1990). The lower net mineralization in clay soils is assumed to be caused by a greater physical protection of soil organic matter and microbial biomass in clay soils than in sandy soils. This physical protection is assumed to be the result of adsorption of organic material by layers in swelled clays, adsorption on the surface of silt and clay particles, or aggregate formation, which renders the organic material less accessible to decomposition by microorganisms.

Initially, soil organic matter is divided into three major fractions: microbial biomass, active organic matter and stabilized organic matter (SOM, an old inactive fraction stabilized by unspecified mechanisms), each of which has a particular decomposition rate (Van Veen *et al.*, 1984). In order to describe differences in physical protection, active organic matter is divided into a physically protected component (POM) and a nonprotected component (NPOM) (Verberne *et al.*, 1990). Soil microbial biomass is also divided into a protected (BIOMP) and a nonprotected component (BIOMN). Decomposing material is distributed between the nonprotected and the protected microbial biomass. The actual distribution is governed by a soil-type specific parameter α, which denotes the proportion of decomposed material retained near protected biomass. The value of α increases proportionally with the clay content of the soil.

Whether a transformation causes nitrogen mineralization or immobilization depends on the values of the carbon to nitrogen ratio of the substrate and product, and of the efficiency factor.

References

Frissel, M.J. & J.A. Van Veen (1981) Simulation model for nitrogen immobilization and mineralization. In: *Modelling wastewater renovation* (ed: I.K. Iskandar), pp359-381. John Wiley & Sons, New York.

Hassink, J., Scholefield, D. & P. Blantern (1990) Nitrogen mineralization in grassland soils. In: *Soil-Grassland-Animal Relationship*. Volume I (ed: Gaborcik, N., Krajcovic, V. & M. Zimkova)., pp206-210. Proceedings 13th General Meeting of the European Grassland Federation, Banska Bystrica.

Jenkinson, D.S., (1977) Studies on the decomposition of plant material in soils. IV. The effect of rate of addition. *Journal of Soil Science* **28**: 417-423.

Johnsson, I.R. (1983) Nitrate uptake and respiration in roots and shoots: a model. *Physiologia Plantarum* **58**: 145-147.

Johnsson, I.R. & J.H.M. Thornley (1983) Vegetative crop growth model incorporating leaf area expansion and senescence, and applied to grass. *Plant, Cell and Environment* **6**: 721-729.

Johnsson, I.R. & J.H.M. Thornley (1985) Dynamic model of the response of a vegetative grass crop to light, temperature and nitrogen. *Plant, Cell and Environment* **8**: 485-499.

McGrath, D. (1988) Seasonal variation in the Water-Soluble Carbohydrates of Perennial and Italian Ryegrass under cutting conditions. *Irish Journal of Agricultural Research* **27**: 131-139.

Parton, W.J., Schimmel, D.S., Cole, C.V. & D.S. Ojima (1987) Analysis of factors controlling soil organic matter levels in Great Plain grasslands. *Soil Science Society of Americal Journal* **51**: 1173-1179.

Spitters, C.J.T., Van Keulen, H. & D.W.G. Van Kraalingen (1989) A simple and universal crop growth simulator: SUCROS87. Simulation and systems management in crop production. In: *Simulation monographs*. (ed: Rabbinge, R., Ward, S.A. & H.H. Van Laar), PUDOC, Wageningen, pp147-181.

Thornley, J.H.M. & E.L.J. Verberne (1989) A model of nitrogen flows in grassland. *Plant, Cell and Environment* **12**: 863-886.

Van Keulen, H. & N.G. Seligman (1987) Simulation of water use, nitrogen nutrition and growth of a spring wheat crop. *Simulation monographs*, PUDOC, Wageningen, 310p.

Van Veen, J.A. & E.A. Paul (1981) Organic carbon dynamics in grassland soils. I.Background information and computer simulation. *Canadian Journal of Soil Science* **61**: 185-201.

Van Veen, J.A., Ladd, J.N. & M.J. Frissel (1984) Modelling C and N turnover through the microbial biomass in soil. *Plant and Soil* **76**: 257-274.

Verberne, E.L.J., Hassink, J., De Willigen, P., Groot, J.J.R. & J.A. Van Veen (1990) Modelling soil organic matter dynamics in different soils. *Netherlands Journal of Agricultural Science* **38**: 221-238.

Verberne, E.L.J. (1992) Simulation of the nitrogen and water balance in a system of grassland and soil. *Internal nota 258*, Research Institute for Agrobiologie and Soil Fertility, AB-DLO Haren, 1992.

The DNDC Model

Changsheng Li
Complex Systems Research Center
Institute for the Study of Earth, Oceans, and Space
University of New Hampshire
Durham, New Hampshire, USA

The denitrification-decomposition (DNDC) model is a process-oriented simulation model of soil carbon (C) and nitrogen (N) biogeochemistry. The model is intended to be used for estimating effects of climate change, land use, agricultural management, soil properties, and atmospheric nitrogen deposition on soil C and N dynamics. DNDC is a site model with simulated time span from several days to centuries.

DNDC was initiated in the Bruce Company, a Washington D. C.-based consulting firm, in 1989 and further developed by a biogeochemical modeling group headed by Dr. Robert Harriss in the University of New Hampshire. Drs. Changsheng Li, Steve Frolking and Robert Harriss were involved in the model development. Some concepts, formulas, or parameters used in DNDC were adopted from other models. The structure of soil organic matter pools was adopted from the NCSOIL model developed by Molina *et al.* (1983), the denitrification sequence from a closed system model developed by Leffelaar *et al.* (1988), and the hydraulic conductivity equations from the models developed by Mualem (1986) and Clapp *et al.* (1978).

DNDC contains four interacting submodels of soil climate, decomposition, denitrification, and plant growth (see Figure 1). The soil climate submodel calculates soil temperature and moisture profiles, and soil water flow over time based on soil physical properties, air temperature, precipitation, and water uptake by plants. This

NATO ASI Series, Vol. I 38
Evaluation of Soil Organic Matter Models
Edited by David S. Powlson, Pete Smith and Jo U. Smith

Figure 1. *The DNDC Model*

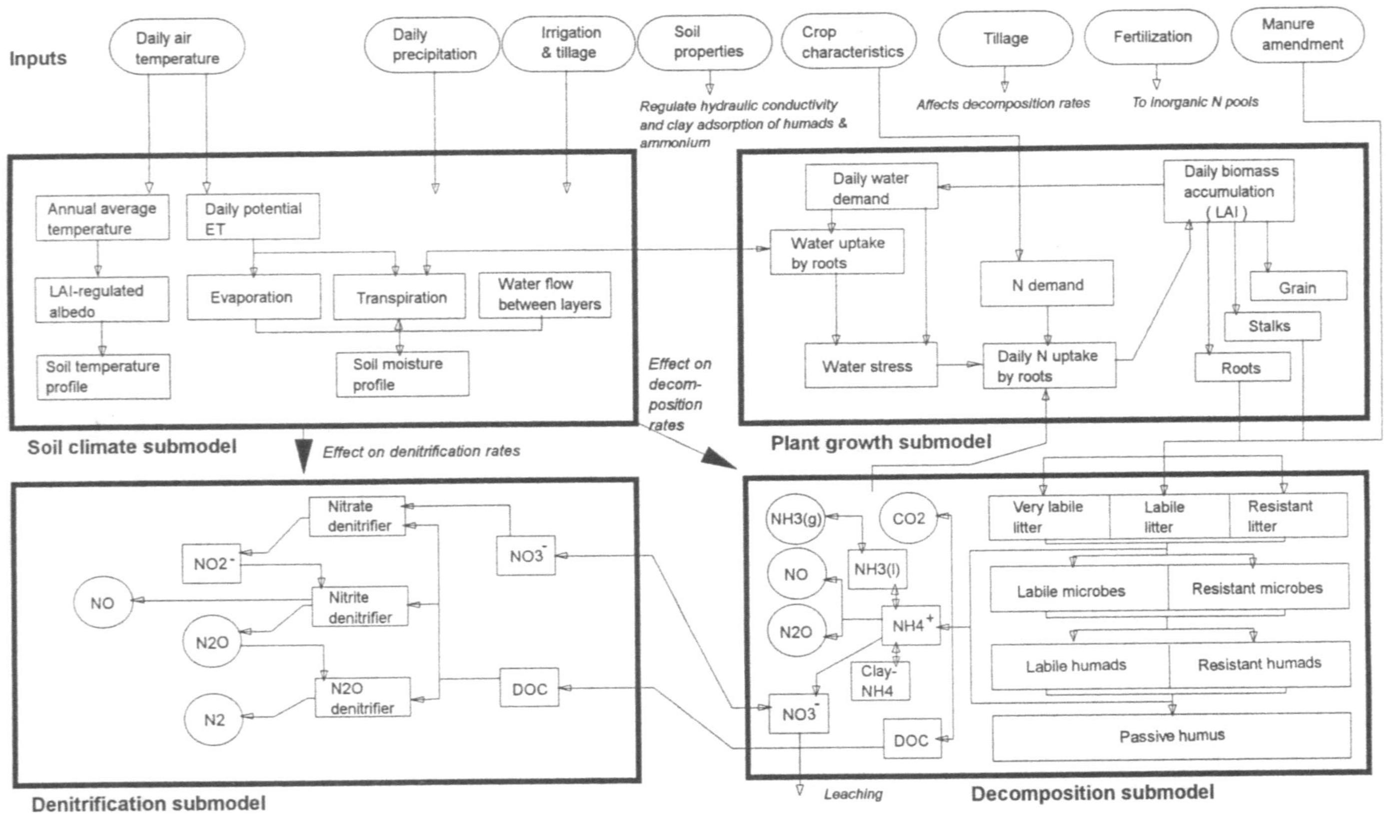

information is fed into either a decomposition or a denitrification submodel. The decomposition submodel calculates daily decomposition, nitrification, movement of nitrate and ammonium along soil profile, ammonia (NH_3) volatilization, and carbon dioxide (CO_2) and other trace gases emissions. The denitrification submodel calculates hourly denitrification processes and nitrous oxide (N_2O), nitric oxide (NO) and dinitrogen (N_2) production during wet periods. The plant growth submodel calculates daily water- and N-uptake by plants, plant growth, root respiration, and biomass partitioning to grain, stalks and roots. Effects of cropping practices including tillage, fertilization, manure application, and irrigation, are incorporated into the model. The time step is hourly for the soil climate and denitrification submodels, and daily for the decomposition and plant growth submodels.

Input parameters required by DNDC are climate (daily air temperature and precipitation), soil properties (texture, organic matter content, bulk density, and *p*H), and cropping practices (rotation, crop characteristics, tillage, fertilization, manure application, and irrigation). The outputs include daily or annual changes in soil C and N pools, emissions of CO_2 by soil and root respiration, emissions of N_2O, NO, N_2 and NH_3 from soil, methane (CH_4) uptake by soil, nitrate available for leaching, and crop biomass (roots, leaves/stalks, and grain) production.

In the DNDC model, soil organic matter (SOM) dynamics is regulated by SOM turnover and litter/manure input. SOM turnover rate is affected by a variety of factors in DNDC, including size of SOM pools, C/N ratio in SOM pools, temperature, moisture, available N content, and soil disturbance by tillage *et al.* In DNDC, SOM pool consists of several sub-pools of labile/resistant litter, microbes, humads, and humus. Each sub-pool has its own C/N ratio. The C/N ratios for the microbes, humads, and humus sub-pools are fixed, but the C/N ratios for the litter sub-pools are allowed to vary. The size and C/N ratio of the litter pools will be recalculated if fresh crop residue or manure is once added into the soil. A specific decomposition rate is set for each of the SOM sub-pools. The size and quality (C/N ratio) of the SOM pools dominate the overall decomposition rate. When SOM decomposes from one sub-pool to another, available N may release from or be required by the decomposing process because of the deference in C/N ratio between the two sub-pools. Decomposition rates will be retarded if the required N is not met by the available N (ammonium and nitrate) pools. The CO_2 flux produced by soil respiration is an indicator for the overall SOM decomposition rate. Soil litter comes from crop roots and the above-ground residue which is left in the field after harvest and finally incorporated into the soil by tillage. The magnitude of annual litter input depends on crop biomass yield and percent of residue returned to the soil after harvest. In DNDC, crop yield is regulated by several factors including user-defined or default optimum yield, water stress, and N availability during the growing season. The percent of residue left in the field after harvest is defined by users according to the practical situation in their fields.

DNDC has been previously evaluated against the field measurements in various croplands in America, Brazil, China, Costa Rica, Denmark, England, Germany, and Venezuela, either on SOM dynamics or on trace gas emissions. Part of the verification results have been published in several papers (Li *et al.*, 1992b, 1994a, and 1994b). DNDC has been used for the studies on alternative agricultural practices in soil carbon sequestration and mitigating greenhouse gases (Li, 1995), and also for estimation of regional trace gas emissions (Li *et al.*, 1995).

References

Clapp, R.B. & Hornberger, G.M. (1978) Empirical equations for some soil hydraulic properties. *Water Resour. Res.* **14**: 601-604.

Leffelar, P.A.& Wessel, W.W. (1988) Denitrification in a homogeneous, closed system: experiment and simulation. *Soil Sci.* **146**: 335-349.

Li, C., Narayanan V. & Harriss R.C. (1995) Nitrous oxide emissions in 1990 from agricultural lands in the United States. (in preparation)

Li, C. (1995) Modeling impact of agricultural practices on soil C and N_2O emissions. In: *Advances in Soil Science: Soil Management and Greenhouse Effect* (ed. R. Lai, J. Kimble, E. Levine, and B.A. Stewart), Lewis Publishers, London: 101-112.

Li, C., Frolking, S. & Harriss, R.C. (1994a) Modeling nitrous oxide emissions from agriculture: A Florida case study. *Chemosphere* 28: 1401-1415.

Li, C., Frolking, S. & Harriss, R.C. (1994b) Modeling carbon biogeochemistry in agricultural soils. *Global Biogeochemical Cycles* **8**: 237-254.

Li, C., Frolking, S. & Frolking, T.A. (1992a) A model of nitrous oxide evolution from soil driven by rainfall events: 1. Model structure and sensitivity. *Journal of Geophysical Research* **97**: 9759-9776.

Li, C., Frolking, S. & Frolking, T.A. (1992b) A model of nitrous oxide evolution from soil driven by rainfall events: 2. Applications. *Journal of Geophysical Research* **97**: 9777-9783.

Molina, J.A.E., Clapp, C.E., Shaffer, M.J., Chichester, F.W. & Larson, W.E. (1983) NCSOIL, a model of nitrogen and carbon transformations in soil: description, calibration, and behavior. *Soil Sci. Soc. Am. J.* **47**: 85-91.

Mualem, Y. (1986) Hydraulic conductivity of unsaturated soils: prediction and formulas. In: *Methods of Soil Analysis, Part 1: Physical and Mineralogical Methods* (ed. Cambell *et al.*) American Society of Agronomy, Inc. & Soil Science Society of America, Inc. Publisher, Madison, Wisconsin, USA: 799-823.

Description of the model NCSOIL

J.A.E. Molina
Department of Soil, Water, and Climate
University of Minnesota
St. Paul, MN (USA)

Introduction

NCSOIL simulates C and N (total and tracers) transformations in soils. It is the module of a larger model which computes C and N flows in the soil-water-air-plant system (NCSWAP). As a stand-alone model, NCSOIL was initiated in 1980 and has been validated against several data sets (Barak *et al.*, 1990 a; Barak *et al.*, 1990 b; Hadas *et al.*, 1993 a; Hadas *et al.*, 1993 b; Hadas *et al.*, 1987; Hadas *et al.*, 1992 a; Hadas *et al.*, 1992 b; Hetier *et al.*, 1989-1990; Houot *et al.*, 1989; Molina *et al.*, 1994; Molina *et al.*, 1980; Molina *et al.*, 1983; Molina *et al.*, 1990; Nicolardot *et al.*, 1994 a; Nicolardot *et al.*, 1994 b).

Structure of NCSOIL

C flows

C from residues is stabilized into two distinct organic pools, Pool II with a half-life of 115 days and the more stable Pool III (Figure 1). The decomposition of residues, Pool II, and Pool III is driven by the activity of the primary decomposers which are in turn cannibalized by other microbes to form microbial successions. Two percent of C carried through the microbial successions goes into Pool II. The same percentage of decomposed C from Pool II is stabilized into Pool III. The microbial efficiency (fraction of decayed C recycled into the biomass) is set to 0.6. The microbial efficiency of residue decomposition is reduced (increased organic C loss by CO_2 evolution) to adjust for N stress. Each residue can be represented by up to 3 pools decaying independently of each other.

NATO ASI Series, Vol. I 38
Evaluation of Soil Organic Matter Models
Edited by David S. Powlson, Pete Smith and Jo U. Smith

The mathematical description of C flow represented by Figure 1 calls for the determination of too many initial conditions and dynamic parameters. The structure represented by Figure 1 was simplified. It was assumed that the microbial successions have the same dynamics of C flow, whatever the primary decomposers; thus, the three microbial successions were grouped into three pools, each with a C feedback loop (Figure 2-A); and finally into one pool: Pool I of NCSOIL (Figure 2-B). C flow through Pool I is dominated by the dynamics of the cannibalistic microbes. Since these microbes have a rapid turnover, the dynamics of Pool I can be simulated by the decomposition of two fractions: Pool I labile, and Pool I resistant, with half-lives of 2 and 17 days, respectively; fifty six percent of C input to Pool I is directed into the labile fraction (Nelson *et al.*, 1979).

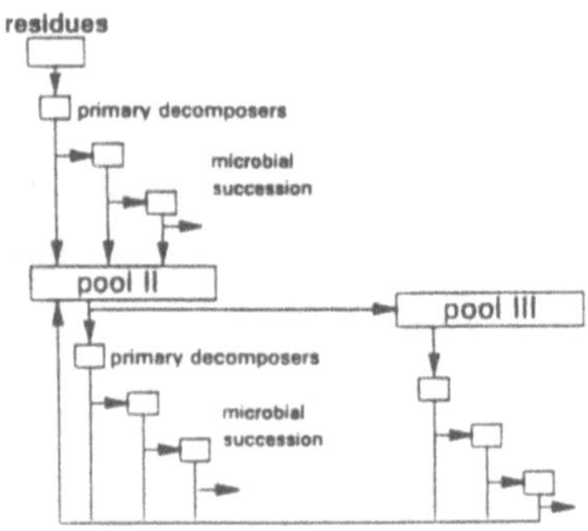

Figure 1 *Organic C flows in NCSOIL.*

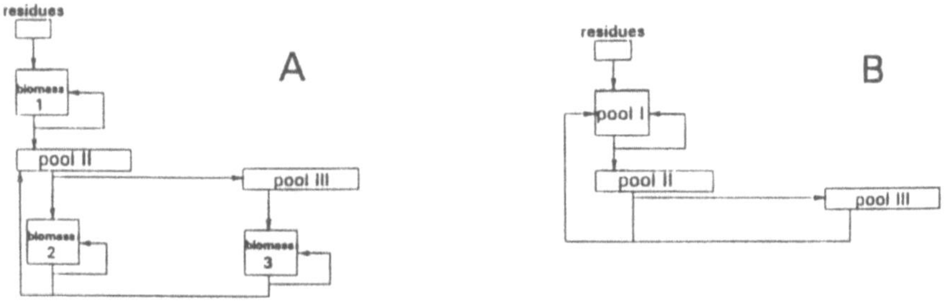

Figure 2 *Organic C flows in NCSOIL: (A) simplification by grouping each microbial succession into one biomass pool with C feedback loop; (B) final form of the model after collapsing the microbial pools into Pool I.*

The kinetics of Pool I (labile plus resistant) has been correlated to that of the microbial biomass as measured by the fumigation method (Nicolardot *et al.*, 1994 b). Pool I plus Pool II represents the potentially mineralizable N (N_o of Stanford and Smith, 1972); and Pool III the remaining soil organic matter. The C/N ratio of Pool I is set to 6; and that of Pool II to 12 when not obtained by calibration.

N flows

N flow is represented by 2 schemes: (1) the mineralization-immobilization turnover (MIT) which is driven by the decomposition of the soil organic matter (Pool I, Pool II, and Pool III); and (2) residues decomposition which is adjusted to mineral N availability after completion of the MIT. The soil organic pools are the site of the MIT process which always results in net N mineralization (Figure 3). The flux of N from Pool I, Pool II, or Pool III

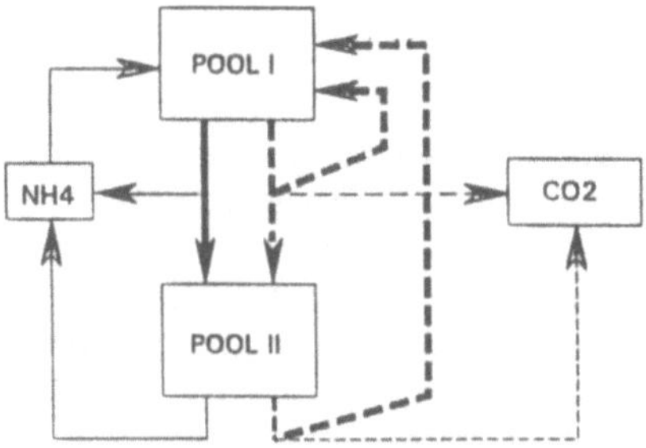

Figure 3 *C and N flows in NCSOIL between Pool I and Pool II. The same structure would apply for flows from Pool II to Pool III and back to Pool I. MIT between these pools always results in net N mineralization. C flow: discontinued line; N flow: solid line; inorganic flow: narrow line; organic flow: large line.*

Figure 4 *C and N flows between residues and Pool I. N flow is carried by: (A) inorganic N; or (B) inorganic and organic N. Scheme B produces net N mineralization (flow 2 suppressed); or net N immobilization (flow 1 suppressed). The user must select between structure A and B. C flow: discontinued line; N flow: solid line; inorganic flow: narrow line; organic flow: large line.*

into Pool I is mediated exclusively by NH_4^+. By contrast, microbes feeding on residues receive their N from inorganic N (Figure 4-A), or inorganic plus residue N. (Figure 4-B). Nitrate is assimilated by the biomass when NH_4^+ has been depleted. One scheme has to be selected by the user.

Kinetic rates

The integration time step is one day or less. Residues and soil organic pools decay with first order rates with respect to their C concentrations. Users have the option to select for residue decomposition a Monod kinetics. Rates of denitrification are proportional to the rates of C decay from residues and soil organic pools. Rates are modified by soil temperature, water content (Doran *et al.*, 1987), and clay content. They are N independent at the MIT site; but are controlled by N availability for residue decomposition (Figure 5).

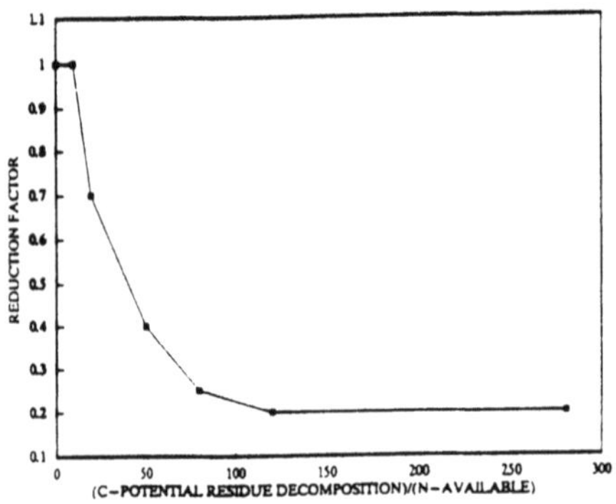

Figure 5 *Influence of N availability on residue decay. Potential residue decomposition corresponds to no N stress. N-available is computed as the sum of inorganic N plus N mineralized from residues.*

One unresolved issue: determination of inital levels of Pool II and Pool III

The initial value for Pool I plus Pool II is accessible by the soil incubation technique. This method, however, is tedious and requires a model (e.g. the simple or double exponential equation) to analyze the kinetics of CO_2 and inorganic N accumulation; thus making it dependent on a decay rate constant. A routine method has to be devised to obtain independent values for the initial levels of Pool I plus Pool II; and the C/N ratio of Pool II which has been found to range from 9 to 30 when determined by calibration against kinetics of CO_2 and inorganic N accumulation during soil incubation. Similarly, a routine method to assess the level of the stable soil organic matter (half life in the order of thousands of years) is needed. This stable organic matter would be substracted from Pool III which would have a shorter half-life than the one presently obtained by calibration (in the order of 200 years).

References

Barak, P., Molina, J.A.E., Hadas, Aviva & Clapp, C.E. (1990 a) Optimization of an Ecological Model with the Marquardt Algorythm. Ecological Modeling **51**:251-263.

Barak, P., Molina, J.A.E., Hadas, Aviva & Clapp, C.E. (1990 b) Mineralization of Amino acids and Evidence of Direct Assimilation of Organic Nitrogen. Soil Sci. Soc. Am. J. **54**:769-774.

Doran, J.W. & Smith, M.S. (1987) Organic matter management and utilization of soil and fertilizer nutrients. *In: Soil Fertility and Organic Matter as Critical Components of Production Systems.* SSSA Spec. Pub. No. **19**: 53-71.

Hadas, Aviva, Feigenbaum, S., Sofer, M., Molina, J.A.E. & Clapp, C.E. (1993 a) Decomposition rates of 15N-labeled wheat and cellulose in soil - Simulation modeling of tracer dynamics. Soil Sci. Soc. Am. J. **57**:996-1001.

Hadas, Aviva & Molina, J.A.E. (1993 b) Simulation of nitrogen assimilation by heterotrophic soil microbial biomass. Physiol. Plant. **87**:528-534.

Hadas, Aviva, Molina, J.A.E., Feigenbaum, S. & Clapp, C.E. (1987) Simulation of 15N immobilization by the model NCSOIL. Soil Sci. Soc. Amer. J. **51**:102-106.

Hadas, Aviva, Molina, J.A.E., Feigenbaum, S. & Clapp, C.E. (1992 a) Factors affecting nitrogen immobilization in soil as estimated by simulation models. Sci. Soc. Am. J. **56**:1481-1486.

Hadas, Aviva, Sofer, M., Molina, J.A.E., Barak, P.,& Clapp, C.E. (1992 b) Assimilation of nitrogen by soil microbial population: NH4 vs. organic N. Soil Biol. Biochem. **24**:137-143.

Hetier, J.M., Zuvia, M., Houot, S. & Thiery, J-M. (1989-1990) Comparaison de trois modeles choisis pour la simulation du cycle de l'azote dans les agro-systemes tropicaux. Cah. ORSTOM, ser. Pedol. **25**:443-451.

Houot, S., Molina, J.A.E., Chaussod, R. & Clapp, C.E. (1989) Simulation by NCSOIL of net mineralization in soils from the "Deherain" and "36 parcelles" fields at Grignon. Soil Sci. Soc. Am. J. **53**:451-455.

Molina, J.A.E., Cheng, H.H., Nicolardot, B., Chauusod, R. & Houot, S. (1994) Biologically active soil organics: A case of double identity. *In*: *Defining Soil Quality for a Sustainable Environment.* (ed: Doran, J.W., Coleman, D.C., Bezdicek, D.F. & Stewart, B.A.) SSSA Special Publication No. **35**: 169-177.

Molina, J.A.E., Clapp, C.E. & Larson, W.E. (1980) Potentially mineralizable nitrogen in soil: A simple exponential model does not apply for the first 12 weeks of incubation. Soil Sci. Soc. Am. J. **44**:442-443.

Molina, J.A.E., Clapp, C.E., Shaffer, M.J., Chichester, F.W. & Larson, W.E. (1983) NCSOIL, a model of nitrogen and carbon transformations in soil: Description, calibration and behaviour. Soil Sci. Soc. Am. J. **47**:85-91.

Molina, J.A.E., Hadas, Aviva & Clapp, C.E. (1990) Computer Simulation of Nitrogen Turnover in Soil and Priming Effect. Soil Biology and Biochemistry, **22**:349-353.

Nelson, D.W., Martin, J.P. & Ervin, J.O. (1979). Decomposition of Microbial Cells and Components in Soil and their Stabilization through Complexing with Model Humic Acid-type Phenolic Polymers. Soil Sci. Soc. Am. J. **43**:84-88.

Nicolardot, B. & Molina, J.A.E. (1994 a) C and N fluxes between pools of soil organic matter: Model calibration with long-term field experimental data. Soil Biol. Biochem. **26**:245-251.

Nicolardot, B., Molina, J.A.E. & Allard, M.R. (1994 b) C and N fluxes between pools of soil organic matter: Model calibration with long-term incubation data. Soil Biol. Biochem. **26**:235-243.

Stanford, G. & Smith S.J (1972) Nitrogen mineralization potentials of soils. Soil Sci. Soc. Am. Proc. **36**:465-472.

Simulating soil carbon and nitrogen dynamics with the soil-plant-atmosphere system model DAISY

T. Mueller, L.S. Jensen, S. Hansen and N. E. Nielsen
Soil, Water and Plant Nutrition,
The Royal Veterinary and Agricultural University,
Thorvaldsensvej 40, DK-1871
Frederiksberg C, Copenhagen,
Denmark

Introduction

DAISY is a deterministic model that simulates the C and N-fluxes in the soil-plant-atmosphere system (Hansen *et al.*, 1990; Hansen *et al.*, 1991b). It consists of submodels for *soil water (incl. solute movement), soil temperature, soil organic matter (incl. microb. biomass)*, soil *mineral N, crop growth* and *system management*. The model in its present version is adapted to the wet temperate climate of North-Western Europe. Beside its use in scientific system studies, it has been used as a management tool at the field scale (Jensen *et al.* 1994b), and in regional-scale studies (Styczen & Storm 1993). It has been validated under various conditions (Hansen *et al.*, 1991a, b, c; Jensen *et al.*, 1994a; Petersen *et al.*, 1995; Svendsen *et al.*, 1995; Vereecken *et al.*, 1991).

Model description for the soil part of DAISY

The model has a *one dimensional structure* allowing for 5 different soil horizons and a number of soil layers represented by node points used in the numeric solutions. Hence, the usefulness for different horizontal spatial scales depends on the spatial variability of the soil input parameters in the field. However, there are no scale limitations in general. DAISY

NATO ASI Series, Vol. I 38
Evaluation of Soil Organic Matter Models
Edited by David S. Powlson, Pete Smith and Jo U. Smith

works with different time scales in the various submodels. The soil submodels run with a basic time step of 1 h. The crop submodel runs with time steps of 1 day.

Soil water dynamics are simulated by a numeric solution of the *Richards equation*. The upper boundary is defined by an evapotranspiration (ET) model, whereas the lower boundary is user defined. The ET model is based on the concept of potential ET and simulates infiltration and water loss rates from soil and crop. *Solute movement* is simulated by a numeric solution to the *convection-dispersion equation*. The considered solutes are ammonium and nitrate. *Water and nitrogen uptake* by roots are modelled as a function of the root density and a single-root uptake model. The assumed mechanisms are Darcian flow for water movement, and mass-flow and diffusion for nitrogen movement.

The soil temperature model simulates heat flow due to conduction and convection. Furthermore the heat flow equation is expanded to include frost and thaw processes.

The *soil-organic-matter* submodel (Figure 1) considers three organic pools, viz. added organic matter (AOM), soil microbial biomass (SMB), and soil organic matter (SOM = total org. C - SMB-C - AOM-C). After crop harvest, turnover of dead root organic matter (ROM) is simulated like AOM once soil tillage has been performed. If several incorporations of AOM occur, these are simulated as separate and identifiable AOM pools (AOMa, AOMb, ...).

Each of the above mentioned organic pools are divided into two subpools: one with a slow turnover (i.e. SOM1) and one with a fast turnover (i.e. SOM2). This discrete division is used as a very simple approximation of the corresponding continuum in nature and facilitates the description of the turnover of all the organic matter pools by first-order reaction kinetics. In order to determine the actual turnover rate coefficients [s^{-1}] of the organic pools in every time step, the standard turnover rate coefficients are multiplied with three or two modifiers. The *three* standard coefficient modifiers of the *turnover rates* of SOM1 and SOM2 as well as those of the *death rates* and of the *maintenance respiration rates* of SMB1 and SMB2 are functions (1) of the actual soil temperature (F_m^T), (2) of the actual soil water potential (F_m^{ψ}) and (3) of the soil clay content (F_m^c) (see Figure 2). *Turnover rates* of AOM1 and AOM2 are calculated using only F_m^T and F_m^{ψ}. The partition of the C fluxes between organic matter pools

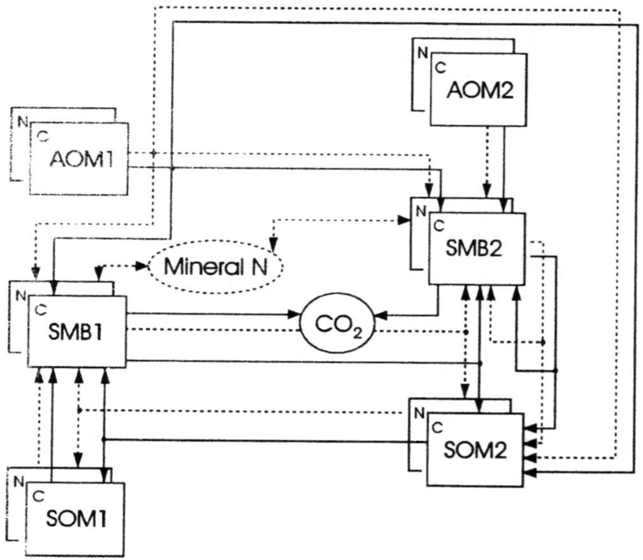

Figure 1 *C and N fluxes between the different pools and subpools of organic matter, mineral N and CO_2 in soil. AOM =Added Organic Matter; SMB = Soil Microbial Biomass; SOM = Soil organic matter.*

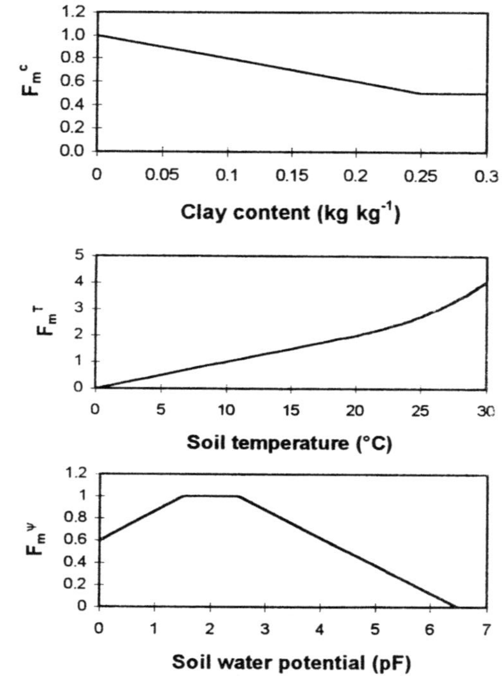

Figure 2 *Functional forms of the turnover rate modifiers of organic matter pools in soil. Clay content modifier (F_m^c), temperature modifier (F_m^T) and water potential modifier (F_m^ψ)*

are defined by partitioning coefficients. C-fluxes into the microbial biomass are multiplied with the substrate utilization efficiency. The remaining substrate C is mineralized to CO_2.

The corresponding N-pools are calculated from the amount of C in the pools using a fixed C/N ratio for each pool. Net N-mineralization or N-immobilization are derived from the N-balance. The simulated growth of SMB1 and SMB2 is limited by mineral N in the soil if N-immobilization occurs. In that case, available NH_4^+ is utilized in preference to NO_3^-.

Nitrification is described by *Michaelis-Menten kinetics*. Abiotic factors influencing the nitrification rate are soil temperature and soil water potential. *Denitrification* is modelled by a simple "capacity approach" and occurs only at more than 80% water-filled pore space. Denitrification is then determined either by the transport of NO_3^- to anaerobic microsites or, in the case of ample supply of NO_3^-, by the actual microbial activity at these sites. The anaerobic soil volume is simulated as a function of the relative soil water content and of soil respiration as an index of substrate availablility.

Model input requirements

Global radiation (R_G), *air temperature* 2 m over ground (T_{2m}) and *precipitation* (P) are ***driving variables*** and have to be available as daily means (R_G, T_{2m}) or daily totals (P). The lower boundary of the system may be defined as free drainage or as a constant or fluctuating groundwater table. The following ***constants*** must be defined: (a) soil physical properties (e.g. water release characteristics, hydraulic conductivity), (b) a soil balance depth in order to calculate the N-balance, (c) C/N-ratios of the different subpools of soil organic matter, (d) standard turnover rate coefficients of the pools of organic matter, (e) partitioning coefficients of the C-fluxes, and (f) nitrification and denitrification parameters. For (c) - (f) a default set of parameters has been calibrated, see Table 1. The following ***initial conditions*** must be defined: (a) soil water potential, volumetric ice content and soil temperature for the defined node points, (b) NO_3-N and NH_4-N content of the soil, (c) total organic C content and total organic C/N-ratio of the soil, (d) fraction of total organic C in SMB1 and SMB2, and (e) division of SOM-C in SOM1 and SOM2. For (d) and (e) default values have been calibrated for temperate arable systems, but the initial division of total SOM-C into SOM1 and SOM2 is an essential calibration parameter when adapting the model to different climatic areas or

Table 1 *Default C/N-ratios, potential turnover rate and partitioning coefficients*

Pool	C/N - ratio	Standard turnover rate coefficient for:	day^{-1}	Partitioning coefficients from	to	
SOM1	11	decay rate	$2.7 \cdot 10^{-6}$	SOM2	SOM1	0.1
SOM2	* [1]	decay rate	$1.4 \cdot 10^{-4}$	SOM2	SOM1	0.9
SMB1	6	death rate	0.001	SMB1	SMB2	0.6
		mainten. respir.rate	0.01	SMB1	SOM2	0.4
SMB2	10	death rate	0.01	SMB2	SMB2	0.6
		mainten. respir. rate	0.01	SMB2	SOM2	0.4
ROM1[2]	100	decay rate	0.007	AOM1	SMB1	0.5
ROM2[2]	12	decay rate	0.07	AOM1	SMB2	0.5
AOM1[2]	90	decay rate	0.005	ROM1	SMB1	0.5
AOM2[2]	15	decay rate	0.05	ROM1	SMB2	0.5
Efficiency of use of organic matter:			0.6			

[1] C/N ratio of SOM2 is calculated from initial total organic-C, -N and fraction in SOM1 (default fraction=0.8)

[2] Depends on crop type and AOM type, listed default values are for cereal roots and straw.

management histories. Every ***application of crop residue or organic fertilizer*** must be characterized by the (a) applied amount of wet weight, (b) dry matter (DM) content, (c) content of organic C, organic N and NH_4-N in DM, (d) percent of NH_4-N lost by volatilization, (e) fraction of AOM-C in AOM1 and AOM2, (f) C/N ratio of AOM1, (g) standard turnover rate coefficients of AOM1 and AOM2, and (h) partition coefficients. Information is also required on irrigation, climatic parameters, atmospheric N deposition, fertilization, soil tillage, sowing and harvest. Currently 10 different crop definition files are available: winter wheat, grass (non-legume), fodder beet, sugar beet, spring rape, winter rape, potatoes, peas and spring barley with and without undersown grass.

Data output

DAISY first simulates soil water (water content, water flow, pressure potential, ice formation), soil temperature and water-limited crop dry-matter production. Secondly, the turnover of soil organic matter and soil mineral N, as well as the nitrogen-limited crop dry-matter production is simulated and four output files are produced as follows. The first is the *Soil Mineral Nitrogen* output file containing total NO_3-N and NH_4-N for each day in every specified layer. The second is the *Water Soluble Nitrogen* output file containing the concentrations of NO_3^--N in the soil water for each day in every specified layer. The third is the *Nitrogen Balance* output file containing several variables, calculated as daily totals from the soil surface to the balance depth: NO_3-N, NH_4-N, net mineralized NH_4-N, nitrified NO_3-N, denitrified NO_3-N, immobilized NO_3-N and leached NO_3-N. The daily totals of atmospheric NO_3- and NH_4-N deposition, NH_4-N volatilization, NH_4- and NO_3-N plant uptake are also given in this file. The *Nitrogen Balance* output file also contains the daily means of the variables of the crop production (top and root DM, total crop nitrogen), harvested yield DM, and the components of the water balance. The fourth output file, the *Soil Organic Matter* output file contains the C- and N-contents in soil of SOM1, SOM2, SMB1, SMB2, AOM, and ROM from the soil surface down to the balance depth at user-defined intervals. If several AOM or ROM applications were performed, these pools are listed separately.

References

Hansen S., Jensen N.E., Nielsen N.E. & Svendsen H. (1990) DAISY-Soil Plant Atmosphere System Model. *NPo-forskning fra Miljøstyrelsen*, **A10**: 272 pages.

Hansen, S., Jensen H.E., Nielsen N.E. & Svendsen H., (1991a) Simulation of nitrogen dynamics in the soil plant system using the Danish simulation model Daisy. In: *Hydrological Interactions Between Atmosphere, Soil and Vegetation* (eds.: Kienitz, G., P.C.D. Milly, M.Th. van Genuchten, D. Rosbjerg & W.J. Shuttleworth) IAHS Publication No. **204**:185-195.

Hansen S., Jensen N.E., Nielsen N.E. & Svendsen H. (1991b) Simulation of nitrogen dynamics and biomass production in winter wheat using the Danish simulation model Daisy. *Fertilizer Research* **27**: 245-259.

Hansen S., Jensen H.E., Nielsen N.E. & Svendsen H. (1991c) Simulation of biomass production, nitrogen uptake and nitrogen leaching by using the Daisy model. In: *Soil and Groundwater Research Report II: Nitrate in Soil.* Final Report on Contracts EV4V-0098-NL and EV4V-00107-C. DG XII. Commission of the European Communities, 300-309.

Jensen, C., Stougaard B. & Olsen P. (1994a) Simulation of nitrogen dynamics at three Danish locations by use of the DAISY model. *Acta Agric. Scand., Sect. B, Soil and Plant Science* **44**: 75-83.

Jensen, C., Stougaard B. & H.S. Østergaard (1994b) Simulation of the nitrogen dynamics in farm land areas in Denmark (1989-1993). *Soil Use and Management* **10**: 111-118.
Petersen, C.T., Jørgensen U., Svendsen H., Hansen S., Jensen H.E. & Nielsen N.E. (1995) Parameter assessment for simulation of biomass production and nitrogen uptake in winter rape. *Eur. J. Agron.* **4**: 77-89.
Styczen, M. & Storm B. (1993) Modelling of N-movement on catchment scale - a tool for Analysis and Decision Making. 2. A Case Study. *Fert. Res.* **36**:7-17.
Svendsen, H., Hansen S. & Jensen H.E. (1995) Simulation of crop production, water and nitrogen balances in two German agro-ecosystems using the Daisy model. Ecological Modelling (in press).
Vereecken, H., Jansen E.J., Hack-ten Broeke M.J.D., Swerts M., Engelke R., Fabrewiz F. & Hansen S. (1991) Comparison of simulation results of five nitrogen models using different data sets. In: *Soil and Groundwater Research Report II: Nitrate in Soils*. Final Report on Contracts EV4V-0098-NL and EV4V-00107-C. DG XII. Commission of the European Communities, 321-338.

Acknowledgements

This work was financed by the Danish Environmental Research Programme. Torsten Mueller was a collaborator via a fellowship under the OECD Project on Biological Resource Management and via a research fellowship of the German Research Society (DFG).

The CENTURY model

W.J. Parton
Natural Resource Ecology Laboratory
Colorado State University
Fort Collins
Colorado 80523
USA

The CENTURY model simulates the dynamics of carbon (C), nitrogen (N), phosphorus (P), and sulfur (S) for different plant-soil systems. The model can simulate the dynamics of grassland systems, agricultural crop systems, forest systems, and savanna systems. The grassland/crop and forest systems have different plant production submodels which are linked to a common soil organic matter submodel. The soil organic matter submodel simulates the flow of C, N, P, and S through plant litter and the different inorganic and organic pools in the soil. The model runs using a monthly time step and the major input variables for the model include: (1) monthly average maximum and minimum air temperature, (2) monthly precipitation, (3) lignin N, P, and S content of plant material, (4) soil texture, and (5) atmospheric and soil N inputs.

Soil Organic Matter Submodel

The SOM submodel is based on multiple compartments for SOM and is similar to other models of SOM dynamics (Jenkinson and Rayner 1977, Jenkinson 1990, van Veen and Paul 1981). The pools and flows of SOM (C, N, P) are illustrated in Figure 1. The model includes three soil organic matter pools (active, slow and passive) with different potential

NATO ASI Series, Vol. I 38
Evaluation of Soil Organic Matter Models
Edited by David S. Powlson, Pete Smith and Jo U. Smith

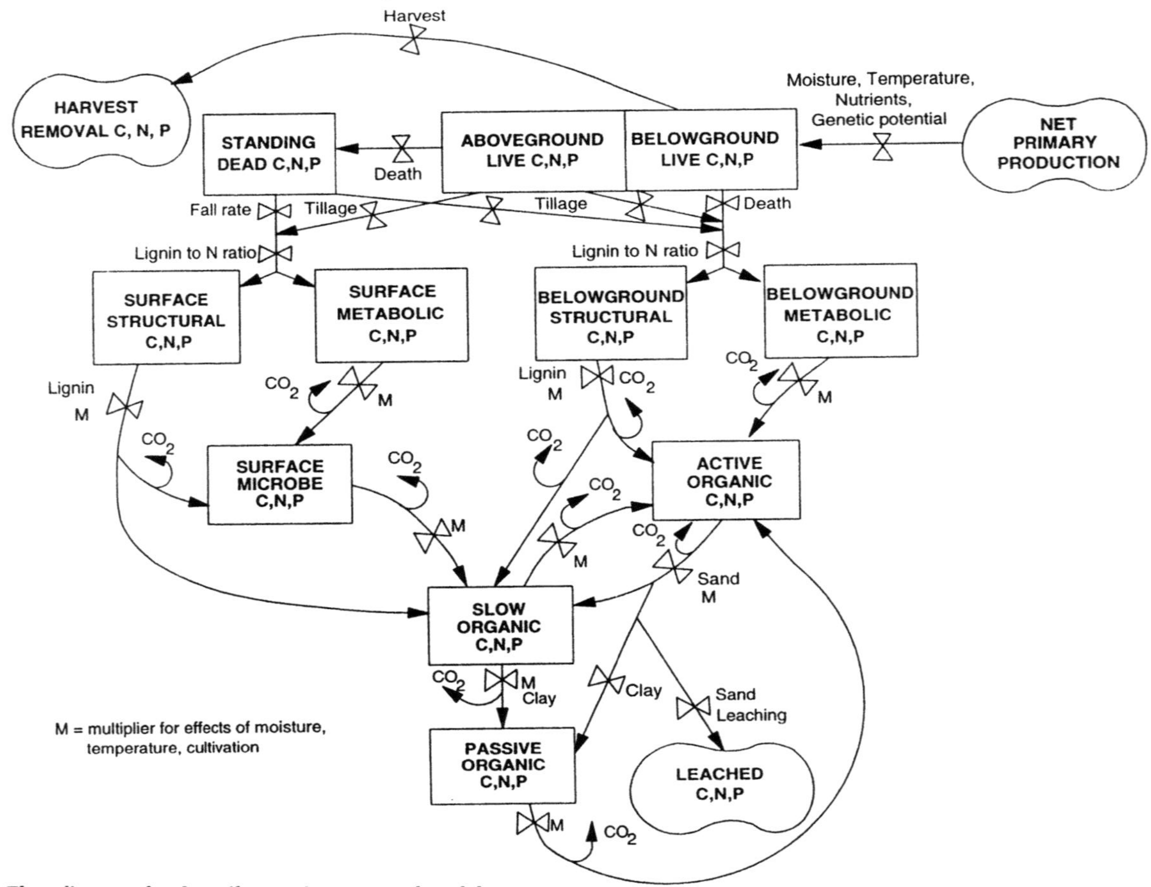

Figure 1 *Flow diagram for the soil organic matter submodel.*

decomposition rates, above and below-ground litter pools and a surface microbial pool which is associated with decomposing surface litter.

Above and below-ground plant residues and organic animal excreta are partitioned into structural and metabolic pools as a function of the lignin to N ratio in the residue. With increases in the ratio, more of the residue is partitioned to the structural pools which have much slower decay rates than the metabolic pools. The structural pools contain all of the plant lignin.

The decomposition of both plant residues and SOM are assumed to be microbially mediated with an associated loss of CO_2 as a result of microbial respiration. In line with long-term soil decomposition studies which show higher microbial respiration, the loss of CO_2 on decomposition of the active pool increases with increasing soil sand content. Decomposition products flow into a surface microbe pool or one of three SOM pools, each characterized by different maximum decomposition rates. The potential decomposition rate is reduced by multiplicative functions of soil moisture and soil temperature and may be increased as an effect of cultivation. Average monthly soil temperature near the soil surface (STEMP) is the input for the temperature function while the moisture function uses the ratio of stored soil water plus current month's precipitation (RAIN) to potential evapotranspiration. The decomposition rate of the structural material is a function of the fraction of the structural material that is lignin. The lignin fraction of the plant material does not go through the surface microbe or active pools but is assumed to go directly to the slow C pool as the structural plant material decomposes.

The active pool represents soil microbes and microbial products (total active pool is ~2 to 3 times the live microbial biomass level) and has a turnover time of months to a few years depending on the environment and sand content. The soil texture influences the turnover rate of the active soil SOM (higher rates for sandy soils) and the efficiency of stabilizing active SOM into slow SOM (higher stabilization rates for clay soils). The surface microbial pool turnover rate is independent of soil texture, and it transfers material directly into the slow SOM pool. The slow pool includes resistant plant material derived from the structural pool and soil-stabilized microbial products derived from the active and surface microbe pools. It has a turnover time of 20 to 50 years. The passive pool is very resistant to decomposition and includes physically and chemically stabilized SOM and has a turnover time of 400 to 2000 years. The proportions of the decomposition products which enter the passive pool from the slow and active pools increase with increasing soil clay content.

A fraction of the products from the decomposition of the active pool is lost as leached organic matter. Leaching of organic matter is a function of the decay rate for active SOM, the clay content of the soil (less loss for clay soils; the model assumes that clay minerals adsorb and stabalize organic matter that leaches down the profile), and only occurs if there is drainage of water below the 30 cm soil depth.

The model has N, P, and S pools analogous to all of the C pools. Each SOM pool has an allowable range of C to element ratios based on the conceptual model of McGill and Cole (1981). Reflecting the concept that N is stabilized in direct association with C, C to N ratios are constrained within narrow ranges, while the ester bonds of P and S allow C to P and C to S ratios to vary widely. The ratios in the structural pool are fixed at high values, while the ratio in the metabolic pool is allowed to float in concert with the nutrient content of the plant residues. The actual ratios for material entering each SOM pool are linear functions of the quantities of each element in the labile inorganic mineral pools in the surface soil layers. Low nutrient levels in the labile pools result in high C to element ratios in the various SOM pools. The N, P, and S flows between SOM pools are related to the C flows. The quantity of each element flowing out of a particular pool equals the product of the C flow and the element to C ratio of the pool. Mineralization or immobilization of N, P, and S occurs as is necessary to maintain the ratios discussed above. Thus, mineralization of N, P, and S occurs as C is lost in the form of CO_2 and as C flows from pools with low ratios, such as the active pool, to those with higher ratios, such as the slow pool. Immobilization occurs when C flows from pools with high ratios, such as the structural pool, to those with lower ratios, such as the active pool. The decomposition rate is reduced if the quantity of any element is insufficient to meet the immobilization demand.

Soil Water and Temperature Submodel

The CENTURY model includes a simplified water budget model which calculates monthly evaporation and transpiration water loss, water content of the soil layers, snow water content, and saturated flow of water between soil layers. If the average air temperature is less than 0°C monthly precipitation occurs as snow. Sublimation and evaporation of water from the snow pack occurs at a rate equal to the potential evapotranspiration rate. Snow melt occurs if the average air temperature is greater than -8°C and is a linear function of the average air temperature.

The potential evapotranspiration rate (PET) is calculated as a function of the average monthly maximum and minimum air temperature using the equations developed by Linacre (1977). Bare soil water loss is a function of standing dead and litter biomass (lower for high biomass levels), rainfall and PET. Interception water loss is a function of above-ground biomass (increases with biomass level), rainfall and PET. Potential transpiration water loss is a function of the live leaf biomass and PET. Interception and bare soil water losses are calculated as fractions of the monthly precipitation and are subtracted from the total monthly precipitation, with the remainder of the water added to the soil.

Water is distributed to the different layers by adding the water to the top layer and then draining excess water (water above field capacity) to the next layer. Transpiration water loss occurs after the water was added to the soil. Water loss occurs first as interception, followed by bare soil evaporation and transpiration (the sum does not exceed the PET rate). The maximum monthly evapotranspiration water loss rate is equal to PET.

Average monthly soil temperature near the soil surface is calculated using equations developed by Parton (1984). These equations calculate maximum soil temperature as a function of the maximum air temperature and the canopy biomass (lower for high biomass) while the minimum soil temperature is a function of the minimum air temperature and canopy biomass (higher for higher biomass). The actual soil temperature used for decomposition and plant growth rate functions is the average of the minimum and maximum soil temperatures.

Plant Production Submodel

The CENTURY model is set up to simulate the dynamics of grasslands, agricultural crops, forests, and savanna (tree-grass) systems. The grassland/crop production model simulates plant production for different herbaceous crops and plant communities (e.g., warm or cool season grasslands, wheat and corn). Existing crop options may be altered to suit particular varieties or environments or new options created. Harvest, grazing, fire and cultivation all directly affect above-ground biomass, while grazing and fire may also impact root to shoot ratios and nutrient content. The forest model simulates the growth of deciduous or evergreen forests in juvenile and mature phases. Fire, large scale disturbances (e.g., hurricanes), and tree harvest practices may impact forest production. The savanna system is simulated as a tree-grass system, essentially using the existing tree and grassland/crop submodels with the two subsystems interacting through shading effects and nitrogen competition.

Both plant production models assume that the monthly maximum plant production is controlled by moisture and temperature and that maximum plant production rates are decreased if there are insufficient nutrient supplies (the most limiting nutrient constrains production). The fraction of the mineralized pools that are available for plant growth is a function of the root biomass with the fraction of nutrients available for uptake increasing exponentially as live root biomass increases from 20 to 300 gm^{-2}. Most forest or grassland/crop systems are limited by nutrient availability and generally respond to the addition of N and P.

The model can simulate a wide variety of crops and grasslands by altering a number of crop specific parameters. The grass/crop production model (Figure 2) has pools for live shoots and roots, and standing dead plant material. Potential production (g C m^{-2} $month^{-1}$) is a function of a genetic maximum defined for each crop and 0-1 scalars depending on soil temperature, moisture status, shading by dead vegetation, and seedling growth.

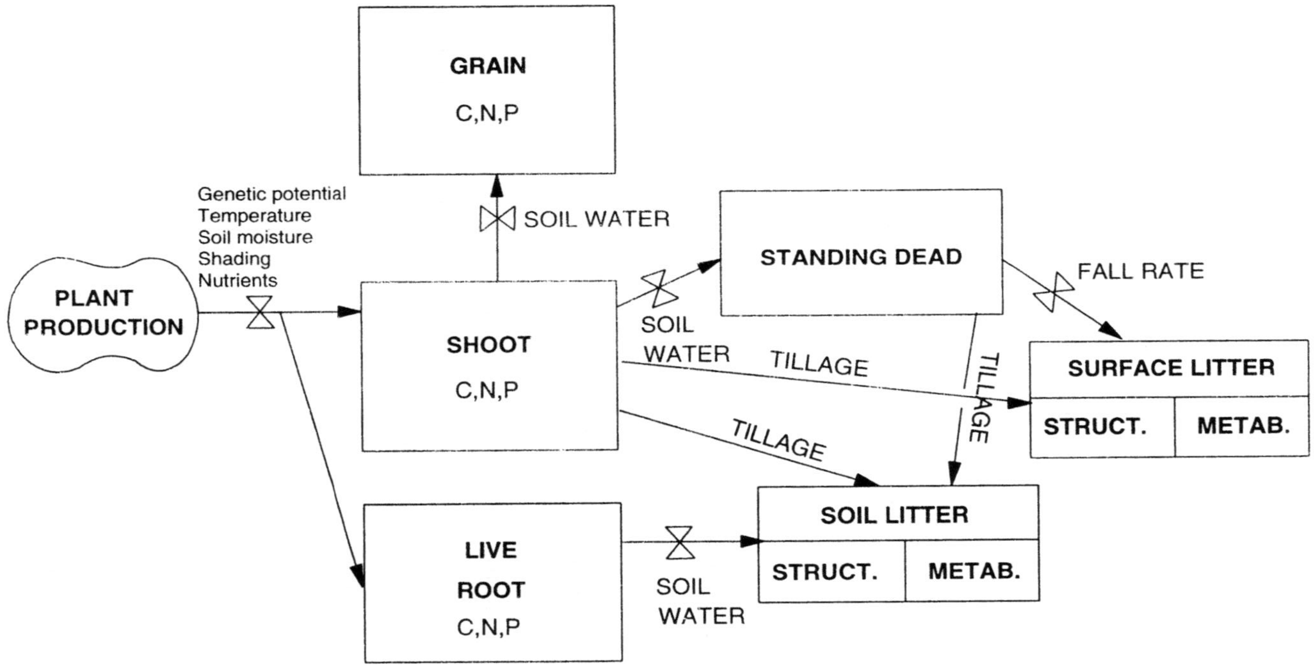

Figure 2 *Flow diagram for the plant production submodel.*

The effects of grazing and fire on plant production are represented in the model by using data from Holland et al. (1992) and Ojima et al. (1990). The major impact of fire is to increase the root to shoot ratio, increase the C:N ratio of live shoots and roots, remove vegetation and return nutrients during the years when fire occurs (Ojima et al. 1990). Grazing removes vegetation, returns nutrients to the soil, alters the root to shoot ratio, and increases the N content of live shoots and roots (Holland et al. 1992).

A detailed description of the CENTURY model is presented in a number of recent publications (Parton et al. 1987, Parton et al. 1988, Parton et al. 1993, Parton et al. 1994) and a users manual for version 4.0 is available (Metherell et al. 1994). The model has been tested using extensive grassland data sets and data sets from long-term agricultural field plots. The grassland version was evaluated using observed plant production and soils data sets from 11 temperate and tropical grasslands around the world. The model correctly predicted the effect of grazing, burning, fertilization and irrigation on plant production and seasonal live biomass patterns.

The CENTURY model has been tested using data from several long-term SOM experiments and used as a scientific tool for integrating and interpreting data. The 55-year winter wheat plots from Pendleton, Oregon (Parton and Rasmussen 1994) were used to look at the impact of different types and amounts of soil organic matter additions (pea vine, manure), straw burning, and fertilizer levels on plant production, and soil C and N levels for a wheat fallow system. Data from a 30-year experiment in Sweden (Paustian et al. 1992) was used to evaluate the impact of adding different types of organic matter (sawdust, straw, green manure, and manure) and fertilizer levels on plant production and nutrient uptake, N budgets, and soil C and N levels. Observations (1970-present) from the Sidney, Nebraska site (Fenster and Peterson 1979) were used to test the simulated impact of different tillage practices (no-till, stubble mulch, and plow) and fertilizer levels on soil water budgets, soil temperature, soil C and N levels, and plant production for a wheat-fallow system (Metherell et al. 1994). Data from continuous wheat system (1970-present) in Australia has recently been used to test CENTURY's ability to simulate tillage management (no-till and conventional tillage), fertilizer levels (0, 40, and 80 kg N ha^{-1} y^{-1}), and straw burning on soil

C and N levels, plant production, soil N mineralization rates and soil water storage (Probert et al., in press). Carter et al. (1993) also recently used CENTURY to simulate changes in soil C and N levels under wheat and pasture rotations at five sites in southern Australia.

References

Carter, M.R., Parton, W.J., Rowland, I.C., Schultz, J.E. & Steed, G.R. (1993) Simulation of soil organic carbon and nitrogen changes in cereal and pasture systems of southern Australia. *Aust. J. Soil Res.* **31**: 481-491.

Holland, E.A., Parton, W.J., Detling, J.K. & Coppock, D.L. (1992) Physiological responses of plant populations to herbivory and their consequences for ecosystem nutrient flows. *Am. Nat.* **140(4)**: 685-706.

Jenkinson, D.S. & Rayner, J.H. (1977) The turnover of soil organic matter in some of the Rothamsted classical experiments. *Soil Sci.* **123**: 298-305.

Jenkinson, D.S. (1990) The turnover of organic carbon and nitrogen in soil. *Philosophical Transactions of the Royal Society, London B* **329**: 361-368.

Linacre, E.T. (1977) A simple formula for estimating evaporation rates in various climates, using temperature data alone. *Agric. Meteorol.* **18**: 409-424.

Metherell, A.K., Parton W.J., Cambardella, C.A., Peterson, G.A., Harding, L.A. & Cole, C.V. (1994) Simulation of soil organic matter dynamics in dryland wheat-fallow cropping systems. *Advances in Soil Science* (in press).

Ojima, D.S., Parton, W.J., Schimel, D.S. & Owensby, C.E. (1990) Simulated impacts of annual burning on prairie ecosystems. In: *Fire in North American Prairies.* (ed: S.L. Collins and L. Wallace), Univ. of Oklahoma Press, Norman, OK, USA.

Parton, W.J. (1984) Predicting soil temperatures in a shortgrass steppe. *Soil Sci.* **138**: 93-101.

Parton, W.J., Schimel, D.S., Cole, C.V. & Ojima, D.S. (1987) Analysis of factors controlling soil organic levels of grasslands in the Great Plains. *Soil Sci. Soc. Am. J.* **51**:1173-1179.

Parton, W.J., Stewart, J.W.B. & Cole, C.V. (1988) Dynamics of C, N, P, and S in grassland soils: A model. *Biogeochemistry* **5**: 109-131.

Parton, W.J., Scurlock, J.M.O., Ojima, D.S., Gilmanov, T.G., Scholes, R.J., Schimel, D.S., Kirchner, T., Menaut, J.-C., Seastedt, T., Garcia Moya, E., Kamnalrut, A. & Kinyamario, J.L. (1993) Observations and modeling of biomass and soil organic matter dynamics for the grasslands biome world-wide. *Global Biochemical Cycles* **7**: 785-809.

Parton, W.J. & Rasmussen, P.E. (1994) Long-term effects of crop management in wheat-fallow: II. CENTURY model simulations. *Soil Sci. Soc. Am. J.* **58**: 530-536.

Paustian, K., Parton, W.J. & Persson, J. (1992) Modeling soil organic matter in organic-amended and nitrogen-fertilized long-term plots. *Soil Sci. Soc. Am. J.* **56**: 476-488.

Probert, M., Parton, W.J. & Keating, B. (in press) Soil organic matter dynamics for the Hermitage Plots. *Aust. J. Soil Res.* (in press).

van Veen, J.A. & Paul, E.A. (1981) Organic carbon dynamics in grassland soils. 1. Background information and computer simulation. *Canadian Journal of Soil Science* **61**: 185-201.

LONG-TERM EXPERIMENTS WITH SOIL ORGANIC MATTER MEASUREMENTS

Soil organic matter dynamics in Sanborn Field (North America)

G.A. Buyanovsky, J.R. Brown and G.H. Wagner
Department of Soil and Atmospheric Sciences
University of Missouri-Columbia
144 Mumford Hall
Columbia, MO 65211
USA

The site of Sanborn Field, when selected in 1888, was an undisturbed area southeast of the campus of the University of Missouri in Columbia. The vegetation on the site was described in early handwritten notes as grass with scattered woody species. The field consists mostly of Mexico silt loam (fine, montmorillonitic, mesic Udollic Ochraqualfs) grading at the east into Lindley loam (fine-loamy, mixed, mesic Typic Hapludalfs). The Mexico soils are somewhat poorly drained, formed in loess over glacial till under tall prairie grasses and the Lindley soils are well-drained, formed in glacial till under hardwood forest. The east series of plots on the field are actually on the boundary between the two soils. The soils of Sanborn Field are in the claypan resources area for which climate has been one of the most influential soil forming factors. The field is 80 m (N→S) x 251 m (E→W), or 2.0 ha in area. The individual plots are 30.6 x 9.5 m in area and are separated by a 1.5 m wide grass border.

Columbia is located at 93°20' west longitude and 38°57' north latitude (North-Central USA). The area is subjected to a continental climate. Potential evapotranspiration increases rapidly from April to July and periods of severe moisture stress are expected from mid-June through early September. Forty-six percent of the 916 mm of annual precipitation falls between 1 May and 30 September. Weather measurements are from the closest meteorological station, now located 24 km south of the field. Observations are available beginning 1889, for daily min-max air temperatures and precipitations. Additional measurements, available since 1940, include wind speed, sunshine hours, relative humidity.

NATO ASI Series, Vol. I 38
Evaluation of Soil Organic Matter Models
Edited by David S. Powlson, Pete Smith and Jo U. Smith

Emphasis of the experiment initiated in 1888 was to evaluate practical approaches in crop production that could be readily adopted by the farmers. Nine cropping practices, using corn (*Zea mays* L.), oats (*Avena sativa* L.), winter wheat (*Triticum aestivum* L.), red clover (*Trifolium pratense* L.) and timothy (*Phleum pratense* L.) under three soil treatments, were originally used in the experiment. The practices were continuous cropping systems of each of the five crops and rotational cropping sequences of two, three, four and six years. Winter wheat, corn, and oats were the major grain crops, and clover and timothy were the main forage components of rotations around 1888. Wheat and corn have been grown continuously and in seven systems of crop rotation. Only one plot with continuous wheat was designed to receive "complete fertilizer for maximum crop," applied in the form of inorganic chemicals. The fertilizer used on other plots was barnyard manure (Upchurch *et al.*, 1985). The continuum of management on Sanborn Field reflects the history of agriculture in the central region of the USA. Records for more than 100 years, complemented with contemporary studies, give a unique opportunity to examine, almost year by year, changes in organic carbon content in the soils of typical agroecosystems.

Periodically during the past 100 years the experimental plan for some plots was revised as certain practices passed out of vogue. Evaluations of management of the field that led to revisions occurred in 1914, 1928, 1940, 1950 and 1989. Historically, the changes in Sanborn Field management paralleled the development of understanding of the nature and properties of soils, the progress in the use of mineral fertilizers, the introduction of improved varieties of crops and the development of new cropping technologies (Brown, 1994).

There were no records of sampling of soils on the field prior to 1914. Since then detailed soil samples have been taken about every 25 years. Past managers took soil samples from each plot before changes were made. The existence of stored units of soil from these various samplings adds significant value of the field for any contemporary studies. Table 1 represents an index of soil samples collected from Sanborn Field now in storage. The historical samples provide an opportunity for assessing long-term trends in the level of SOM, including the impact of mineral fertilizers. For instance, three plots in corn-soybean-wheat rotation since 1928 have had full treatment (N, P, K), with minus P and minus K assigned to separate plots. This series provides resources to evaluate effect of mineral nutrient stress upon SOC balance. The amounts of P and K removed from the soil were found to be much higher than the soil available storage pools and as the pools declined, productivity decreased dramatically .

Table 1. Index of archived soil samples from Sanborn Field† and reference point*

Years sampled	Sampling depth, cm
1914	0-10, 10-20, 20-30, 30-46, 46-61, 61-91, 91-122
1928	0-20, 20-46, 46-61
1938	0-10, 10-20, 20-30, 30-46, 46-61, 61-91, 91-122
1949	0-10, 10-20
1962	Same as 1938, also Ap, A_1, A_2, & top 30 cm of B_2
1988	a) to 120 cm by 10 cm increments b) by diagnostic horizons to 120 cm

† In addition, there has been several sets of plow layer samples taken for testing to provide a basis for modification of P, K and limestone treatment.

* Tucker Prairie is a mixed-grass that is managed to maintain a vegetation diversity similar to native prairie. Soil is Putnam-Mexico silt loam, similar to that of the original Sanborn Field site.

There have been no comprehensive analyses of the soil samples collected on the different dates using the same methodology. The methods of analysis were those in vogue at the time of the sampling. The most consistent methodology has been the use of the Kjeldahl total N analysis. Total C, when done, has been by some form of dry combustion in a purified stream of oxygen. Soil organic matter content has been estimated by quick-test chromic acid methods. Sampling of plant material and grain has been sporadic. Prior to 1950, crop residues were removed from the plots but no analyses were made. No plant samples were stored before 1989. Currently the residues of grain crops are returned to the soil as is the common practice.

Soil organic matter (SOM) level has been significantly influenced by the crop and soil management practices applied to Sanborn Field during the last 100 years (Buyanovsky et al., 1990; 1995). When cultivation was introduced on the field, SOM declined from the virgin level. The rate of loss of SOM was initially rapid and progressive until 25 to 45 years later, at which time a new equilibrium level was reached. The new level was characteristic of the crop and the management practices followed on each respective plot. Annual input of carbonaceous residues, their humification, and the yearly mineralization of the organic matter existing in the soil were factors defining the equilibrium level. Of particular interest are plots in which the annual residue input increased after 1950 because straw and stalks were incorporated into the soil rather than removed (Buyanovsky and Wagner, 1995).

There are no direct data on crop residue inputs into plots under different cropping systems on Sanborn Field. The available data are those of grain and forage yields. However, reliable data on crop residue yields, both above- and belowground, for corn, wheat and soybean have been accumulated in small-scale field experiments (Buyanovsky and Wagner, 1986). These data have been used to calculate representative annual inputs of carbon from plant residues into soil for three different crops (Table 2).

Table 2. Carbon content in the plant residues and input into the soil with post-harvest remnants under contemporary cropping approaches.

	Wheat		Soybean		Corn	
	Tops	Roots	Tops	Roots	Tops	Roots
Carbon content (%)	37.0	29.6	40.5	26.2	40.9	26.1
Mean annual input, Mg ha^{-1}	2.18	1.54	2.17	1.26	5.81	3.40
Tops plus roots Mg ha^{-1}	3.72		3.43		9.21	

To approximate the quantities of residues for the diverse management systems in long-term Sanborn Field experiments, one can use grain and forage yields and data on above- and underground biomass obtained in the small-scale field experiments, as well as relationships between grain yield and total harvested biomass (Buyanovsky *et al.*, 1990)

Continuous wheat, chemically fertilized to optimize yields, demonstrated an equilibrium level of soil C at 40 years of about 0.9% (plot 2). This was characteristic of the management system where straw was removed from the plot each year, so that the annual input of residues was only that of stubble and roots. From 1950 onward, the straw was added back and incorporated into the soil of this plot. The additional residue input resulted in a significant increase in soil C over the next 40 years after which a probable equilibrium level is developing of at least 1.2% C, the level in 1986 (Fig. 1).

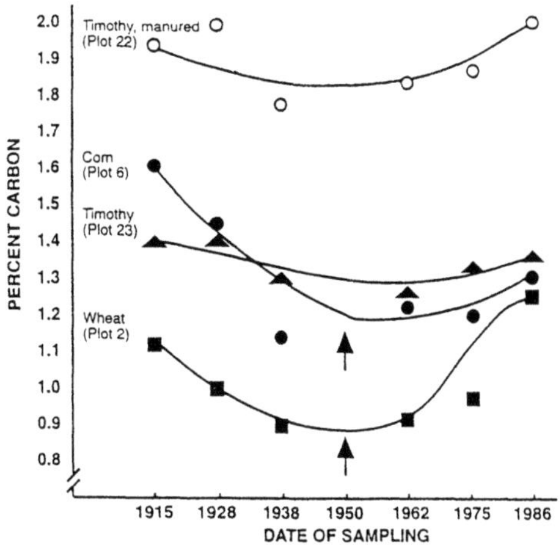

Figure 1. *Changes in organic carbon content of soil in several plots of Sanborn Field during cropping of wheat, corn, and timothy. Arrows indicate date after which all above-ground residues on particular plots were returned to the soil.*

Natural abundance of ^{13}C has been used to distinguish the origin of soil C (Balesdent et al., 1988). Total C and ^{13}C data demonstrated that under wheat culture, much SOM of prairie origin was rapidly lost and very slowly replaced in part by new SOM of crop origin, as illustrated in Fig. 2. It was estimated that labile fraction had a half-life of 11 years, a sizable pool of SOM was estimated to have a turnover time approaching 1000 years. After 100 years of cultivation, the resistant pool of prairie origin still comprised at least 50% of current SOM under all systems of cultivation studied.

Changes in total soil N over time for corn, wheat and timothy plots with and without manure are shown in Fig. 3. Without treatment, the soil demonstrated a progressive decline in N throughout the 100 years of experimentation for both corn and wheat. In the case of timothy, soil N was stable after 1915. Scattered legumes in the timothy plot or the occurrence of non-symbiotic N-fixation is probably keeping N at a uniform level and compensating for the annual harvest of N in the forage removed equaling 11 kg ha^{-1} yr^{-1}.

The timothy plot receiving an annual manure treatment to supply an estimated 110+ kg N ha^{-1} yr^{-1} had a soil N level over time that was 0.05 percentage points higher than that for the unmanured plot. Differences in total soil N between manured and non-manured timothy appears to be particularly significant at the 1986 sampling. Efforts after 1980 to be more precise in defining the N added in manure may have been the reason for the latter observation. The N level in the corn and wheat plots receiving manure annually appears to be unchanged during the past 75 years (Wagner, 1990).

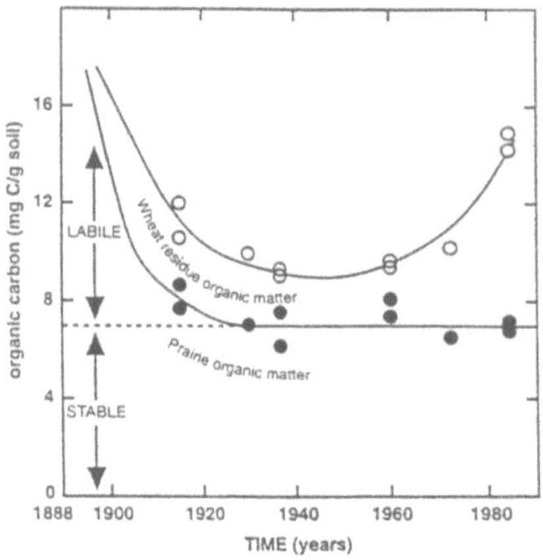

Figure 2. *Changes in amount and origin of soil organic carbon with cropping to winter wheat on formerly virgin prairie soil. Open circles indicate total soil carbon; closed circles indicate that from prairie vegetation. Each data set includes observations for 0-10 and for 10-20 cm sampling depths.*

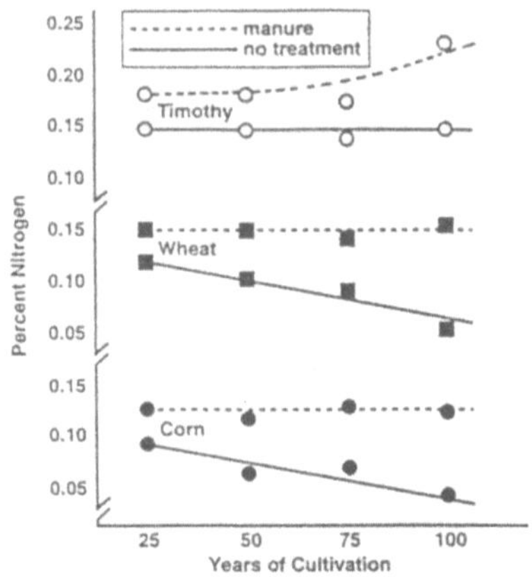

Figure 3. *Changes in total soil nitrogen over time for plots of Sanborn Field cultivated continuously for 100 years to various crops without treatment and with 13.4 Mg ha^{-1} of manure annually.*

References

Balesdent, J., Wagner, G.H., and Mariotti A. (1988). Soil organic matter turnover in long-term field experiments as revealed by the ^{13}C natural abundance tracer technique. *Soil Sci. Soc. Am J.* **52**: 118-124.

Brown, J.R. (1994). The Saborn Field experiment. In: *Long-term Experiments in Agricultural and Ecological Sciences* (ed: R.A. Leigh and A.E. Johnston). CAB International: 39-52.

Buyanovsky, G.A, Brown, J.R., and Wagner, G.H. (1995). Sanborn Field: Effect of one hundred years of cropping on soil parameters influencing productivity. In: *Organic Matter in the Temperate Agroecosystems: Long-Term Experiments in North America* (ed: E.A. Paul, K. Paustian, E.T. Elliott and C.V. Cole), Lewis Publ. (In press).

Buyanovsky, G.A., Nelson, C.J., and Breight, S. (1990). 100-year old cropping systems: Yields and soil properties. *Proceedings of Sanborn Field Centennial*, University of Missouri, Columbia **SR-415**: 140-153.

Buyanovsky, G.A. and Wagner, G.H. (1995). Crop residue input to SOM on Sanborn Field. In: *Organic Matter in the Temperate Agroecosystems: Long-term Experiments in North America* (ed: E.A. Paul, K. Paustian, E.T. Elliott and C.V. Cole), Lewis Publ. (In press).

Upchurch, W.J., Kinder, R.J., Brown, J.R. and Wagner, G.H. (1985). Sanborn Field. Historical perspective. *Missouri Agr. Exp. Stn. Res. Bull.* **1054**. Columbia, MO, 49 pp.

Wagner, G.H. (1990). Lessons in soil organic matter from Sanborn Field. *Proceedings of Sanborn Field Centennial*, University of Missouri, Columbia **SR-415**: 64-70.

The Askov Long-Term Experiments on Animal Manure and Mineral Fertilizers

Bent T. Christensen
Danish Institute of Plant and Soil Science
Research Centre Foulum
P.O. Box 23
DK-8830 Tjele, Denmark

Sites and treatments

Askov Experimental Station is located in the south of Jutland (55°28'N, 09°07'E; elevation 63 m a.s.l.). Annual average precipitation and potential evapotranspiration is 862 mm and 543 mm, respectively, and mean annual temperature is 7.7°C (1961-1990 averages). Monthly mean temperatures for January and July are -0.2 and 15.5°C, respectively. Global radiation and number of hours with sunshine are 3385 MJ m^{-2} and 1539 hours per year.

The long-term experiments were started in 1894 and are placed on two sites with contrasting texture, a coarse sand (Sandmarken) and a sandy loam (Lermarken). The 0-20 cm soil (<2 mm) of Sandmarken holds 4% clay (<2 μm), 4% silt (2-20 μm), 34% fine sand (20-200 μm) and 57% coarse sand (200-2000 μm); the corresponding values for Lermarken are: 12% clay, 13% silt, 38% fine sand, and 37% coarse sand. According to the Soil Taxonomy System, Sandmarken is classified as Inseptisol (Orchrept) and Lermarken as Alfisol (Typic Hapludalf). Lermarken had been cultivated for nearly a century when the experiments started, while the cropping history of Sandmarken dated back at least two hundred years. The two experimental sites are located < 1 km apart and experience the same climatic conditions. A fully automated climate station is located next to Lermarken.

NATO ASI Series, Vol. I 38
Evaluation of Soil Organic Matter Models
Edited by David S. Powlson, Pete Smith and Jo U. Smith

On each soil type, four fields (blocks) were employed with plot sizes ranging from 55 m^2 on Sandmarken (field G1 to G4) to 110 m^2 on Lermarken (field B2 to B5).

The experiments grow a four-course crop rotation of winter cereals (wheat or rye), root crops (mangolds, sugar beets, turnips, or potatoes), spring cereals (barley or oats) and a clover/grass mixture (on Sandmarken, lupins or peas were grown from 1944 onwards). Table 1 summarizes the cropping history. Every rotation element is represented every year.

Table 1 *Crops included in the four-course rotation of the Askov long-term experiments.*

Rotation element	Site	Period	Crop
Winter cereals	Sandmarken	1894-1995	Rye
	Lermarken	1894-1931	Rye
		1932-1995	Wheat
Root crops	Sandmarken	1894-1906	Four subplots*
		1907-1922	Mangolds and potatoes**
		1923-1948	Turnips and potatoes**
		1949-1995	Turnips or potatoes
	Lermarken	1894-1922	Mangolds and potatoes**
		1923-1943	Mangolds
		1944-1948	Mangolds and turnips**
		1949-1995	Sugar beets or turnips
Spring cereals	Sandmarken	1894-1972	Oats
		1973-1995	Barley
	Lermarken	1894-1931	Oats
		1932-1995	Barley
Clover/grass (legumes)	Sandmarken	1894-1943	Clover/grass mixture
		1944-1967	Lupines
		1968-1995	Peas
	Lermarken	1894-1995	Clover/grass mixture

* Plots subdivided into four parts growing mangolds, turnips, potatoes and carrots each year.

** Plots subdivided into two parts growing both crops each year.

Table 2 lists the treatments, the year in which the treatments were initiated, and the total number of replicates. Altogether, there are 96 plots on Sandmarken and 174 plots on Lermarken. The main treatments of the experiments are different levels of nitrogen, phosphorus and potassium (0, ½, 1, 1½, 2) given in mineral fertilizers (NPK) or animal manure (AM). Also included are plots receiving N, P or K alone or in combinations of two, and on Sandmarken AM supplemented with P and K. Table 3 shows the amount of nitrogen, phosphorus and potassium given in 1 NPK and 1 AM. From 1894 to 1972, AM was farmyard manure (9-10 t ha^{-1} yr^{-1}) which during 1923 to 1972 was supplemented with liquid manure (4 t ha^{-1} yr^{-1}). After 1973 cattle slurry (25 t ha^{-1} yr^{-1}) has been used in the AM treatments.

Table 2 *The treatments included in the Askov long-term experiments, their year of establishment, and the number of replicates. N, P and K denote additions in mineral fertilizers, and AM is animal manure treatments.*

	Sandmarken		Lermarken	
Treatment	Year	No. of replicates	Year	No. of replicates
0, unmanured	1893	8	1893	21
1/2 AM		0	1894	11
1 AM	1894	12	1894	21
1 1/2 AM		0	1894	17
2 AM		0	1923	3
1/2 NPK	1923	8	1923	15
1 NPK	1894	12	1894	20
1 1/2 NPK		0	1923	13
2 NPK		0	1923	4
1 AM + 1/2 P	1894	8		0
1 AM + 1/2 K	1894	8		0
1 AM + 1/2 PK	1908	8		0
1 1/2 N + 3PK		0	1923	4
1 PK	1894	4	1934	13
1 NK	1949	8	1934	10
1 NP	1894	8	1894	10
1 N	1894	4	1894	4
1 P	1894	4	1894	4
1 K	1894	4	1894	4

The amount of nutrients added was changed in 1907, 1923, 1949 and 1973, but during each period almost equivalent dressings of nitrogen, phosphorus and potassium were applied in NPK and in AM.

Table 3 *Amounts of nitrogen, phosphorus and potassium (annual mean of crop rotation) applied in the 1 NPK and 1 AM. The dressings were changed in 1907, 1923, 1949 and 1973.*

Period	kg nutrient ha^{-1} $year^{-1}$					
	1 AM			1 NPK		
	N	P	K	N	P	K
1894-1906	40	13	29	39	12	28
1907-1922	42	13	33	42	13	32
1923-1948	72	16	65	70	17	70
1949-1972	93*	19	58	70	18	66
1973-1995	98	19	83	100	19	83

* AM supplemented with N in mineral fertilizer

The experimental layout, the historical changes in treatments, present field operations and yield results have been presented by Christensen *et al.* (1994). A directory of data obtained from the experiments has been established. At present, climatic variables have not been incorporated into the experimental database. Tillage and drilling, weed control and crop protection, cultivars, and liming practices have followed the general development in agriculture. Accordingly field operations have undergone continuous change with time. Sampling of soil to 20 cm depth takes place every fourth year. Samples are analyzed for C, N, P and K content and archived in an air dried state. Crop yields are determined every year; every fourth year pooled samples of individual crops are analyzed for nutrient content. Animal manure is analyzed before application.

Examples of results

During the past 100 years, a large amount of diverse results have been extracted from the long-term experiments. Crop yields have been recorded right from the start of the experiments while regular soil sampling was initiated in 1923. Analyses of crops for nitrogen, phosphorus and potassium content have been carried out on a regular basis since 1949, before that occasional determinations were made. Determinations of total carbon and nitrogen, and extractable potassium and phosphorus in the soil date back to 1912 (on selected plots).

Besides that, a large number of specific research projects have collected diverse information on the crops and soils of the long-term experiments. Most of the results obtained from the

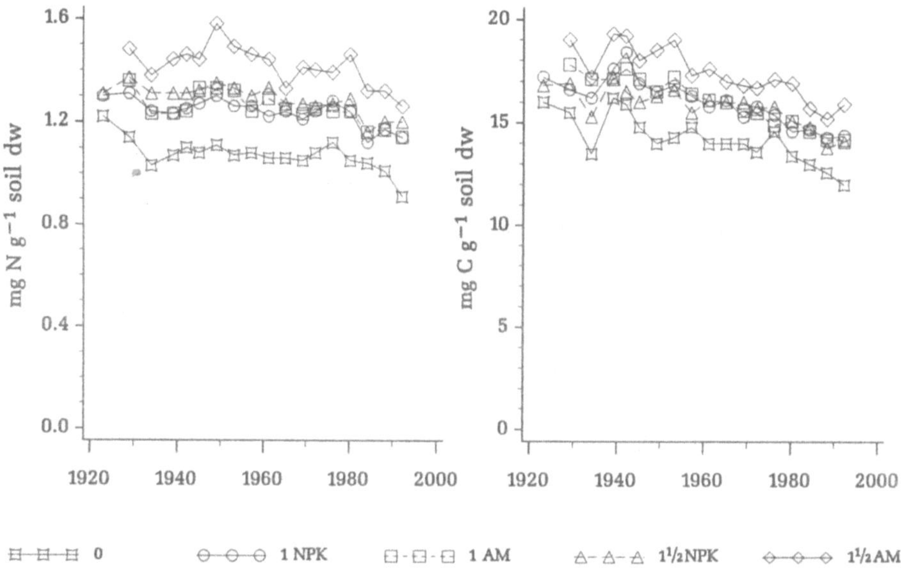

Figure 1 *The C and N content in the B_3-field on Lermarken.*

experiments have already been published through various channels. Results from more resent research are presented here. Figure 1 shows the content of C and N in the plough layer (0-20 cm) in the B_3-field on Lermarken. SOM has declined in all soils, and the decreases are somewhat similar in all treatments. This could be ascribed to a continuing decline in the pool of "original" SOM, derived from the vegetation that preceded cultivation. Differences between treatments had been established by 1923 and those differences have hardly changed since then suggesting that the equilibrium between build-up and decomposition of SOM in each treatment was reached within 30 years. Almost similar contents of SOM have been attained in soils treated with 1 AM, 1 NPK and 1½ NPK. In contrast, unmanured and 1½ AM plots have lower and higher SOM contents, respectively. The higher SOM level in fertilized than in unmanured plots is ascribed to a greater crop productivity and hence a greater return of crop residues (rhizodeposits and stubble) following fertilization. Crop yields have been generally 2-3.5 times higher on 1 NPK and 1 AM treated plots than on unmanured ones, the greater yields usually being obtained with NPK. Similar amounts of plant nutrients have been added in 1 NPK and 1 AM, but the moderate extra organic matter input provided with animal manure has not raised the SOM level compared to 1 NPK. Similarly, SOM contents have not responded to increased rates of mineral fertilizers despite the significantly larger yields of the harvested crops obtained on 1½ NPK plots compared to 1 NPK (Christensen *et al.*, 1994). In contrast, plots getting 1½ AM have higher SOM levels. Taken together, the results suggest that animal manure does possess a potential for raising SOM levels on this light textured soil but only when large amounts are applied. It should be recalled that animal slurry has been used since 1973, and that well managed, straw rich FYM may be more beneficial to SOM build-up than slurry low in bedding materials and stored anaerobically.

Whole soils and size separates have been subject to solid-state ^{13}C-NMR analysis to chemically characterize the nature of SOM in differently fertilized soils (Randall *et al.*, 1995). The relative abundance of major functional SOM groups as calculated from NMR spectra is presented for whole soils in Table 4. Aromatics, O-alkyls and alkyls are the dominant functional groups. The relative distribution of the different functional groups were very similar for unmanured, 1½ NPK and 1½ AM.

Table 4 *Relative abundance of major functional groups calculated from ^{13}C-NMR spectra taken on whole soils. Results from Randall et al. (1995).*

Major functional group	Spectral region (ppm)	0	1½ NPK	1½ AM
Alkyl	0-45	18.9	17.7	19.9
N-alkyl	45-65	13.9	14.7	14.5
O-alkyl	65-90	23.0	23.2	24.9
Acetal	90-110	13.6	12.6	11.7
Aromatics	110-160	25.0	25.2	23.2
Carboxyl	160-190	9.4	10.4	9.7

Clay contains a smaller proportion of aromatics than silt sized separates (Fig. 2) and clay contains a larger proportion of alkyls. Silt associated SOM has a higher proportion of acetals than clay SOM. The relative distribution of functional groups in silt associated SOM is little affected by fertilization regimes. Animal manure tends to increase the proportion of N- and O-alkyls but decrease alkyls and carboxyls compared to the unmanured treatment. SOM in clay sized organomineral complexes appeared to change slightly more with differences in fertilization regime. Animal manure increases the proportion of alkyls and N-alkyl compared to unmanured, while the presence of aromatics and carboxyls is reduced. The overall conclusion for clay is however the same as for silt, namely that fertilization does not markedly alter the distribution of functional groups.

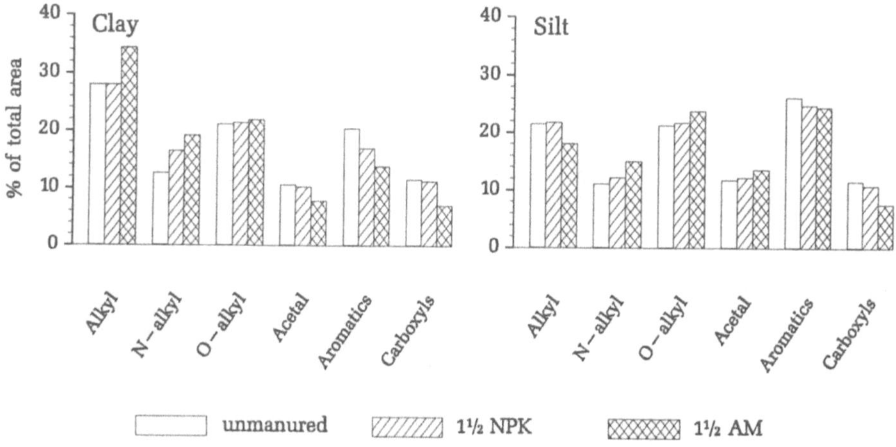

Figure 2 *Relative abundance of functional groups (Table 4) in SOM isolated with clay and silt sized organomineral separates from the B_2-field of Lermarken. Data from Randall et al. (1995).*

Table 5 shows the C and N content in clay, silt and sand sized primary organomineral complexes isolated from unmanured, 1½ NPK and 1½ AM plots in the B_2-field on Lermarken. Clay associated SOM accounts for 57-60% of the soil carbon, whilst silt and sand hold 29-31 and 9-10%, respectively. The C-to-N ratio of clay SOM is lower, and that of silt higher, than the C-to-N ratio of the whole soil. Contents of C and N in clay, silt and sand increased in the order: unmanured, 1½ NPK and 1½ AM. Although some trends are suggested by the data in Table 5 (e.g. in clay and whole soil C-to-N ratios), significant effects of the different fertilization regimes are not evident. If the SOM contents in clay and silt from unmanured plots are used as baseline, then increases in carbon and nitrogen caused by applications of NPK and AM suggest that the AM derived SOM contributes similarly to clay and silt SOM whereas SOM derived solely from crop residues resulting from the use of NPK contributes relatively more to SOM in silt than to SOM in clay. Considering the greater stability of silt than of clay associated organic matter (Christensen, 1992, 1995), it may be speculated that in a soil where similar total contents of organic matter have been reached either by the use of mineral fertilizer or animal manure, the manure treated soil will contain a larger proportion of more labile organic matter than that of the fertilizer treated counterpart.

Table 5 *Distribution of carbon and nitrogen across particle size separates from unmanured, 1½ NPK and 1½ AM plots in the B_2-field, Lermarken. Data from Christensen (1988).*

	Unmanured		1½ NPK		1½ AM	
	C	N	C	N	C	N
$mg\ g^{-1}$ whole soil						
Clay	6.83	0.633	7.67	0.725	9.14	0.879
Silt	3.48	0.244	4.17	0.296	4.72	0.331
Sand	1.10	n.d.[a]	1.39	n.d.	1.43	n.d.
Relative distribution, %[b]						
Clay	58	67	57	63	60	66
Silt	29	26	31	26	31	25
Sand	9	n.d.	10	n.d.	9	n.d.
C-to-N ratio						
Clay	10.8		10.6		10.4	
Silt	14.4		14.1		14.3	
Whole soil	12.7		11.7		11.5	
Relative increase, %[c]						
Clay			12.3	14.5	33.8	38.9
Silt			19.8	21.3	35.6	35.7

[a] n.d. = not determined
[b] % of whole soil SOM
[c] % increase above that in unmanured

Soil physical properties have been examined in unmanured, 1½ NPK and 1½ AM plots on Lermarken (Schjønning *et al.*, 1994). Samples from the B_5-field, show an increasing trend in soil porosity: unmanured < mineral fertilizer < animal manure (Table 6). Compared to 1½ NPK, 1½ AM increased the volume of < 30 μm pores whereas the volume of larger pores was decreased. The associated increase in the amount of plant available water retained in 0.2-30 μm pores in the 0-20 cm layer corresponds however to only 3-4 mm water.

Table 6 *Soil pore volume in unmanured, 1½ NPK and 1½ AM soil samples from the 0-20 cm of the B_5-field, Lermarken. Data from Schjønning et al. (1994).*

	Pore volume, % of total sample volume		
Pore size class	Unmanured	1½ NPK	1½ AM
<0.2 μm	7.4	7.8	8.0
0.2-30 μm	17.7	18.5	19.7
<30 μm	25.1	26.3	27.7
>30 μm	12.9	13.6	12.6
Total	38.0	39.9	40.3

Wet stable aggregates in the range 2 to 8 mm were more abundant in fertilized than in unmanured plots, but for this physical soil property the beneficial effect of the extra SOM formed through dressings of animal manure was distinct (Table 7). The largest increase of 1½ AM over 1½ NPK was seen for the largest aggregates where 1½ AM caused 61% of the soil to be in 6-8 mm aggregate while the corresponding values for 1½ NPK and unmanured were 43 and 32%, respectively. Also the stability of wet-stable aggregates was significantly improved following 1½ AM.

Table 7 *Wet-stable large aggregates in the 2-5 cm layer of unmanured plots and plots receiving 1½ NPK and AM. B_3-field, Lermarken. Data from Schjønning (1985).*

Aggregate stability	Unmanured	1½ NPK	1½ AM
Size class stability, %			
6-8 mm[a]	32	43	61
4-8 mm[a]	40	48	64
2-8 mm[a]	42	49	65
1-8 mm[b]	49	56	71
Δ MWD, mm[c]	2.96	2.47	1.75

[a)] Percent of aggregates in a given size class not broken down after 5 min. of wet sieving.
[b)] Percent of 2-8 mm aggregates not broken down to <1 mm after 5 min. of wet seiving.
[c)] Δ MWD = change in meanweight diameter of aggregate classes.

Soil consistency have been examined using drop-cone measurements on samples from the B_5-field (Schjønning *et al.*, 1994). Plots taking 1½ AM were found to have plastic as well as liquid limits at higher soil water contents than unmanured, plots with 1½ NPK taking an intermediate position (Fig. 3). The plastic limit signifies the water content at which the soil goes from a plastic to a friable state suitable for soil tillage. Thus for 1½ AM, the transition of the soil to a workable state occurs at a water content of 12.4% (w/w), whereas the water content of the unmanured soil has to be reduced to 9.8% before being workable. Soils taking mineral fertilizer and animal manure had a greater strength than unmanured soils at comparable water contents. Differentiation into cohesive and frictional soil strength components showed that unmanured soil increased steeply in internal friction upon drying while the cohesion component remained constant. In contrast, soils taking mineral fertilizer or animal manure increased in cohesion.

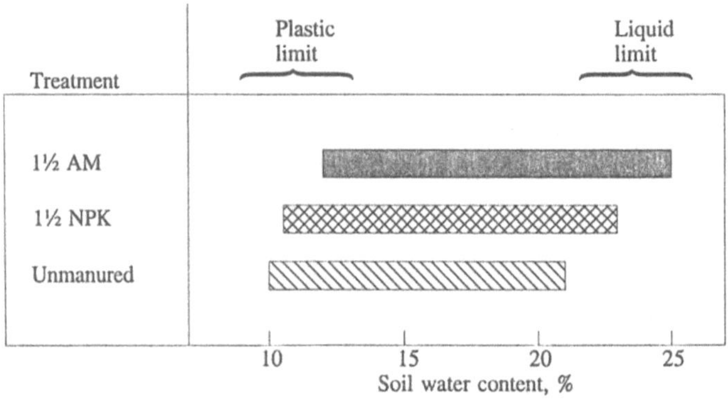

Figure 3 *Soil consistency limits for samples from the 0-20 cm of the B_5-field, Lermarken. Data from Schjønning et al. (1994).*

Rubaek and Sibbesen (1994) determined a range of inorganic and organic P fractions in soils from the B_5-field on Lermarken. Since 1½ NPK and 1½ AM plots have received the same quantity of P during the years, the study allows both a quantitative and qualitative assessment of the P distribution in soils with different SOM contents. Fertilized plots had similar amounts of total-P while unmanured plots were 40% lower in total-P (Table 8). The proportion of total-P present as organically bound P differed little between fertilized treatments (48 and 44%, respectively) whereas 62% of the total-P in unmanured plots were organically bound. Values for plant available inorganic P were rather similar for 1½ NPK and 1½ AM plots while these pools were much lower in unmanured soil. Although the various organic P fractions tended to be slightly higher following addition of animal manure, the overall conclusion is that the amounts and distribution of P forms are very similar for 1½ NPK and

1½ AM. In contrast, unmanured soil is significantly lower especially in inorganic P fractions, and the relative distribution of the total P among the chemical fractions differs clearly from the manured plots except for Hd-HCl-P_i and Hd-HCO_3-P_0.

Table 8 *Distribution of P among different chemically defined soil fractions from unmanured, 1½ NPK and 1½ AM plots on B_5-field, Lermarken. Results from Rubaek and Sibbesen (1995).*

	mg P kg^{-1} soil			% of total-P		
P-fraction	Unmanured	1½ NPK	1½ AM	Unmanured	1½ NPK	1½ AM
Total-P	318	553	535	100	100	100
Total-P_i	122	309	279	38	56	52
Total-P_o	196	244	256	62	44	48
Olsen-P	4.9	26	25	1.5	4.7	4.6
Mp-resin P_i	3.5	34	34	1.1	6.1	6.3
Hd-resin-P_i	2.0	27	27	0.6	5.0	5.0
Hd-HCO_3-P_i	9	57	52	2.8	10	11
Hd-NaOH-P_i	55	162	141	17	29	26
Hd-HCl-P_i	26	46	38	8.1	8.4	7.2
Mp-resin P_o	0.6	2.5	2.8	0.19	0.45	0.52
Hd-HCO_3-P_o	20	23	29	6.3	4.2	5.4
Hd-NaOH-P_o	164	205	214	52	37	40
Hd-residual-P	66	80	84	21	15	16

Mp = macroporous resin method (Rubaek and Sibbesen, 1993).
Hd = Hedley fractions (Hedley *et al.*, 1982).

References

Christensen, B.T. (1988) Effects of animal manure and mineral fertilizer on the total carbon and nitrogen contents of soil size fractions. Biology and Fertility of Soils 5, 304-307.

Christensen, B.T. (1992) Physical fractionation of soil and organic matter in primary particle size and density separates. *Advances in Soil Science* **20**: 1-90.

Christensen, B.T. (1995) Carbon in primary and secondary organomineral complexes. *Advances in Soil Science* (in press).

Christensen, B.T., Petersen, J., Kjellerup, V. & Trentemøller, U. (1994) The Askov Long-Term Experiments on Animal Manure and Mineral Fertilizers: 1894-1994. *SP-report no. 43. Danish Institute of Plant and Soil Science, Copenhagen.*

Hedley, M.J., Stewart, J.W.B. & Chauhan, B.S. (1982) Changes in inorganic and organic soil phosphorus fractions induced by cultivation practices and by laboratory incubations. *Soil Science Society of America Journal* **46**: 970-976.

Randall, E.W., Mathieu, N., Powlson, D.S. & Christensen, B.T. (1995) Fertilization effects on organic matter in physically fractionated soils as studied by ^{13}C-NMR: Results from two long-term field experiments. *European Journal of Soil Science* (in press).

Rubaek, G. & Sibbesen, E. (1993) Resin extraction of labile, soil organic phosphorus. *Journal of Soil Science* **44**: 467-478.

Rubaek, G. & Sibbesen, E. (1995) Soil phosphorus dynamics in a long-term field experiment at Askov. *Biology and Fertility of Soils* **20**: 86-92.

Schjønning, P. (1985) *Unpublished internal report. Danish Institute of Plant and Soil Science, Research Centre Foulum.*

Schjønning, P., Christensen, B.T. & Carstensen, B. (1994) Physical and chemical properties of a sandy loam receiving animal manure, mineral fertilizer or no fertilizer for 90 years. *European Journal of Soil Science* **45**: 257-268.

The Tamworth Legume/Cereal Rotation

Graham J. Crocker and Ian C.R. Holford
NSW Agriculture
Tamworth Centre for Crop Improvement
Tamworth, NSW 2340
Australia

Introduction

Cereal cropping is a major component of farming enterprises in northern NSW, Australia, but research has shown dramatic reductions in soil fertility, expressed in grain protein, with continuous cropping (Dalal *et al.*, 1992). Rotation trials were established in 1966 at Tamworth (latitude 31 degrees 6 minutes S and longitude 150 degrees 56 minutes E, 430 metres elevation) on the eastern edge of the cereal belt. The region has a warm, temperate, subtropical climate with an average annual rainfall of 675 mm; 60% of this falls during the summer (October to March). The rainfall is very unreliable and has ranged from 360 mm in 1965 to 1100 mm in 1955. Evaporation from a free water surface is 1865 mm/year, with the mean minimum and maximum temperatures being 10.5 and 23.8 degrees C.

Site Details

The site was originally a dry sclerophyll forest with Eucalyptus tree species dominant. Area is grazing and cropping country, having been cultivated for around 100 years and was badly eroded by 1958. For the five years prior to the trial commencing in 1966, the soil was used for wheat, fallow, sorghum, wheat and fallow.

The site consists of trials on adjacent Chromic and Pellic Vertisol soils. A description of the two soils is given in Holford & Doyle (1978). Only the Pellic Vertisol or Black Earth will be described here as it was the soil for which organic carbon was modelled. The self

NATO ASI Series, Vol. I 38
Evaluation of Soil Organic Matter Models
Edited by David S. Powlson, Pete Smith and Jo U. Smith

mulching soil is derived from tertiary basalt having a strong blocky structure with $CaCO_3$ present throughout the profile. It had a pH of 8.4 (in water), a bulk density of 1.4 and no impermeable layer in the top two metres. The soil contains 50% clay, stores 300 mm available water to a depth of two metres, had a soil organic matter (SOM) content of 2% and an approximate C:N ratio of 10.

The trials were established to determine the effects of different cropping systems on the long term sustainability of cereal, mainly wheat production. Rotations are imposed for 3-4 years, then cereals were grown to measure any induced differences. There have been three phases, the first from 1966 to 1978 where wheat was the test crop, the second from 1979 to 1987 when sorghum was the test crop, and the third phase from 1988 to 1994 with wheat being the test crop. There were six treatments (rotations) that consisted of three durations of lucerne (Medicago sativa) during 1966-70, alternate crops of grain legumes and wheat, long fallow and continuous wheat. Only two rotations will be described here as these were the two that were modelled at the Workshop.

Treatments

Rotation one consisted of lucerne from 1966 to 1969, wheat from 1970 to 1978, lucerne from 1979 to 1983, sorghum from 1983 to 1987, wheat in 1987, subterranean clover from 1988-1990 and wheat from 1990 to 1994. The other rotation consisted of long fallow from 1966 to 1969, wheat from 1970 to 1979, fallow every second year with wheat in 1981, sorghum sown in 1983 and 1985 and then back to wheat in 1987, 89, 91, 93 and 1994.

The original plot size was 30.5 x 11.3 metres, and this was split to 30.5 x 2.6 m in 1970 to allow two rates of nitrogen application (0 and 45 kg ha^{-1}) in 1970-75, and three rates (0, 60, and 120 kg ha^{-1}) in 1991 and 1993. Phosphorus was applied as a basal application to all plots at rates from 5 to 40 kg ha^{-1} in most years.

Measurements

Vegetative cuts of lucerne and cereals were taken at flowering and grain yield of all cereal crops. This plant material was analysed for N and P content. Soil measurements included total organic carbon and total nitrogen which were measured yearly from 1970 to 1981, and then every second year. Soil moisture and nitrate were also measured to 1.2 metres most years, while soil P, ammonium N, bulk density and pH were recorded less frequently. Total N was determined by autoanalysis after micro-Kjeldahl digestion in concentrated sulphuric acid, organic carbon by the Walkley-Black procedure, soil moisture gravimetrically and mineral N in 1 M potassium chloride (Holford, 1990).

Climatic data is available since 1965 from the Gunnedah Meteorological Station which is about 65 km west of the trial site, a difference in longitude of 40 minutes, with similar elevation. This provides daily maximum and minimum temperatures, soil temperatures at 10 cm, sunshine hours, total wind run, humidity, dewpoint and evaporation plus Tamworth trial site rainfall. This information except for sunshine hours is available from the trial site only for 1987 onwards.

Wheat stubble was burnt in December or January up to 1976, after which it was grazed by sheep prior to being incorporated in March or April. Lucerne was grazed heavily, leaving less than 1000 kg dry matter ha^{-1}, prior to being ploughed in. It was estimated that sheep would have eaten about 75% of wheat stubble present at harvest, but less than 50% of sorghum stubble.

Results - *Organic Carbon*

The site had an original organic carbon (OC) level of 1.14% in 1966 and this rose to 1.21% in 1970 following the lucerne phase (Figure 1). This level gradually declined during the wheat cropping phase from 1970-1978 except for 1976-1977. The higher level was due to ploughing in of heavy weed growth in 1976, caused by well above average rainfall (534mm compared to 228 mm) during the first three months of the fallow. Soils were sampled for OC measurements in June or July. The second lucerne phase from 1979-1983 had little effect on OC as well below average rain occurred in three of the four years. Four sorghum crops from 1983-87 lifted the OC level, before it stabilised under subterranean clover. The increase in organic carbon levels under sorghum is attributed to the incorporation of sorghum residues, which were much heavier than wheat residues (Holford, 1990).

Figure 1 *Effect of crops and long fallow on total organic carbon*

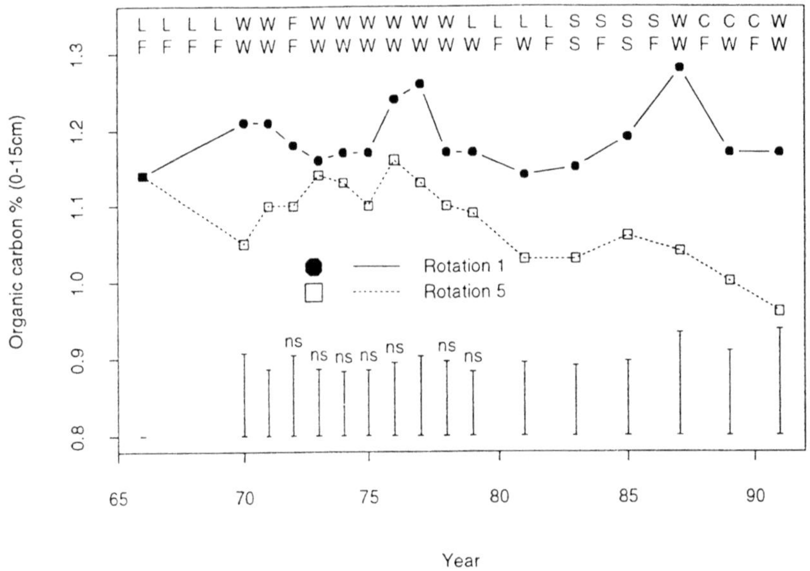

L=lucerne W=wheat S=sorghum F=fallow C=clover
Vertical bars represent LSD ($P<0.05$)

Continuous fallow from 1966 to 1970 caused the OC level to fall to 1.05% which was significantly lower than rotation one in 1970 and 1971. There was no significant difference during the rest of the wheat growing phase, except for 1977. Organic carbon levels, for rotation five, have continued to decline since 1976 except during the sorghum phase (Figure 1). Organic carbon levels have maintained a steady trend except for the increase following the sorghum phase, thus giving significant differences in OC levels between the two rotations.

Lucerne significantly increased the organic carbon levels above those of long fallow or continuous wheat growing in most years (Holford, 1981). The increase in organic carbon was proportional to the length of the lucerne ley. During the sorghum phase from 1983-1987, two sorghum crops in four years maintained organic carbon while four sorghum crops increased the levels (Holford, 1990). Organic carbon levels tended to decrease with a wheat crop every second year (long fallow 1979-1983 and 1988-1991).

Total Soil Nitrogen

Lucerne had a larger effect on total soil nitrogen than on OC, supplying 140 kg N ha^{-1} y^{-1} for each year of lucerne growth from 1966-70 and 103 kg N/ha from 1979-83 (Holford, 1990). The larger input of N during the first lucerne phase was due to higher rainfall. Total soil nitrogen remained significantly greater than for the fallow rotation during the 9 wheat crops from 1970-78. This trend continued during the sorghum phase and at the end of this period the long fallow rotation was the only one to have lower total soil nitrogen levels during the 8 years of the sorghum phase (Holford, 1990).

Atmospheric nitrogen deposition has not been measured at this site, but is thought to be low due to lack of industry in the area and the semi-arid environment. However, for the use of this data in models it was decided to use 10 kg N ha^{-1} y^{-1} for atmospheric N deposition. Figures from European data suggest 50 kg ha^{-1} y^{-1} is not uncommon in that region.

References

Dalal, R.C., Strong, W.M., Weston, E.J., Cahill, M.J., King, A.J., Cooper, J.E. & Lehane, K.J. (1992) Restoration of fertility-depleted soils: The Warra experiment. *Rural Science* 1992: 30-35

Holford, I.C.R. (1980) Effects of duration of grazed lucerne on long-term yields and nitrogen uptake of subsequent wheat. *Australian Journal of Agricultural Research* **31:** 239-250.

Holford, I.C.R. (1981) Changes in nitrogen and organic carbon of wheat growing soils after various periods of grazed lucerne, extended fallowining and continuous wheat. *Australian Journal of Soil Research.* **19:** 239-249.

Holford, I.C.R. (1989) Yield and nitrogen uptake of grain sorghum in various rotations, including lucerne, annual legume and long fallow. *Australian Journal of Agricultural Research.* **40:** 255-260.

Holford, I.C.R. (1990) Effects of 8-year rotations of grain sorghum with lucerne, annual legume, wheat and long fallow on nitrogen and organic carbon in two contrasting soils. *Australian Journal of Soil Research.* **28:** 277-291.

Holford, I.C.R. and Doyle, A.D. (1978) Effect of grazed lucerne on the moisture status of wheat-growing soils. *Australian Journal of Experimental Agriculture and Animal Husbandry* **18:** 112-117.

Long-term fertilization trials in Hungary

K. Debreczeni , I. Sisák, K. Sárdi, and G. Csitári
Department of Agricultural Chemistry
Pannon University of Agricultural Sciences
Keszthely,
Hungary

Introduction

In 1967, according to the proposal of specialists in Hungary, the National Long-Term Fertilization Trials (NLFT) were initiated to establish a uniform network of fertilizer trials representing different agro-ecological regions of the country. The aim of conducting these trials was to study the effects of increasing fertilizer rates on soil nutrients, as well as to predict fertilizer consumption in Hungary. In recent decades environmental pollution in water, soils and the atmosphere has increased; thus investigations on the environmental impact of fertilization (e.g. on soil acidification, nutrient leaching etc.) has become of primary importance. Results obtained in the long-term fertilization trials provide, a) essential information on these processes and b) a reliable database for interpretation of the possible causes of these changes.

The National Long-Term Fertilization Trials in Hungary have been continued at 9 experimental sites (Figure 1), including sites at several universities and research institutions. The whole program is supervised by the Pannon University of Agricultural Sciences in Keszthely which is the founder of the trials and is responsible for directing and coordinating the maintenance of the sites, the analysis of soil and plant samples, the processing of data, and the interpretation of results. Unfortunately, current funding difficulties are having a negative impact on both the analysis of samples and the processing of data.

NATO ASI Series, Vol. I 38
Evaluation of Soil Organic Matter Models
Edited by David S. Powlson, Pete Smith and Jo U. Smith

The experimental site from the network described in this paper is Keszthely, the location of Pannon University of Argricultural Sciences. The dataset from this trial is available in the prototype standard data format for the European GCTE SOMNET.

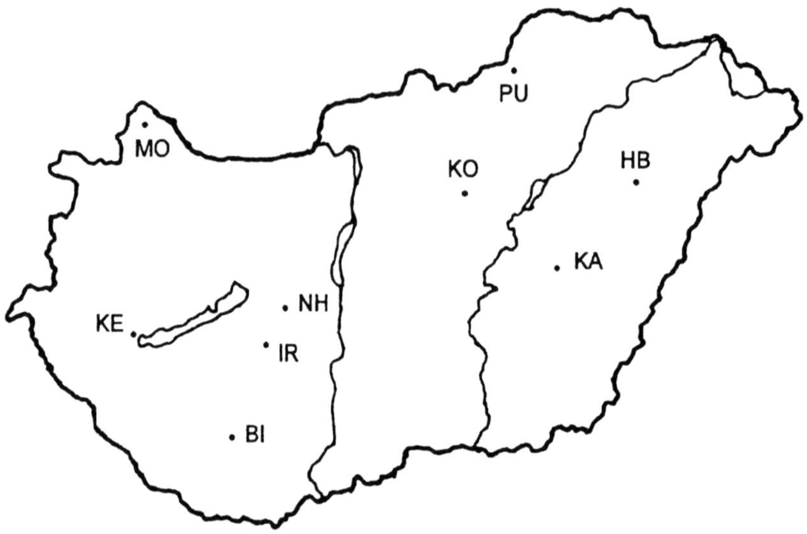

Figure 1 *Sites of the Long-Term Fertilization Field Trials in Hungary. KE = Keszthely, KA = Karcag, KO = Kompolt, MO = Mosonmagyaróvár, NH = Nagyhörcsök, BI = Bicsérd, HB = Hajdúböszörmény, IR = Iregszemcse, PU = Putnok*

Experimental site description

Keszthely is in Hungary, Europe and has a cool temperate continental climate with some sub-mediterranean influences. The latitude and longitude of the site are 46.77° N and 17.246° respectively. The site is 132 m above sea level. Before the trials began, winter wheat was grown on the experimental plots.

Main soil characteristics:

The soil is a loam formed on glacial, alluvial and sandy loess deposits during the Pleistocene. It is a brown forest soil categorised by FAO as a Eutric cambisol and by the USDA as an Orthic Eutrochrept. The main soil characteristics of the control plot are as follows: Organic matter content is 1.3 %, pH_{KCl} is 5.9-6.3 (in Exp. I, II and III), ammonium lactate (AL) extracted P_2O_5 is 20-45 mg kg^{-1} soil, AL extracted K_2O is 116-125 mg kg^{-1} soil and the C:N ratio is 8:1. Further details of soil and other parameters are available in Debreczeni (1991), Debreczeni & Debreczeni (1994) and Debreczeni, K,(1994).

Description of land-use (crop rotations):

Four-year crop rotations were initiated over a successive four year period after 1967. The three experiments at each site therefore began in 1967, 1968 and 1969 (experiments I, II and III respectively). These experiments were initiated on plots of 50-70 m^2 with one of two crop rotations and a corn monoculture. Each has 20 treatments with 4 replicates of each treatment. The crop rotations are as follows:

Crop rotation A - Winter wheat-Corn-Corn-Pea	at 9 sites
Crop rotation B - Winter wheat-Corn-Corn-Winter wheat	at 9 sites
Crop rotation C - Corn in monoculture	at 4 sites

Corn monoculture is found only in the first experiment. Code numbers used in the experiments are as follows:

Year	Exp. Number	Code number
1967	I	17
1968	II	18
1969	III	19

The numbers of years after the experiment was initiated is given after the code-numbers i.e. results of experiment obtained in the 20th year in experiment No.17 and rotation A is numbered as 1720A.

Fertilizer treatments used in the experiments:

The treatments used in each of the experiments are as follows:

- Unfertilized control
- N, P and K fertilizer rates:
- N rates: gradually increasing by 50 kg ha^{-1} N from 50 to 250
- P rates: gradually increasing by 50 kg ha^{-1} P_2O_5 from 0 to 200
- K rates: 0, 100 and 200 kg ha^{-1} K_2O

Code numbers indicating the fertilizer rates are given in the order of NPK.

Measurements made during the trials and their frequency:

Soil samples from a depth of 0-20 cm were taken in the last year of the crop rotations and soil pH and available soil nutrient status i.e. P and K content of soils in the AL-extract were determined. Ammonium lactate (pH = 3.7) extractant is used in Hungary for routine soil tests (see Egner, Riehm & Domingo, 1960). Plant samples have been taken for NPK analysis from 5 treatments of Experiment No. I (fertilizer code numbers 000, 121, 221, 321, 431 respectively). Because of economic constraints, chemical analyses other than routine soil testing and plant analysis are made only in specific cases, e.g. to determine nitrate leaching or heavy metal content etc.

Meteorological data:

The meteorological station is situated 250m from the experimental plots. The 50-year average annual precipitation at the site was 716 mm between 1901 and 1950 and 709 mm between 1951 and 1980. Between 1 October and 31 March, an average of 250-300 mm of rain falls whilst between 1 April and 30 September, 400-450 mm falls. The 50-year average daily air temperature is 10.0 °C (from data between 1901 and 1950). In January the average air temperature is - 1 °C whilst in July it is + 21 °C. The average air temperature during the period of vegetation growth is 17 °C. Average solar radiation is 1800 hours per year (between 1951-1980, Varga-Haszonits1987). Annual potential evapotranspiration has been estimated at 860 mm. The monthly average soil moisture content in upper 20 cm layer are as follows for each decade since the experiment began (all values in mm):

Decade	Jan	Feb	March	Apr	May	June	July	Aug	Sep	Oct	Nov	Dec
1.	37	35	31	28	26	25	23	22	24	27	32	36
2.	37	35	30	25	24	26	25	23	24	28	36	37
3.	36	33	30	26	25	24	22	25	26	28	36	37

Cropping details:

The effects of previous crop on the main crop in the rotations have been studied and the results can be found in Debreczeni, K. (1993).

Methods of determining C and N and changes over the years

C and N were determined at the time soil samples were taken, i.e. in the last year of the crop rotations; in every fourth year. Soil organic matter content was calculated according to Fekete, Hargitai & Zsoldos (1967), after a digestion with chromium-sulphuric acid.

Trend in SOM or C over the course of the experiment

Kemenesy (1959) found the following values for root residues from old crops after 60 days in the brown forest soils at Keszthely: in the upper 0-20 cm soil layer of plots having grown maize there is an average of 2.5 t ha^{-1} of roots, consisting of 1.15% N with a C:N ratio of 45.5:1. Winter wheat root residues however, contribute only 1.2 t ha^{-1} to the SOM with an N content of 1.32 % and a C:N ratio of 75:1. Peas leave an average of 0.57 t ha^{-1} of root residues with an N content of 2.18 % and a C:N ratio of 22:1.

When the experiments began, the average total carbon content of soil exceeded 1% whilst by the mid 1980's, these values had decreased below 1%. However, due to the relatively small number of measurements and the large standard deviation, no definite statements can be made about this trend.

References

Debreczeni, B. & Debreczeni, Bné (Eds) (1994) Fertilization Research 1960-1990. Akadémiai Kiadó, Hungary, 411 p.

Debreczeni, B. (1991) State-wide Long-Term Fertilization Trials as a Research Activity Representing National Wealth. Proceedings of the XXXIII. Georgikon Days, p. 20-27. Debreczeni, K. (1993) Effect of crop rotation on the fertility of different soil types within long-term fertilization trials. Proceedings of the XXXV. Georgikon Days. p.509-516.

Debreczeni, K. (1994) Country Report 5. The Potassium Resources and Efficiency of Potassium Fertilizers in Hungary. International Potash Institute, Basel. 116 p.

Egner, H., Riehm, H. and Domingo, W.R. (1960) Untersuchungen Uber die chemische Bodenanalyse als Grundlage fur die Beurteilung des Nahrstoffzustandes der Boden II. *K. Lantbr. Hogsk. Ann.* **26**:199.

Fekete, Z., Hargitai, L. & Zsoldos, L. (1967) Talajtan és Agrokémia. (Soil Science and Agrochemistry. Mezőgazdasági Kiadó, Hungary.

Kemenesy, E. (1959) Talajerőgazdálkodás. (Soil Fertility and Management) Akadémiai Kiadó, Hungary, 308 p.

Varga-Haszonits, Z. (1987) Agrometeorológiai információk és hasznositásuk. Mezőgazdasági Kiadó Budapest. 248 p.

Acknowledgements

The authors acknowledge the assistance of colleagues in conducting the experiments and in the processing of data.

Essai permanent plots, Gembloux

Marc Frankinet and Yves Raimond
Station de Phytotechnie - C.R.A.
Chemin de Liroux, 11 - 5030 Gembloux
Belgium

Summary

In 1959, a long term trial was started in order to assess the effects of organic matter (farmyard manure, pig slurry, crop residues, green manures, etc.) and calcic amendments on soil physical structure, fertility and crop production.

Changes in soil organic matter (SOM) content showed that the addition of farmyard manure raised the SOM content of the soil (1.9 % humus) whilst the SOM content remained stable when crop residues were ploughed in (1.6 % humus). Without the addition of organic matter, SOM content decreased (1.4 % humus).

Treatments combining slurry and slaked lime were more effective over the last 10 years. Aggregate stability declined for all treatments with the lowest values occurring where no organic matter is added. When the input / output balance of NPK mineral nutrients is maintained, crop yields are unaffected by the observed differences in SOM content and in soil physical properties.

Keywords : Long term trial - Organic matter - Calcic amendment - Physical fertility - C and N content - Crop yields.

NATO ASI Series, Vol. I 38
Evaluation of Soil Organic Matter Models
Edited by David S. Powlson, Pete Smith and Jo U. Smith

Introduction

In 1959, a trial was established on the Essai Permanent Plots to study the effects of substituting farmyard manure with crop residues, green manures and other organic amendments.

Site description

The experimental site at Gembloux , Belgium, is located 50 km southwest of Brussels. The soil is a leached brown soil developed on a deep loess deposit. The texture is : clay 15 % (< 2 µm), silt 80 % (< 50 µm) and sand 5 % (> 50 µm). This soil is sensitive to sealing and crusting especially in spring. Chemical parameters are determined by the *Station de Chimie et de Physique agricoles - C.R.A. Gembloux* and are listed in Table 1.

Table 1 *Chemical parameters*

pH (water)	6.8 (6.3 to 7.8)
Soil organic matter (C x 1.724)	1.3 to 1.8 %
Organic N	0.87 to 1.01 %
P	150 to 200 mg kg^{-1} soil
K	200 to 250 mg kg^{-1} soil
Ca	2000 to 3000 mg kg^{-1} soil

These measurements are made every year on samples taken just after harvest. The climatic data are colected at the meteorological station of the *Station de Chimie et de Physique agricoles* which is located at Gembloux (2000 m from the experimental site). These data are available on request from the *Bureau d'Informatique et de Statistique appliquées - C.R.A. Gembloux*. The mean values of temperature and rainfall are presented in Table 2. In the datafiles, the climatic data are expressed per 10 days (in French : décades). Annual precipitation is 806 mm, occuring on average 212 days per year.

Table 2 *Monthly average temperature (°C) and precipitation (mm)*

Months	J	F	M	A	M	J	J	A	S	O	N	D
Temperature	1.7	2.0	5.0	7.8	11.9	14.9	16.6	16.5	13.9	10.1	5.5	3.3
Precipitation	65.5	56.7	65.6	53.5	69.0	73.0	71.7	75.2	62.8	65.7	75.0	72.1

Experimental design

The experimental design is a latin square with six treatments. Each treatment covers a total of 0.46 ha. The six treatments are :

Treatment 1 = Remaoval of crop residues (1959-1994)

Treatment 2 = Ploughing in of crop residues - Pig-slurry and slaked-lime

a) 1959-1974 : - removal of crop residues
- 2 t ha^{-1} Krillium in 1959 only

b) 1975 : - ploughing in of crop residues
- 45 t ha^{-1} waste lime every 2 or 3 rotatations
- 40 t ha^{-1} pig slurry every rotation

Treatment 3 = Removal of crop residues - Pig slurry and slaked-lime

a) 1959-1974 : - removal of crop residues
- 45 t ha^{-1} waste lime every 2 or 3 rotations

b) 1975-1992 : - removal of crop residues
- 45 t ha^{-1} waste lime every 2 or 3 rotations
- 40 t ha^{-1} pig slurry every rotation

c) 1993 : - one year of fallow with ryegrass and clover

d) 1994 : - treatment 3 becomes the same as treatment 2

Treatment 4 = Removal of crop residues - Farmyard manure (1959-1994)

- removal of crop residues
- 40 t ha^{-1} farmyard manure (from cattle) every rotation

Treatment 5 = Removal of beet leaves - Ploughing in of straw

a) 1959-1993 : - removal of beet leaves

- ploughing in of straw

b) 1994 : - treatment 5 becomes the same as treatment 1

Treatment 6 = Ploughing in of crop residues - green manure (1959-1994)

- ploughing in of crop residues
- one green manure (*Vicia* sp.) every rotation

The crop rotations are listed in Table 3.

According to general practice, during the first two four-course rotations (1959-1966), large amounts of P and K fertilizers were applied. During the third and fourth rotations some adjustments were made to reduce applications of P and K for the treatments with FYM and waste lime applications.

Since 1975, a three-course rotation has been usedand the P and K dressings have been calculated, rotation by rotation, based on the input / output balance of P and K removed by harvest or in crop residues, and the input from organic or calcic amendments.

For nitrogen up to 1974, this value was estimated in several different field experiments by comparing increasing amounts of mineral nitrogen fertilizer with an equivalent effect of organic matter ploughed in. From 1975 to 1990, soil samples from each treatment were analysed and the nitrogen dressing was estimated according to the analytical results. In 1984 and during the 7 years following, all plots were split into four sub-plots with four levels of nitrogen dressing, the average values being calculated for the dataset used in the modelling exercise.

Table 3 *Crop rotations, nitrogen applications (kg ha^{-1}) and yields (grain or sugar) (t ha^{-1})*

Year	Crop	Treat. 1		Treat. 2		Treat. 3		Treat. 4		Treat. 5		Treat. 6	
		N	Y	N	Y	N	Y	N	Y	N	Y	N	Y
59	S.B.	200	4.9	200	5.8	160	4.8	180	5.1	200	5.0	200	4.8
60	S.O.	70	5.3	70	5.5	70	5.6	70	5.5	70	5.6	50	5.5
61	F.B.	20	3.7	20	3.6	20	3.8	20	3.7	20	3.7	20	3.8
62	W.W.	67	7.1	67	7.0	67	7.0	67	7.1	67	7.2	67	7.3
63	S.B.	180	9.4	180	9.4	180	9.2	143	9.4	180	9.4	180	9.6
64	W.W.	85	5.6	85	5.7	85	5.5	65	5.5	75	5.5	42	6.0
65	S.O.	60	3.3	60	3.7	60	3.8	60	3.4	60	3.5	54	3.9
66	W.B.	100	2.2	100	2.5	100	2.4	100	2.1	100	2.5	100	2.9
67	S.B.	160	10.7	160	10.3	120	10.6	100	11.2	180	11.1	160	11.0
68	S.W.	140	4.5	140	4.5	140	4.3	120	4.5	130	4.5	80	4.5
69	F.B.	20	1.7	20	1.7	20	1.7	20	1.6	20	1.7	20	1.7
70	W.W.	76	6.0	76	6.0	76	6.0	76	6.2	70	6.1	70	6.2
71	S.B.	180	11.3	180	11.4	180	11.4	150	11.7	170	11.5	130	11.8
72	W.W.	120	6.1	120	6.1	120	5.9	110	6.1	110	6.2	65	5.9
73	W.B.	115	5.2	115	5.2	115	5.7	99	6.0	131	5.4	107	7.5
74	S.O.	86	5.3	76	5.2	86	5.2	76	5.5	76	5.3	76	5.3
75	S.B.	160	5.9	30	4.9	30	5.7	120	5.4	150	5.9	140	6.5
76	W.W.	120	5.6	105	5.6	130	5.3	120	5.8	120	5.8	90	7.8
77	W.B.	104	5.6	94	5.4	104	5.4	104	5.3	94	5.5	84	5.4
78	S.B.	170	8.7	135	8.9	115	9.1	135	8.9	160	8.5	135	8.9
79	F.B.	-	4.7	-	4.8	-	4.8	-	4.6	-	4.4	-	4.6
80	W.W.	86	6.3	76	6.3	76	6.1	76	6.6	86	6.3	76	6.4
81	S.B.	170	7.3	100	7.9	100	8.0	130	6.7	170	6.7	155	7.0
82	S.O.	48	6.8	48	7.1	48	6.8	48	6.8	48	6.8	48	7.0
83	W.W.	116	7.1	81	7.2	97	7.2	92	7.0	108	7.2	92	7.1
84*	S.B.	113	6.5	83	9.0	68	8.0	78	6.2	113	6.2	106	6.3
85*	W.W.	140	4.8	138	7.1	138	6.6	139	5.8	140	4.8	126	4.9
86*	W.B.	176	7.5	166	8.2	171	7.7	171	7.8	176	7.6	171	7.9
87*	S.B.	195	6.4	135	7.6	135	7.4	155	6.7	195	6.2	155	6.7
88*	F.B.	-	4.0	-	3.9	-	4.1	-	4.0	-	4.2	-	4.0
89*	W.W.	145	7.7	116	8.2	116	7.8	131	7.8	145	7.8	131	8.0
90*	W.B.	185	7.4	155	7.8	155	7.4	170	7.8	185	7.7	170	7.8
91	S.B.	131	7.5	131	8.9	131	9.2	131	8.7	131	8.0	131	8.5
92	W.W.	170	8.0	170	8.0	170	8.1	170	8.2	170	8.1	170	8.2
93	W.B.	160	5.4	160	6.6	-	-	160	5.8	-	-	160	6.2
94	S.B.	140	8.0	140	8.8	140	8.1	140	8.2	140	7.3	140	7.6

* Mean values : all plots were split into 4 sub-plots for N applications

N = Nitrogen; Y = Yield; S.B. = sugar beet; W.W. = winter wheat;

W.B. = winter barley; S.O. = spring oats; F.B. = field beans.

In 1993, a one year fallow (ryegrass and clover) was grown on the plots of treatments 3 and 5 and in 1994, treatment 5 was modified and became the same as treatment 1 (removal of crop residues) and treatment 3 became the same as treatment 2.

The organic and calcic amendments are applied during late summer or in early autumn before the sugar beet crop. Green manure is also sown before sugar beet. All of the agronomic data for amendments, green manure, yields, crop residues are expressed in t ha^{-1} of fresh material. The mean values of the dry matter content of these products are listed in Table 4.

Table 4 *Average dry matter (DM) content*

Products		DM %
Farmyard manure		27.59
Pig slurry		7.43
Grain	from 1959 to 1974	84.00
	from 1975 to present	85.00
Roots of sugar beet		20.00
Green manure	mustard	21.99
	Vicia sp.	14.87
Straw		84.04
Top and leaves of sugar beet		17.34

Experiment results

Measurement of soil organic matter content confirmed the results obtained from other similar long-term trials, that farmyard manure applied regularly raised the total organic matter content in the soil. Soil organic matter content remained stable when crop residues were ploughed in. Without organic amendments soil organic matter content decreased. The organic matter content in the soil was modified from 1.6 % at the beginning of the trial to 1.4 % with maximum residue removal. Addition of farmyard manure increased organic matter levels to 1.9%.

Aggregate stability and the mean diameter of aggregates declined in all treatments. This drop was not as pronounced for treatments where organic matter (slurry, farmyard manure, crop residues and waste lime) was added. The poorest results were observed where no organic matter was added. However, when the input / output balance of NPK mineral nutrients were maintained, yields of sugar beet, winter wheat and winter barley were unaffected by such differences in soil physical properties. Treatments involving the addition of slurry and waste lime from sugar factories appear to have yielded more over the last 10 years. This tendency has yet to be confirmed.

Sources of additional information

Crohain, A. and Rixhon, L. (1967) La valeur fertilisante pratique des feuilles et collets de betterave sucrière. *Bull. Rech. Agron. Gembloux* **2(3):** 397-428.

Crohain, A. and Rixhon, L. (1968) La valeur fertilisante pratique d'un engrais vert de vesces, d'une paille de céréale et d'écumes de sucrerie. *Bull. Rech. Agron. Gembloux* **3(3):** 392-431.

Crohain, A. (1975) Fumure phosphopotassique et amendements organiques. *Rev. Agric.* **28(5):** 1133-1148.

Darcheville, M. (1972) Optimisation de la fumure minérale de quelques plantes de grande culture. In: *VIIe Congrès Mondial des Fertilisants*, Baden, 317-324.

Darcheville, M., Destain, J.P. and Droeven, G. (1986) La fumure minérale des plantes de grande culture. Essais de longue durée en région limoneuse (1967-1984). *Rev. Agric.* **39(1):** 5-40.

Destain, J.P. and Raimond, Y. (1983) La composition chimique du lisier, ses facteurs de variation et ses conséquences agronomiques. *Rev. Agric.* **36(1):** 39-49.

Dumortier, G. and Droeven, G. (1954) Appareil permettant la détermination de la stabilité de la structure des sols. *Bull. Inst. Agron. Stn. Rech. Gembloux* **T 22(3/4):** 213-215.

Frankinet, M., Raimond, Y., Destain, J.P., Roisin, C. and Grévy, L. (1993) Organic matter management and calcic amendments in order to maintain or improve soil fertility. In: *Soil Biota, Nutrient Cycling, and Farming Systems* - Paoletti, M.G., Foissner, W. and Coleman, D. 27-39.

Guiot, J. (1971) Evolution de l'azote minéral dans le sol et fumure azotée du blé d'hiver. *Bull. Rech. Agron. Gembloux* **6(3-4):** 280-326.

Guiot, J. (1975) Estimation des réserves azotées du sol par détermination de l'azote minéral. *Rev. Agric.* **28(5):** 1117-1132.

The Waite Permanent Rotation Trial

Peter R. Grace

Cooperative Research Centre for Soil and Land Management and University of Adelaide. Adelaide, South Australia

Site description

The Waite Permanent Rotation trial is located at the Waite Agricultural Research Institute in suburban Urrbrae, approximately 7 km from the centre of Adelaide. The climate is typically mediterranean with a mean annual rainfall of 626 mm, of which 487 mm falls within the growing season of April-October. Mean annual maximum and minimum temperatures are 21°C and 12°C respectively.

The fine sandy loam found on this site is a prime example of one of the most important soil types developed for wheat production in Australia. It is described in the Handbook of Australian Soils (Stace *et al.*, 1968) as a Red-brown earth of the non-sodic Urrbrae series. It is more commonly known in Soil Taxonomy (Soil Survey Staff, 1975) as an Alfisol, specifically a Rhodoxeralf. The upper 10 cm contains 18% clay, increasing to 32% in the A2 horizon, with illite and kaolin being the main clay minerals. Soil pH (in water) is typically 5.9 with a negligible amount of calcium carbonate. Soil samples taken from the site in 1924 indicate an initial soil organic carbon (SOC) value of 1.82% in the top 22.5 cm, which equates to 2.75% in the top 10 cm. The inert fraction in the top 10 cm as measured by ultraviolet photo-oxidation (Skjemstad *et al.*, 1993) is 0.66%. The site has daily rainfall records since 1925 with daily maximum and minimum temperature, pan evaporation, sun hours, solar radiation and wind run being recorded since 1973.

NATO ASI Series, Vol. I 38
Evaluation of Soil Organic Matter Models
Edited by David S. Powlson, Pete Smith and Jo U. Smith

Site history

The trial was initiated 1925 on a site previously under native grassland. The original 10 rotations were wheat (W), wheat-fallow (WF), wheat-peas (WPe), wheat-oats-fallow (WOF), wheat-peas-fallow (WPeF), wheat-barley-fallow (WBF), wheat-oats-peas (WOPe), wheat-barley-peas (WBPe), wheat-oats-pasture-fallow (WOPF) and wheat-pasture (WP). The treatments were arranged on thirty-four 0.06 ha plots (reduced to 0.05 ha, in 1938) in an unreplicated design (except for WF, WPe and W). For all rotations each phase has been represented each year in adjacent plots.

Seven of the original rotations have remained in an unbroken sequence since 1925 (W, WF, WPe, WPF, WOF, WBPe and WOPF). In 1948, the WPeF and WBF rotations were converted to 2 years wheat followed by a 4 year pasture phase (WWPPPP) and the WOPe rotation converted to wheat followed by a 2 year pasture phase (WPP). The replicated WPe and W plots (numbers 29, 30 and 31 respectively in the sequential design) as well as the WP treatment (plots 32 and 33) were changed to a wheat-2 years pasture-fallow (WPPF) rotation in 1948. Plot 29 was subsequently converted to permanent pasture (PP) in 1951. In 1985 the WF plots at the western end of the trial were converted to permanent pasture to study changes in soil properties when converting from cereal to ley farming.

The wheat cultivar sown has only changed once since the trials inception in 1925 with Seewari (seeding rate of 90-101 kg ha^{-1}) being preferred until replaced by Molineaux in 1990. In the WOPF and WPF rotations, wimmera rye grass is sown with grazing oats. In the WPP, WPPF and WWPPPP rotations, rye grass is sown with subterranean clover. On the PP plot, phalaris, subterranean clover, lucerne and wimmera rye grass was sown.

Annual fertilizer inputs have been restricted to a basal dressing of superphosphate of 209 kg ha^{-1} between 1926 and 1943. This amount was subsequently halved in later years. Herbicide use was discontinued from 1978-82 to enable research into the influence of rotations on weed populations. Initial fallow workings since 1973 have been with either a chisel plough, to prevent erosion, or a standard spring release cultivator in preference to the mould board plough. Cereal crops since 1975 have been reaped with a self-propelled plot harvester in preference to the binder and thresher used since the experiment commenced. Cages were used to exclude grazing in pasture phases to estimate aboveground pasture production and the collection of plant composition data. Grazing of pasture plots and stubbles was discontinued in 1992. Pasture plots have been mown since that date but with little removal of product. Grain and straw of all crops are weighed at harvest and removed from the site.

Yields

Mean decadal yields of the 10 wheat rotations presently in existence are reported in Table 1. Wheat yields from all plots show considerable seasonal variation, particularly where a fallow does not follow a pasture phase. The inclusion of a short-term pasture phase, whether it be oats for grazing or ryegrass and subterranean clover, has resulted in higher yields of wheat than those obtained from rotations without a pasture phase. All treatments containing fallow phases consistently yielded over 2 t ha^{-1} of grain during the first 20 years of the trial, experiencing a sharp decline between 1945 and 1955. The WPe rotation has consistently yielded low but variable grain yields, but there is no significant trend over 69 years. After an initial rapid decline in yields on the continuous wheat plots they have gradually recovered in the past 20 years to 1930's levels. It is has been suggested (Grace *et al.*, 1995) that this may be due to the gradual build up of resistant light fraction material in this treatment aiding in the maintenance of structure and nutrient availability.

Table 1. *Wheat grain yields (t ha^{-1}) for rotations at the Waite Agricultural Research Institute Permanent Rotation Trial (1925-1993).*

Rotation[1]	1925-34	1935-44	1945-54	1955-64	1965-74	1975-84	1985-93
WOPF	3.00	3.06	2.85	2.53	2.62	2.61	1.57
WPF	2.66	2.72	2.32	1.94	2.14	2.14	1.33
WBPe	2.44	2.25	1.86	1.80	1.81	2.13	1.49
WOF	2.61	2.25	2.16	1.61	2.03	1.96	1.31
WF[2]	2.67	2.51	1.76	1.45	1.53	1.72	1.24
WPe	1.77	1.51	1.24	1.22	1.40	1.62	1.11
W	1.39	0.83	0.78	0.52	0.65	0.92	1.09
WPP[3]	-	-	-	1.88	1.86	2.34	1.79
WPPF[4]	-	-	-	2.30	2.50	2.33	1.88
(W)WPPPP[5]	-	-	-	1.42	1.75	2.04	1.78
W(W)PPPP[5]	-	-	-	1.42	1.60	2.00	1.56

[1]W-Wheat, Pe-Peas, F-Fallow, P-Pasture, O-Oats, B-Barley, Pasture; [2]Plots 34 and 35 only
[3]Commenced 1948, formerly WOPe; [4]Commenced 1948, formerly WPe, W and WP
[5]Commenced 1948, formerly WPeF and WBF. Parentheses indicate first or second W crop

Soil analyses

Soil samples were taken from the trial in 1925, 1946, 1963, 1967, 1973, 1980, 1983 and 1993. The samples were taken at 0-10 cm and 10-22.5 cm depth increments except in 1925 (22.5 cm increments to 45 cm) and 1967 (0-2.5 cm only). Since 1963, sample have been taken prior to sowing, with twenty 5.0 cm diameter soil samples taken randomly along the length of each plot (approximately 90 m) and then bulked. A less intensive sampling regime was used on selected plots in the other years. Total C of stored samples from 1925, 1946 and 1963-1993 has been determined by dry combustion assuming the Urrbrae red brown earth to be entirely organic (based on its known chemical properties).

There has been a significant influence of rotation on the changes in soil organic carbon (SOC) in the surface 10 cm since 1925. Soil organic carbon contents for both the 0-10 and 10-22.5 cm layers are ranked in Table 2 as 1963-1993 average values because the majority of rotations have tended to approach steady state since 1963.

Table 2. *Mean value (1963-1993) of soil organic carbon (%) at two depths in the Waite Agricultural Research Institute Permanent Rotation Trial.*

Rotation[1]	0-10 cm	10-22.5 cm	Difference
PP	2.46	1.28	1.18
WWPPPP	1.92	1.02	0.90
WPP	1.88	1.03	0.85
WPPF	1.60	0.90	0.70
WBPe	1.59	0.88	0.71
W	1.52	0.88	0.64
WOPF	1.45	0.89	0.56
WPe	1.30	0.84	0.46
WOF	1.27	0.83	0.44
WPF	1.11	0.77	0.34
WF	1.04	0.67	0.37

[1]W-Wheat, Pe-Peas, F-Fallow, P-Pasture, O-Oats, B-Barley, PP-permanent pasture

The five rotations with a fallow phase contain lower amounts of SOC in both the 0-10 and 10-22.5 cm layers (1.29% and 0.81% respectively) compared to the remaining six rotations (1.78% and 0.99%). For the eleven treatments currently in the trial, SOC in the top 10 cm has declined from 2.75% in 1925 to a mean value in 1993 of 1.56% (equivalent to 46% of the initial value). The smallest loss (10%) being in the plot which had reverted to PP from a WPe rotation in 1950. Of the seven rotations which have remained intact since 1925, the average SOC content in the surface layer has declined to 1.33% after 68 years of cropping.

There are negative linear relationships between the amount of SOC present in 1993 in both 0-10 cm and 10-22.5 cm layers and increasing the frequency of fallows and cereals in the rotations. Increasing the length of the pasture phase has a positive effect on SOC content. The value of 1.04% for SOC in the WF plots is probably well below the threshold for degradation in these red-brown earths considering that little, if any, aboveground crop residues have been returned to these soils since 1925 thereby promoting rapid degradation. The fact that yields in four (WBPe, WPF, WPe and WOF) of the six original rotations are now converging at 1.54 t ha^{-1} (Grace *et al.*, 1995) would also suggest that the soils have been sufficiently exploited by the rotations and residue removal and their SOC levels (average 1.36% in the surface 10 cm and 0.84% at 10-22.5 cm) are either rapidly approaching or are at threshold levels.

A similar linear relationship to that found for SOC and pasture frequency also exists for total soil N content for a selection of plots sampled in 1993. Total soil N in the continuous wheat plot declined by 62% between 1925 and 1993 (Grace and Oades, 1994), whilst the permanent pasture and WWPPPP plots have been accumulating N since 1963 at 8.5 kg ha^{-1} yr^{-1} (Table 3). It is difficult to assess the effect of rotation on either soil N dynamics or pH in the Permanent Rotation Trial because of the lack of detailed historical data.

Under the original rotations which commenced in 1925, there had been little change in pH (6.3) in the top 20 cm when sampled in 1963 and 1973. Under the newer rotations established post-WWII, in which clover pastures make up a substantial part of the rotation, there has been a decrease in pH in all cases, with values of 6.15 and 5.95 in 1963 and 1973 respectively.

The trial has also been used to study changes in water stable aggregate stability and soil organic chemistry under contrasting rotations. Soil biology has been a field of research neglected on this site until recently where changes in microbial biomass and protozoan populations are now studied. Commencing in 1995, detailed soil nitrate profiles and water balance studies will be undertaken on the wheat phases of each rotation. This will complement *in situ* mineralization studies which commenced in 1994. On harvest of the 1995 crop, residue retention will be imposed on one-half of each cereal plot for the first time since the trial's inception.

Table 3. *Total soil nitrogen (kg ha^{-1}) and its linear trend (kg ha^{-1} yr^{-1}) in 0-10 cm in selected plots at the Waite Agricultural Research Institute Permanent Rotation Trial (1925-1993).*

Rotation[1]	1925[2]	1963[2]	1993[3]	1963-93
W	3974	1840	1495	-11.5
WF	3974	1280	1235	-1.5
WPF	4078	1760	1333	-14.2
WPP	4288	1920	2080	+5.3
WWPPPP	3869	2080	2340	+8.7
PP	3974	2480	2730	+8.3

[1]W-Wheat, F-Fallow, P-Pasture, PP-Permanent pasture
[2]calculated from Greenland (1971) assuming a bulk density of 1.3 Mg m^{-3}
[3]Grace (unpublished)

References

Grace, P.R. and Oades, J.M. (1994) Long-term field trials in Australia. In: *Long-term Ecological Research in Agricultural and Ecological Sciences* (ed: R.A. Leigh and A.E. Johnson), CAB International, Wallingford: 53-81.

Grace, P.R., Oades, J.M, Keith, H. and Hancock, T. (1995) Trends in wheat yields and soil organic carbon in the Permanent Rotation Trial at the Waite Agricultural Research Institute, South Australia. *Aust Journal of Experimental Agriculture* (in press).

Greenland, D.J. (1971) Changes in the nitrogen status and physical condition of soils under pastures, with special reference to the maintenance of the fertility of Australian soils used for growing wheat. *Soils and Fertilizers* **34**: 237-251.

Skjemstad, J.O., Janik, L.J., Head, M.J. and McClure, S.G. (1993) High energy ultraviolet photo-oxidation: a novel technique for studying physically protected organic matter in clay- and silt-sized fractions. *Journal of Soil Science* **44**: 485-499.

Soil Survey Staff (1975) *Soil Taxonomy,* Handbook No. 436, USDA, Washington.

Stace, H.C.T., Hubble, G.D., Brewer, R., Northcote, K.H., Sleeman, J.R., Mulcahy, M.J., Hallsworth, E.G. (eds) (1968) *A Handbook of Australian Soils.* Rellim, Glenside, 435pp.

Woodslee Tile Runoff Experiment: Fertilization Effects on Soil Organic Matter

E.G. Gregorich, C.F. Drury [2], *C.S. Tan* [2] *& B.H. Ellert* [3]
Research Branch
Agriculture and Agri-Food Canada
Ottawa, K1A 0C6
Canada

Introduction

The original objective of this field experiment was to determine the effects of crop rotation and fertilization on soil and water quality by examining water flow and loss of N, P, and K through tile drains in a clay loam soil (Bolton *et al.*, 1970). From this field study several papers have been published on nutrient and atrazine losses from tile drains (Von Stryk and Bolton, 1970), structural changes resulting from soil and crop management systems (McKeague *et al.*, 1987), effects of distance from the tile drain on maize growth over a 10-yr period (Bolton *et al.*, 1982), and the effects of fertilization and crop rotation on maize root disease (Stone *et al.*, 1987). The cropping history and yield data (Drury and Tan, 1995) from this site provide a unique opportunity to evaluate the effects of fertilization on soil organic matter turnover and storage of maize-residue C (Gregorich *et al*, 1996).

Site information

The site is located at Woodslee (42° 13′N 82° 44′W) in southwestern Ontario, Canada, and the soil is a Brookston clay loam (28% sand, 35% silt, and 37% clay) that is poorly drained in its natural state (Orthic Humic Gleysol [Canadian soil classification system];

[2, 3] at Agriculture and Agri-Food Canada, Harrow Ontario and Lethbridge Alberta, respectively.

NATO ASI Series, Vol. I 38
Evaluation of Soil Organic Matter Models
Edited by David S. Powlson, Pete Smith and Jo U. Smith

clayey, mixed, mesic Typic Haplaquolls [U.S. system]). The average annual temperature at Woodslee is 8.7°C, with an average maximum of 13.4°C and minimum of 3.9°C. The average total precipitation is 876 mm, with rainfall accounting for 767 mm. Soybean, maize and wheat are the predominant crops grown in the area around Woodslee.

Cropping history

The land was summer fallowed and levelled in 1954; tile drains (100 mm diameter) were installed in 1955 at a depth of 70 cm and a spacing of 12 m. The plots, 76 m long by 12 m wide, were centred longitudinally above the tile drains. Maize (*Zea mays* L.) was grown on all plots from 1956–1958 to reduce residual effects and obtain uniform data. In 1959, 12 plots were established, consisting of six cropping treatment plots with fertilizer, and six without. The cropping treatments were continuous maize, continuous bluegrass (*Poa pratensis* L.), and four plots of a maize–oat (*Avena sativa* L.)–alfalfa (*Medicago sativa* L.)–alfalfa rotation (each crop in the rotation was grown every year). The fertilized plots received 5–20–10 at 336 kg ha^{-1} (16.8 kg N, 67.2 kg P, and 33.6 kg K ha^{-1}) at planting; maize plots received an additional 112 kg N ha^{-1} of ammonium nitrate side-dressed (incorporated). Maize residues were incorporated (about 15 cm deep) into the soil by moldboard plow each autumn, whereas the oat straw, alfalfa hay, and bluegrass were left on the soil surface after harvest. There is a mixed-hardwood forest adjacent to the experimental plots that has been undisturbed for at least the past 100 years.

Yields and estimated residue inputs

The maize grain yields during 1959–1990 showed considerable year-to-year variability, presumably because of climatic factors (Drury and Tan, 1995). The average annual grain yield (dry weight) on fertilized plots was 5.09 Mg ha^{-1}; on the unfertilized plots it was 1.45 Mg ha^{-1} over the 32-yr period. After 1960 the fertilized plots consistently produced greater maize yields than the unfertilized plots, and Drury and Tan (1995) observed that crop rotation and fertilization increased maize yields and reduced yield variability. Precipitation was the only climatic variable that significantly contributed to grain yield in fertilized maize. They also noted that for unfertilized maize, weather variation did not play a significant role in grain production, and that soil fertility was the factor most limiting to maize production.

The actual amount of maize straw returned to the soil during the experimental period was not measured. The amount of C from maize residues that were returned to the soil was estimated using the grain yields (Drury and Tan, 1995), assuming a carbon content of 45% in plant tissues and a root:shoot ratio of 0.15. The harvest index was assumed to be 45% for the fertilized maize and 30% for the unfertilized maize. A minimum value of 3 Mg ha^{-1} y^{-1} of maize residue was assumed for the unfertilized maize when the crop failed (no grain yield), as occurred in 1983 and 1985, and when the grain yield was less than 1 Mg ha^{-1} y^{-1}. The total amount of maize residue returned to the soil was estimated at 286 Mg ha^{-1} for the fertilized system and 159 Mg ha^{-1} for the unfertilized system (Figure 1).

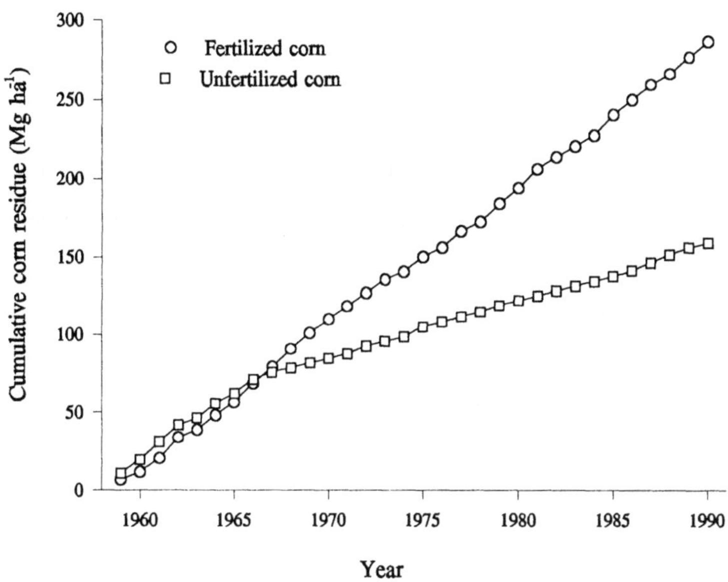

Figure 1. *Estimated cumulative corn residue (stover + roots) returned to the soil under fertilized and unfertilized treatments (from Gregorich et al., 1996).*

Soil sampling - 1991

In May 1991 the fertilized and unfertilized continuous maize, fertilized continuous bluegrass sod, and forest were sampled. Four replicate soil samples were obtained at three depths (0–10, 10–25, 25–45 cm) from each plot before seeding. Soil organic C and N, ^{13}C natural abundance, and microbial biomass C were measured on all soil samples.The light fraction (LF) organic matter from the surface 10 cm of each soil was separated, and organic C and N were measured. Since the soils were free of carbonates to a depth of 60 cm, total soil C is equivalent to soil organic C. Bulk density of the plow layer was determined using soil cores; bulk density values for depths below the plow layer were obtained from published data measured at the same site.

Gregorich *et al.* (1996) reported that, based on total organic C and ^{13}C natural abundance measurements, fertilized soils had more organic C than unfertilized soils, the difference accounted for by a larger amount of C_4-derived C in the fertilized soils. About 22 to 30% of the soil C in the plow layer had turned over and was derived from maize in the fertilized soils, whereas in unfertilized soils only 15 to 20% was derived from maize (Table 1). Assuming first order kinetics, the half-life of C_3-derived C in the surface 10 cm was the same in both soils - about 19 yr. Approximately 10–20% of the added residue C was retained in the soil. Fertilized soils had larger amounts of LF C than unfertilized soils (Table 2). More than 70% of the C in the LF of fertilized soils was derived from maize, but in unfertilized soils only 41% was derived from maize. The estimated half-life of C_3-derived C in the LF was 8 yr. These results indicate that adequate fertilization increased crop yields, which in turn led to greater C storage, and that fertilization did not significantly alter the rate of turnover of native soil organic matter.

Dataset

Data collected over the length of the experiment include grain yield and climate (total precipitation, rainfall, and average daily temperature). Measurements on soil samples collected in 1991 include total organic C and N, ^{13}C natural abundance, microbial biomass C, light fraction C and N, and bulk density for the fertilized and unfertilized continuous maize, continuous bluegrass, and forest soils. In 1994, samples from all cropped and forest soils were collected (to 60 cm depth) and are currently being analyzed for the above properties. Soil samples that were obtained during earlier studies (McKeague *et al.*, 1987) will be analyzed for organic C and N and ^{13}C natural abundance.

Table 1 *$\delta^{13}C$, organic C, C_4-C, C_3-C, and C_4-C/organic C in fertilized and unfertilized maize soils (from Gregorich et al., 1996).*

Crop	Depth	$\delta^{13}C$	Organic C	C_4-C	C_3-C	% C_4-C
	cm	‰	---------- g kg^{-1} ----------			%
Fertilized maize	0-10	-22.7±0.3†	21.8±1.7	6.4	15.4	30
	10-26	-23.1±0.4	21.7±0.9	4.8	16.9	22
	26-42	-25.7	6.5	0.4	6.1	6
Unfertilized maize	0-10	-24.0±0.3	19.5±0.5	3.8	15.7	20
	10-26	-24.1±0.3	19.8±0.6	2.9	16.9	15
	26-41	-25.2	7.4	0.7	6.7	10
Sod	0-10	-26.7±0.3	51.3±3.2	-	51.3	-
	10-22	-26.0±0.1	30.1±3.2	-	30.1	-
	22-40	-26.5	11.1	-	11.1	-

† Mean ±standard deviation

Table 2 *Amounts of total organic C, C_4-C, and C_3-C in the LF organic matter in the surface 10 cm of fertilized and unfertilized maize soils (from Gregorich et al., 1996).*

Crop	$\delta^{13}C$	C_4-C	LF		
	‰	%	-------------- g C kg^{-1} soil ----------		
Fertilized maize	-17.2±0.2†	72	1.11±0.08	0.80	0.31
Unfertilized maize	-21.7±0.5	41	0.48±0.08	0.20	0.28
Sod	-27.7±0.6	-	4.91±2.74	-	4.91

† Mean ±standard deviation

References

Bolton, E.F., Aylesworth, J.W. & Hore, F.R. (1970) Nutrient losses through tile drains under three cropping systems and two fertility levels on a Brookston clay soil. *Can. J. Soil Sci.* **50**: 275-279.

Bolton, E.F., Dirks, V.A. & McDonnell, M.M. (1982) The effect of drainage, rotation and fertilizer on corn yield, plant height, leaf nutrient composition and physical properties of Brookston clay soil in southwestern Ontario. *Can. J. Soil Sci.* **62**: 297-309.

Drury, C.F. & Tan, C.S. (1995) Long-term (35 years) effects of fertilization, rotation and weather on corn yields. *Can. J. Plant Sci.* **75:** 355-362.

Gregorich, E.G., Ellert, B.H., Drury, C.F. & Liang, B.C. (1996) Fertilization effects on soil organic matter turnover and corn residue C storage. *Soil Soc. Am. J.* (In press).

McKeague, J.A., Fox, C.A., Stone, J.A. & Protz R. (1987) Effects of cropping system on structure of Brookston clay loam in long-term experimental plots at Woodlsee, Ontario. *Can. J. Soil Sci.* **67**: 571-584.

Stone, J.A. (1991) Core sampling technique for bulk density and porosity determination on a clay loam soil. *Soil & Tillage Res.* **21**: 377-383.

Stone, J.A., McKeague, J.A.& Protz R. (1987) Corn root distribution in relation to long-term rotations on a poorly drained clay loam soil. *Can. J. Plant Sci.* **67**: 231-234.

Von Stryk, F.G. & Bolton E.F. (1977) Atrazine residues in tile-drain-water from corn plots as affected by cropping practices and fertility levels. *Can. J. Soil Sci.* **57**: 249-253.

Effect of Farm Yard Manure and Fertilizer Nitrogen in Pearl millet-Wheat cropping Sequence

A.P Gupta, R.P. Narwal and R.S. Antil
Department of Soil Science
CCS Haryana Agricultural University
Hisar, 125 004, India

Introduction

Application of farm yard manure (FYM) is known to maintain soil productivity longer than inorganic fertilizers (Flaig, 1975). Farm yard manure contains all the macro- and micronutrients required for plant growth (Hesse, 1984), but its main effect is due to nitrogen, phosphorus, and potassium (Heck, 1931). Maintenance of organic matter in soil is important for the improving nutrient and structural status of soils, particularly under tropical conditions as in India (Acharya et al., 1988; Gattani et al., 1976). Apart from its nutritional role, FYM controls the dynamics of all the macro- and micronutrients (Katyal, 1977; Stevenson, 1982). Therefore, a long-term field experiment was initiated in October, 1967 to study the savings in fertilizer N with the use of FYM in pearl millet-wheat cropping sequence.

Site description

The experimental site is located between 29.16° N latitude and 75.75° E longitude in the North-west part of India. The climate of the experimental area is semi arid with mean annual rainfall of 447 mm and mean annual temperature of 23.9°C. The monthly rainfall and mean temperature is presented in Fig. 1.

NATO ASI Series, Vol. I 38
Evaluation of Soil Organic Matter Models
Edited by David S. Powlson, Pete Smith and Jo U. Smith

Figure 1 *Monthly rainfall and mean temperature distribution during the year,* ● = *rainfall;* + = *temperature*

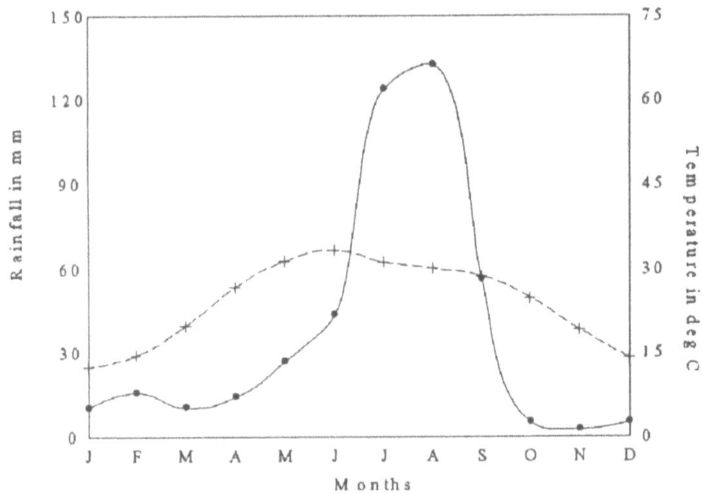

The experimental area was covered with perennial shrubs and xerophytic trees. The area was levelled after removing the shrubs and trees and planted with pearl millet and wheat crops from 1963. The present experiment was established in October 1967. The experimental field has been classified as Fine Loamy Typic Ustochrept. The pH of the soil was 8.2 (1:2), 0.47% organic C, 1.12% calcium carbonate, 100 mg kg^{-1} available N, 13 mg kg^{-1} available P and 249 mg kg^{-1} available K.

Experimental description

The experimental treatment consisted of 3 levels of FYM (15, 30, 45 t ha^{-1}) applied under 3 different modes i.e. application every summer season (June), application every winter season (October), and application in both summer and winter seasons. One control treatment (no FYM) was also included. These 10 treatments were imposed on main plots and each main plot was subdivided into 3 sub plots receiving urea fertilizer N at 0, 60, and 120 kg ha^{-1}. All treatments were replicated four times, each subplot measuring 10x6 m. Farmyard manure was incorporated in the top 15-20 cm layer 3 to 4 weeks before sowing crops. The cropping sequence was pearl millet in summer and wheat in winter. The fertilizer N was applied in two equal splits i.e. half at sowing and the rest 25-30 days after sowing. Both crops were irrigated with canal water. For two years (1974, 1975) the FYM application rates were reduced from 15, 30 and 45 t ha^{-1} to 5, 7.5 and 10 t ha^{-1}. All other cultural practices were common to all the treatments. No chemical fertilizers other than urea have been applied to any treatment since the inception of the experiment. Both crops were cut at about 2-5 cm above ground level. The harvested crop was removed and the roots and stubble were incorporated. Pearl millet is susceptible to temporary water logging due to heavy rains during the growth period when the whole crop can be damaged. During some years, the pearl millet crop could not be sown due to continuous rain during sowing season.

Description of measurements

The grain and straw yield of both the crops has been recorded every season. The yields of the wheat crop are available from 1983 onwards. Prior to this, only the mean yields of the treatments are available. The nutrient (N, P, K) content of the crops were analyzed only once between 1983 and 1995. It was planned that the soil organic C content of the soil would be determined at 5 yearly intervals, but since 1983 this has become more frequent and it has been estimated 4 times. Meterological data is available from 1972. The meterological observatory is located about 900 m from the experimental site.

Method used for determination of C and N

The soil samples were collected from 5 different positions on each plot to a depth of 15cm, mixed thoroughly and a composite sample was taken. The soil samples were dried in the shade and ground to pass through a 2 mm sieve. Soil organic C was estimated by the wet digestion method of Walkley and Black (1934). Available N was distilled using alkaline potassium permanganate. Ammonia was absorbed in boric acid and the amount of ammonia was obtained by titration (Subbiah and Asija, 1956). Total N of the soil was estimated by digesting the soil with sulphuric acid in the presence of a mixture of copper sulphate, mercuric oxide and selenium powder (Jackson, 1962). The 'inert' carbon content of the soil and the contribution of atmospheric N deposition has never been estimated.

Results

The organic C content of the soil increased with the application of FYM at the various rates and under all modes (Figure 2). The pearl millet sown in July 1987 failed because of heavy rains that destroyed the whole crop at the full growth stage. This could explain the apparent increase in soil organic C occurring around 1988.

Figure 2 *Periodic changes in organic carbon under various treatments in 0-15cm soil layer,* ● *= control, + = 240 kg N ha*$^{-1}$*,* ⋆ *= 15 t FYM ha*$^{-1}$*,* □ *= 15 t FYM ha*$^{-1}$ *plus 240 kg N ha*$^{-1}$*.*

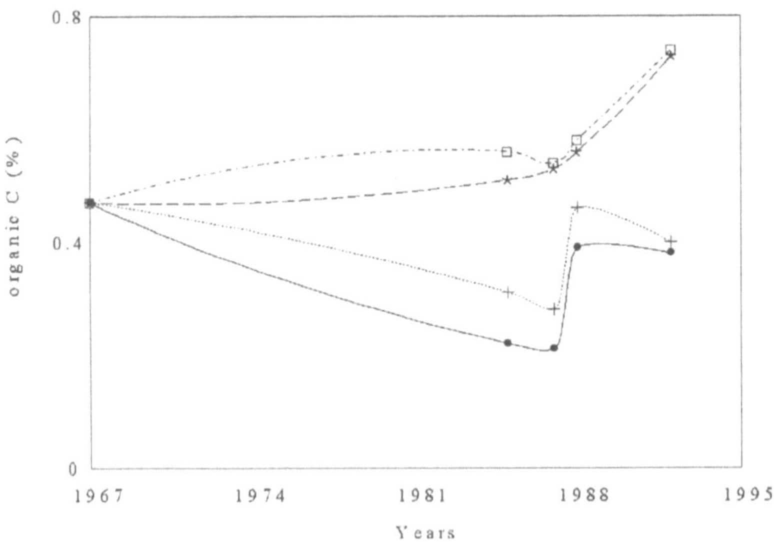

The organic C content of the soil decreased from its original level when no FYM was applied. The highest organic C content of the soil was approaching 1.8% in the plot receiving 45 t FYM ha^{-1} on both crops. Both crops responded to inorganic fertilizer N application in the presence of all levels of FYM. The grain yield of both the crops increased with the application of FYM and N (Fig. 3 a & b).

Figure 3 *Periodic changes in (a) wheat yield and (b) pearl millet yield under various treatments, ● = control, + = 240 kg N ha^{-1}, ⋆ = 15 t FYM ha^{-1}, □ = 15 t FYM ha^{-1} plus 240 kg N ha^{-1}.*

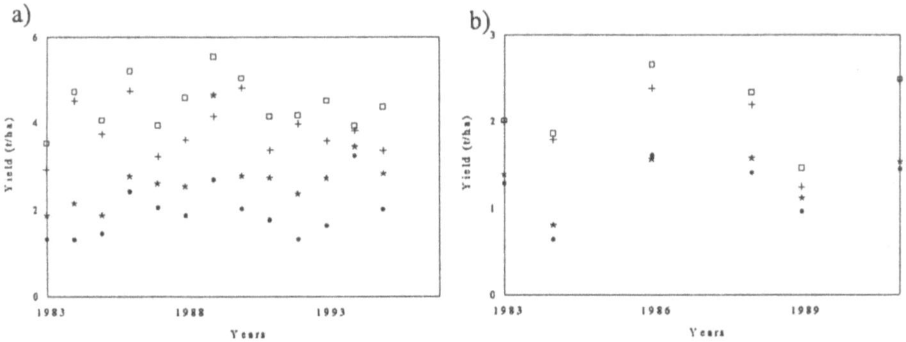

It has also been observed that both the crops responded to N application even in the presence of 45 t FYM ha^{-1}. This may be due to an increase in the potential for crop production. Application of all other chemical fertilizers can be avoided if FYM is applied.

References

Acharya, C.L., Bishnoi, S.K. & Yaduvanashi, H.S. (1988) Effect of long-term application of fertilizers and amendments under continuous cropping on soil physical and chemical properties in Alfisol. *Indian Journal of Agricultural Sciences* **58**:509-516.

Flaig, W. (1975) Specific effects of soil organic matter on the potential of soil productivity. *FAO Bulletin (FAO, Rome)* **27**,. 31-69.

Gattani, P.D., Jain, S.V. & Seth, S.P. (1976) Effect of continuous use of chemical fertilizers and manures on soil physical and chemical properties. *Journal Indian Society of Soil Science* **24**: 284-289.

Gupta, A.P., Antil, R.S. & Narwal, R.P. (1988) Effect of farm yard manure on organic carbon, available N and P content of soil during different periods of wheat growth. *Journal Indian Society of Soil Science* **36**:269-273.

Gupta, A.P., Narwal, R.P., Antil, R.S. & Dev, S. (1992) Sustaining soil fertility with organic C, N, P and K using farmyard manure and fertilizer N in a semi arid zone - A long-term study. *Arid Soil Research and Rehabilitation* **6**:243-251.

Heck, A.F. (1931) Conservation and availability of the nitrogen in farm yard manure. *Soil Science* **31**:335-363.

Hesse, P.R. (1984) Potentials of organic materials for soil improvement. In: *Organic Matter and Rice.* International Rice Research Institute, Los Banos. Philippines, 35-43.

Jackson, M.L. (1962) *Soil chemical analysis.* Asia Publishing House, New Delhi, 183 pp.

Katyal, J.C. (1977) Influence of organic matter on the chemical and electrochemical properties of some flooded soil. *Soil Biology and Biochemistry* **9**:259-266.

Kukreja, K., Mishra, M.M., Dhankar, S.S., Kapoor, K.K. & Gupta, A.P. (1991) Effect of long-term manurial application on microbial biomass. *Journal Indian Society Soil Science* **39**: 685-688.

Nath, J., Sarkar, M.C. & Singh, M. (1972) Dynamics of FYM on physical properties and organic matter build up of soil. *Haryana Agricultural University Journal of Research* **2**:165-168.

Ruhal, D.S. & Shukla, U.C. (1979) Effect of continuous application of farmyard manure and nitrogen and carbon and available N, P and K content of soil. *Indian Journal of Agricultural Chemistry* **12**: 11-18.

Sagwal, O.P. & Ruhal, D.S. (1984) Effect of long-term application of farmyard manure and nitrogen on different forms of nitrogen in soil. *Bulletin Indian Society of Soil Science* **13**: 173-180.

Stevenson, F.J. (1982) *Humus chemistry, genesis, composition, reactions.* Wiley, New York.

Subbiah, B.V. & Asija, C.L. (1956) A rapid procedure for the estimation of available nitrogen in soil. *Current Science* **25**: 172-194.

Walkley, A. & Black, C.A. (1934) An estimation of the method for the determination of soil organic matter and a proposed modification of the chromic acid titration method. Soil Science **37**: 29-38.

The Breton Classical Plots

R. César Izaurralde, Noorallah G. Juma, James A. Robertson, William B. McGill, and Jeffrey T. Thurston
Department of Renewable Resources
University of Alberta
Edmonton, Alberta, T6G 2E3
Canada

Introduction

The Breton Plots research site is owned and operated by the University of Alberta located at latitude 53° 07' N and longitude 114° 28' W near the town of Breton (Alberta, Canada). The research site currently holds five long-term crop rotations and tillage experiments. The oldest plots, the Breton Classical Plots, were initiated in 1930 with the objective of finding "systems of farming suitable for the wooded soil belt" (Wyatt *et al.*, 1936).

The soils of the region are dominantly Gray Luvisolic developed mostly on glacial till parent material under boreal forest vegetation. Climax plant communities of the Breton area correspond to the Lower Foothills Section with a variable mixture of trembling aspen (*Populus tremuloides*), balsam poplar (*Populus balsamifera*), white birch (*Betula papyrifera*) and white spruce (*Picea glauca*). Breton has a long-term annual precipitation of 547 mm, a potential annual evapotranspiration of 732 mm, and a mean annual air temperature of 2.1 °C.

Luvisolic soils represent about 15% of the cultivated area in Alberta. These soils are characterized by low-nutrient status and, depending on parent material, presence of top- and sub-soil acidity (Robertson & McGill, 1983). Typical Luvisolic soil profiles have well developed Ae and Bt horizons, are well to imperfectly drained, and their texture may range from sandy-loam to clay.

The objective of this paper is to describe the Breton Classical Plots experiment and to highlight results on crop yields and soil properties.

NATO ASI Series, Vol. I 38
Evaluation of Soil Organic Matter Models
Edited by David S. Powlson, Pete Smith and Jo U. Smith

Description of Field Study

The Breton Classical Plots experiment, initiated in 1930, compared the productivity of two cropping systems: continuous spring wheat (*Triticum aestivum* L.) and a four-year rotation with three years of cereal grains and one year of legumes. The plots (each 269 m^2) were laid out as an unreplicated experiment with every phase of the crop rotations present every year. In 1938, the continuous wheat system was converted to a wheat-fallow to help control weeds. In 1939, the four-course rotation was changed to a five-year rotation of wheat, oat (*Avena sativa* L.), barley (*Hordeum vulgare* L.), forage, forage. The forage crop has varied over time but has always included a legume. From 1939 to 1954 it was "mixed legumes". From 1955 to 1966 it was a five-crop mixture (alfalfa [*Medicago sativa* L.], red clover [*Trifolium pratense* L.], brome [*Bromus inermis* Leyss.], creeping red fescue [*Festuca rubra* L.], and timothy [*Phleum pratense* L.]). In 1967 the forage mixture was changed to alfalfa and brome because it was the commonly recommended forage mixture for the area. The present setup of the Breton Classical Plots allows for comparisons between a five-year rotation of spring wheat, oat, barley underseeded to alfalfa/brome mixture followed by two more years of the hay mixture (WOBHH); and a two-year rotation of spring wheat and fallow (WF) (McGill *et al.*, 1986; Wani *et al.*, 1994; Juma *et al.*, 1995). Cultivars have been changed over the years to reflect advances in crop improvement (e.g., cereal cultivars with improved resistance to lodging and diseases or with increased harvest index).

The soil amendments included several combinations of the nutrients nitrogen, phosphorus, potassium and sulphur as well as lime and farmyard manure. Fertilizer application methods and rates have varied over the years. There are eight fertility treatments established across the various rotation phases. Three contrasting treatments were selected for this workshop: (i) nil, (ii) commercial fertilizers containing N, P, K, and S, and (iii) farmyard manure calculated to add N at rates similar to those added with commercial fertilizers.

N additions to the Classical Plots has not been constant over the years. N rates applied per rotation cycle varied on average as follows (Izaurralde *et al.*, 1995a): WOBHH fertilizer received N at 43 kg ha^{-1} from 1939 to 1979 and 175 kg ha^{-1} from 1980 onwards; WOBHH manure received N at 299 kg ha^{-1} from 1939 to 1979 and 175 kg ha^{-1} from 1980 onwards; WF fertilizer received N at 8 kg ha^{-1} from 1939 to 1979 and 91 kg ha^{-1} from 1980 onwards, WF manure received N at 305 kg ha^{-1} from 1939 to 1979 and 91 kg ha^{-1} from 1980 onwards. Prior to 1980, cattle manure was added in the fall once per rotation cycle; during the fallow phase in the WF rotation and after plowing down of the alfalfa/brome mixture in the WOBHH rotation. After 1980, the manure application in the latter rotation was split equally into two at the end of the oat crop and the alfalfa/brome mixture. Approximately 90% of all above ground plant biomass is harvested and removed from the plots every year as grain, straw or forage cuts.

Results and Discussion

Update of the Breton Plots Dataset

The completeness of the meteorological information has evolved over the years. From 1968 to 1973 the daily records only contain precipitation data. From 1974 to 1984 the records contain precipitation and average air temperature. The meteorological records have been fairly complete since 1985 thanks to the installation of an automated weather station. Weather data can now be obtained in real time.

Plant biomass data are complete. Estimated nitrogen removals were made from plant N concentration data. Amounts and dates of nutrient inputs are based on historical records. Dates and details on tillage practices were somewhat generalized for the years 1980 through 1994. Data on total N and organic C in surface soil begin in 1936 and sampling continues at between five to ten year intervals. For early sampling times, soil C is derived from total soil N. The dataset contains partial information on microbial biomass C and $\delta^{15}N$ but no information neither on $\delta^{13}C$ nor on $\delta^{14}C$.

Long-term yield trends

Long-term yield trends of the Breton Classical Plots were recently presented (Juma *et al.*, 1995) and compared to yield trends of other long-term experiments in Alberta (Izaurralde *et al.*, 1995b). A fundamental question we asked in these studies was whether yields were sustainable. Wheat yields in the wheat-fallow (WF) rotation with and without fertilizers NPKS (Figure 1) were not much different during the first 40 years but departed substantially after the fertilizer rate revisions made at the end of the 1970s.

The WF - Nil system could be described as a non-sustainable system because of lack of nutrient replenishment and surface cover during half the rotation cycle. Despite rather adverse management, however, yields did not decline over time. What is not known, however, is to what extent a likely declining yield trend caused by this non-sustainable management was compensated for by improved cultural practices and unaccounted inputs from the atmosphere. The addition of nutrients either via commercial fertilizers or manure had a sustained impact on yields of wheat grown after fallow and after forages. Further, the inclusion of forages added, undoubtedly, an important 'rotation effect' to the yields of annual cereals.

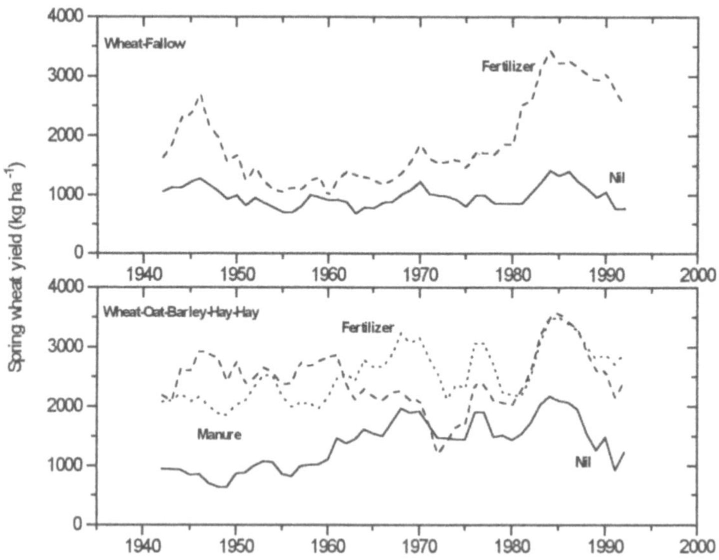

Figure 1 *Long term wheat yields of Wheat-Fallow and Wheat-Oats-Barley-Hay-Hay rotations with and without fertilizers or manure.*

'Sustainability' is a widely used word but a rather vague concept. We attempted to quantify the elements of sustainability by conducting a spatial analysis of the Breton Classical Plots yield data (Feng & Izaurralde, *in preparation*). Datasets were prepared for the three fertility treatments of the five-year rotation (WOBHH) and semivariances of dry matter yields calculated. Semivariograms[1] in Figure 2 suggest that the 'nil' and 'manure' treatments have induced a more prolonged 'rotation effect'[2] on dry matter yields than the 'fertilizer' treatment. 'Fertilizer' semivariances calculated at different lags[3] approach their general variance (σ^2~1,900,000 [kg ha^{-1}]^2) sooner (lag ~7 years) than the other two treatments (σ^2~900,000 [kg ha^{-1}]^2) at lag ~20 years for 'nil' and σ^2~2,100,000 [kg ha^{-1}]^2 at lag ~20 years for 'manure'). We hypothesize from these data that if we were to grow, for example, continuous barley on all these plots without any addition of fertilizers, the 'rotation effect' would be observed for a longer period on either 'nil' or 'manure' plots than on 'fertilizer' plots.

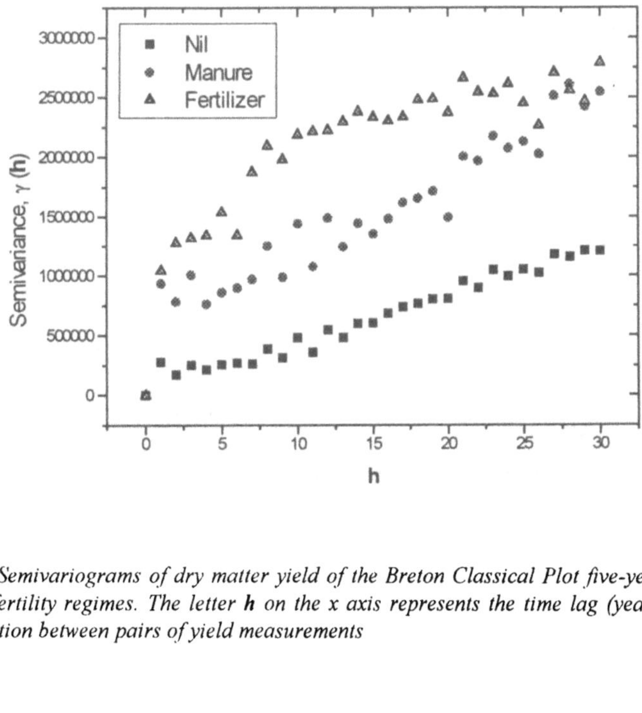

Figure 2 *Semivariograms of dry matter yield of the Breton Classical Plot five-year rotation under three fertility regimes. The letter* ***h*** *on the x axis represents the time lag (years) or distance of separation between pairs of yield measurements*

[1] A semivariogram is a geostatistical term that describes the spatial relation between the property studied and the temporal or spatial 'distance' or 'lag'.
[2] Rotation effect: the influence (e.g., nitrogen release from legume residues or an interuption to an increasing pest or disease problem associated with monoculture) of past crops grown on a field on the growth and development of crops grown in the present.
[3] Lag or **h**: is a vector of specified distance (or time in this case) and direction.

Soil organic matter

Soil organic matter is a key indicator of soil quality. Soil organic matter holds essential nutrients, promotes aggregation, acts as a chemical buffer, and provides protection against soil erosion. Both concentration and mass of soil organic matter can be altered by management. Most cultivated soils in Alberta have experienced substantial decline in organic matter content. The observed reductions have been attributed to the intensity and nature of tillage practices and to associated erosion losses. After some years of decline, soil organic matter tends to find new steady states whereby rates of organic matter decomposition approximately equal those of carbon additions via residues, roots, or manure. Soils, however, are dynamic natural bodies that respond to changes in management. Results from field studies have demonstrated that producers can increase soil organic matter provided they apply management principles that favor productivity and reduce erosion and/or decomposition losses.

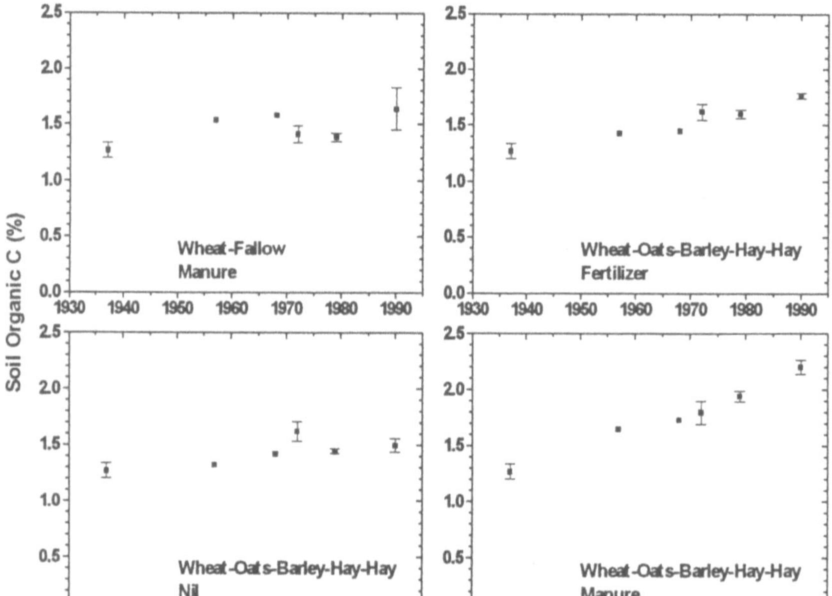

Figure 3 *Soil organic C on four treatments of Breton Classical Plots from 1936 to 1990. Vertical lines are one standard error each above and below mean values. Data for 1957 are from archives while for 1968 are from Khan (1969).*

Recently, historical sets of soil samples from the Breton Classical Plots were analyzed or re-analyzed for soil organic C using a LECO Carbon Determinator CR12 (LECO Corp., St. Joseph, Mich.) (Figure 3) (Izaurralde *et al.*, 1995c). Luvisolic soils brought into cultivation early in this century had rather low quantities of soil organic matter in their surface layers (soil organic C in 1936 in the surface 0.175 m was 1.27%). Therefore any management technique that favored crop productivity in these cool, subhumid environments brought slow but steady increases in soil organic matter. Specific management that favored soil organic matter accumulations in this environment included the addition of nutrients via manure and fertilizers and the inclusion of perennial forages in the crop rotation.

Soil microbial measurements

Microbial biomass C and water-soluble C were measured during two consecutive years (1981 - 1982) on six treatment combinations of the Breton Classical Plots (McGill *et al.*, 1986). The six treatments studied resulted from a factorial combination of the two crop rotations (WF and WOBHH) and three fertility treatments (Nil, Fertilizer NPKS, and Manure).

Average biomass C in the surface 5-cm depth was 46 mg 100 g^{-1} in 1981 and 35 mg 100 g^{-1} in 1982. These quantities corresponded to approximately 3% of the total soil organic C present in the first 5 cm soil depth.

On average, biomass C on plots in the WOBHH rotation was greater than in the WF rotation (Figure 4). As well, surface soil receiving manure consistently contained greater biomass C than soil receiving either no amendment (Nil) or complete fertilization (NPKS) (Figure 4). Estimates made over two years revealed microbial biomass in the WF rotation to turnover 1.5 - 2.0 faster than in the WOBHH rotation.

Recently, Wani *et al.* (1994) used treatments of the Breton Classical Plots and other long-term rotations at the Breton Plots to evaluate crop productivity and resulting soil quality under legume-based cropping systems. Surface soil (from 0 to 15 cm depth) collected in May 1990 from the WOBHH 'Nil' plots had 19 mg C 100 g^{-1} as microbial biomass and respired 6 mg C 100 g^{-1} over a 10-d incubation period. Barley productivity was low mainly due to the position of the barley crop within the crop sequence (third cereal crop after legume/grass mixture). Microbial measurements fell between those from an 8-yr rotation (63% of time occupied by legumes or legume/grass mixtures) on the high end, and those from a continuous barley grain system with annual addition of fertilizer N on the low end.

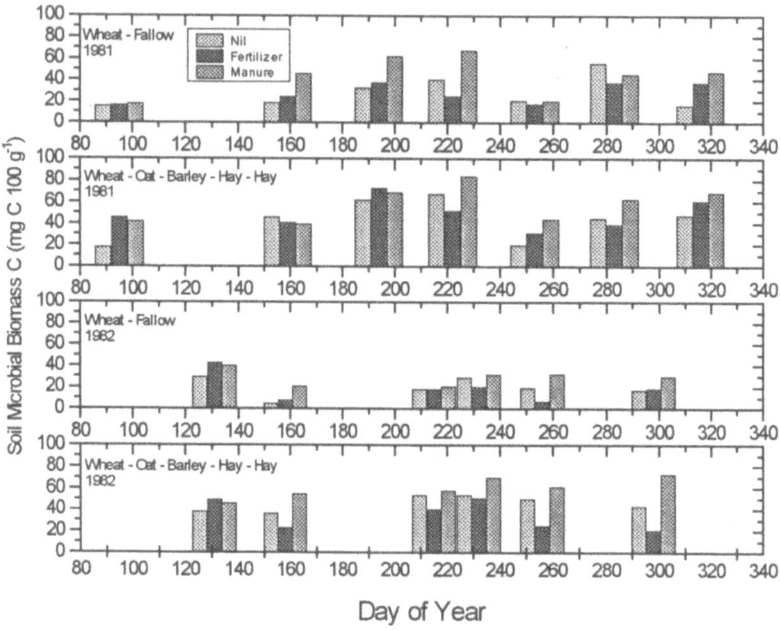

Figure 4 *Dynamics of soil microbial biomass C in six treatments of the Breton Classical Plots during 1981 and 1982.*

Nitrogen cycling and leaching

The extent of nitrate leaching in cultivated soils of Alberta is relatively unknown. Izaurralde *et al.* (1995a) studied the influence of long-term management at the Breton Classical Plots on nitrate leaching. Soil samples were taken in the fall of 1993 from the six treatments discussed in this paper in 0.3-m depth increments from 0 to 3.9 m and analyzed for NO_3-N. Nitrate accumulations below 0.9 m depth of the Breton Classical Plots ranged from 0 to 67 kg N ha^{-1} (Figure 5).

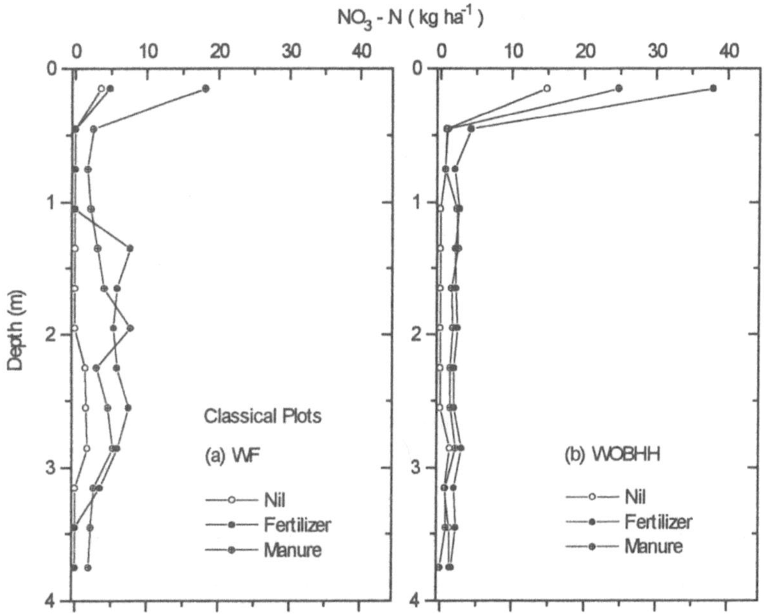

Figure 5 *NO_3-N distribution with soil depth in Classical Plots at Breton (Alberta): (a) WF: 2-yr crop rotation of wheat-fallow, and (b) WOBHH: 5-yr crop rotation of wheat-oat-barley-alfalfa/brome-alfalfa/brome.*

On average, the WF 'fertilizer' and 'manure' rotations had eight times more NO_3-N below 0.9 m depth than the WF nil rotation (39 vs. 5 kg ha^{-1} of NO_3-N, $p<0.06$). The fertilized plots in the WOBHH rotation contained more NO_3-N in the 0-3.9 m depth range than did 'nil' plots of the same rotation ($p<0.04$). Manure did not influence NO_3-N content in the WOBHH rotation irrespective of the depth range considered, although values for the manure treatment tended to exceed those of the control and were generally less than those of the fertlized system.

Although rotation had no significant effect on soil NO_3-N, nutrient-amended plots of the WF rotation tended to have more NO_3-N ($p<0.3$) below 0.9 m than similar plots of the WOBHH rotation. A more continuous plant cover and a more extensive soil exploration by roots of perennial forages may have prevented NO_3^- from accumulating below 0.9 m. By comparison, soil NO_3-N found in the rotation with forages was small compared to the N that cycled during the sixty years through the plant-soil system (Table 1).

Table 1 *N additions and removals on Classical Plots at Breton over 50 years (1940-1989).*[z]

	WF			WOBHH		
Treatment	Added	Removed	'Net' removal	Added	Removed	'Net' removal
			kg ha^{-1}			
Nil	0	773	773	0	1517	1517
Manure	1622	1603	-19	2801	3322	521
Fertilizer	625	1399	774	733	3357	2624

[z] **Source:** Izaurralde *et al.* (1995a).

Summary

- A comprehensive dataset of the Breton Classical Plots at Breton (Alberta, Canada) has been assembled and is now available for use in the testing of soil organic matter simulation models.

- Records of crop management, yields, and soil quality measurements have been invaluable for the understanding of agroecosystems functioning at high latitudes.

- Use of forages in rotation and addition of nutrients via commercial fertilizers or manure, alone or in combination, changed the carbon dynamics of a Gray Luvisol with originally low soil organic carbon content.

References

Feng, Y., & Izaurralde, R.C. Temporal analysis of crop yields of the Breton Classical Plots. (*in preparation*).

Izaurralde, R.C., Feng, Y., Robertson, J.A., & McGill, W.B, & Juma, N.G. (1995a) Long-term influence of cropping systems, tillage methods, and N sources on nitrate leaching. *Canadian Journal of Soil Scence* **75**: (*in press*).

Izaurralde, R.C., Arshad, M.A., Bremer, E., Janzen, H.H., Jensen, T.L., McKenzie, R.H., Robertson, J.A., & VanderPluym, H.P. (1995b) In search of sustainable cropping systems: crop rotation and tillage studies in Alberta. In: *31th Annual Alberta Soil Science Workshop Proceeding*. March. 13-15 1995. Grande Prairie, Alberta: 81-96.

Izaurralde, R.C., McGill, W.B., Robertson, J.A., & Juma, N.G. (1995c) Carbon balance of the Breton Classical Plots over 55 years. *Canadian Soil Science Society Abstracts*: 3.

Juma, N.G., Izaurralde, R.C., Robertson, J.A., & McGill, W.B. (1995) Crop yield and soil organic matter over 60 years in a Typic Cryoboralf at Breton, Alberta. In: *Soil organic matter in temperate agroecosystems: long-term experiments in North America.* (eds: E.A. Paul, K. Paustian, E.T. Elliott, & C.V. Cole, Lewis Publishers, CRC Press. Boca Raton, FLA: (*in press*).

Khan, S.U. (1969) Some carbohydrate fractions of a Gray Wooded soil as influenced by cropping systems and fertilizers. *Canadian Journal of Soil Science* **49**: 219-224.

McGill, W.B., Cannon, K.R., Robertson, J.A., & Cook, F.D. (1986) Dynamics of soil microbial biomass and water-soluble organic C in Breton L after 50 years of cropping to two rotations. *Canadian Journal of Soil Science* **66**: 1-19.
Robertson, J.A., & McGill, W.B. (1983) New directions for the Breton Plots. *University of Alberta Agriculture and Forestry Bulletin* **6**: 36-41.
Wani, S.P., McGill, W.B., Haugen-Kozyra, K.L., Robertson, J.A., & Thurston, J.J. (1994) Improved soil quality and barley yields with fababeans, manure, forages and crop rotations on a Gray Luvisol. *Canadian Journal of Soil Science* **74**: 75-84.
Wyatt, F.A. (1936) Fertilizers for the black and grey soils of central Alberta. *Scientific Agriculture* **16**: 238-240.

Acknowledgments

We sincerely thank KR Cannon, C Nguyen, and MC Quiroga Jakas for helping to prepare the electronic datasets. The Breton Plots are supported by the University of Alberta, the Breton Plots Endowment Fund, and the Canada-Alberta Environmentally Sustainable Agriculture (CAESA) Agreement.

Long-term field experiment Praha - Ruzyně, Czech Republic

J. Klír

Research Institute of Crop Production

Praha - Ruzyně,

Czech Republic

Introduction

The oldest field experiments in the Czech Republic were initiated at five different sites (Praha - Ruzyně, Čáslav, Lukavec, Ivanovice and Pohořelice) in 1955. The Plant Nutrition and Fertilization Experiment at Praha - Ruzyně was started by Dr. Škarda on nine fields at Research Institute of Crop Production. Now, in 1995 there are still five main fields with different organic and mineral fertilization on this area.

Site description

The fields, where the experiments are situated have been under ordinary arable cultivation for several hundred years.

Location

The Plant Nutrition and Fertilization Experiment Praha-Ruzyně is located at the Research Institute of Crop Production in suburban Ruzyně, approximately 10 km west of the centre of Praha in the central region of the Czech Republic (50°05′N, 14°20′E; elevation: 340 m above sea level).

NATO ASI Series, Vol. I 38
Evaluation of Soil Organic Matter Models
Edited by David S. Powlson, Pete Smith and Jo U. Smith

Weather

The annual average precipitation is 464 mm (ranged from 255 mm to 701 mm) and the annual average air temperature is 8.0°C (from 6.4°C to 9.7°C) over the course of the experiment. Daily meteorological data are available for 1956 to present and are in electronic format on disk from 1961 to present, with exception of soil temperature (1976 to present) and radiation (1984 to present). The meteorological station is at Praha - Ruzyně airport and is 3 km from the experimental fields.

Soil characteristics

The soil of the experimental field is brown earth and the parent material is loess mixed with highly weathered chalk. The ground water level is 20 m below field surface. The upper 30 cm contains 27 % clay (particles under 1 μm of diameter), increasing to 40 % clay in the soil layer 30 - 40 cm, and to 49 % clay in soil layer 40 - 50 cm. Soil pH is about 6.5. Soil samples taken from the site in 1955 indicate an initial soil organic carbon value of 1.17 % in the top 20 cm.

Treatments and crops

The experiment consists of five main fields (I, II, III, IV and B). Each field is divided to 96 plots (24 variants x 4 replications), each plot being 12 x 12 m. Three factors are considered in the experiment: crop rotation, organic fertilization and mineral fertilization.

Crop rotations

There are three type of cropping system in this experiment:

1. classical crop rotation (45 % cereals, 33 % root crops, 22 % legumes) 9 years: lucerne, lucerne, winter wheat, sugar beet, spring barley, potato, winter wheat, sugar beet, spring barley with lucerne underseeding (fields II, III, IV)
2. cereal crop rotation (67 % cereals, 11 % root crops, 22 % legumes) 9 years: lucerne, lucerne, winter wheat, winter wheat, spring barley, potato, winter wheat, winter wheat, spring barley with lucerne underseeding (field I)

3. crop rotation without legumes (50 % cereals, 50 % root crops) 2 years: spring wheat, sugar beet (field B)

Organic fertilization

There are four different manuring treatments on each field:

1. no manure, 2. straw + N, 3. straw + pig slurry, 4. straw (field I)
1. no manure, 2. farmyard manure, 3. dung-water + straw, 4. poultry slurry (field II)
1. no manure, 2. farmyard manure, 3. pig slurry + straw, 4. pig slurry (field III)
1. no manure, 2. farmyard manure, 3. cattle slurry + straw, 4. cattle slurry (field IV)
1. no manure, 2. farmyard manure, 3. cattle slurry + straw, 4. compost (field B)

The quantities of organic fertilizer added are as follows:

farmyard manure: 21 t ha^{-1} for sugar beet, 15 t ha^{-1} for potato
cattle slurry: 60 t ha^{-1} for sugar beet, 43 t ha^{-1} for potato
pig slurry: 68 t ha^{-1} for sugar beet, 49 t ha^{-1} for potato, 20 t ha^{-1} for cereals (field I only)
poultry slurry: 45 t ha^{-1} for sugar beet, 32 t ha^{-1} for potato
dung-water: 40 t ha^{-1} for sugar beet, 30 t ha^{-1} for potato
straw: according to the yield of straw after harvesting the plot

Mineral fertilization

All organic fertilization treatments are subdivided into six mineral fertilization treatments: without mineral fertilizers, P_2K_2, $N_1P_1K_1$, $N_2P_2K_2$, $N_3P_3K_3$, $N_4P_2K_2$.

The average nutrients doses are given in Table 1.

Table 1. *Average nutrient treatments (kg ha^{-1} y^{-1})*

	N1	N2	N3	N4	P1	P2	P3	K1	K2	K3
field B	40	70	100	130	9	18	27	25	50	75
field I	40	53	66	80	25	35	-	88	120	-
fields II-IV	40	63	67	91	24	31	-	106	142	-

Soil sampling and analysis

The parameters measured annually by the Research Institute of Crop Production are:

- Total carbon (wet combustion) since 1972
- Total soil nitrogen (Kjeldahl method) since 1972
- Available phosphorus (Egner)
- Available potassium (Schachtschabel)
- Available magnesium (Schachtschabel)

Effect of fertilization on crop yield

The effects of fertilization on the yields of sugar beet and spring wheat are shown in figures 1 and 2.

Figure 1 *Long-term effect of different fertilization on the yields of sugar beet (Praha-Ruzyně, field B, 1966-1993)*

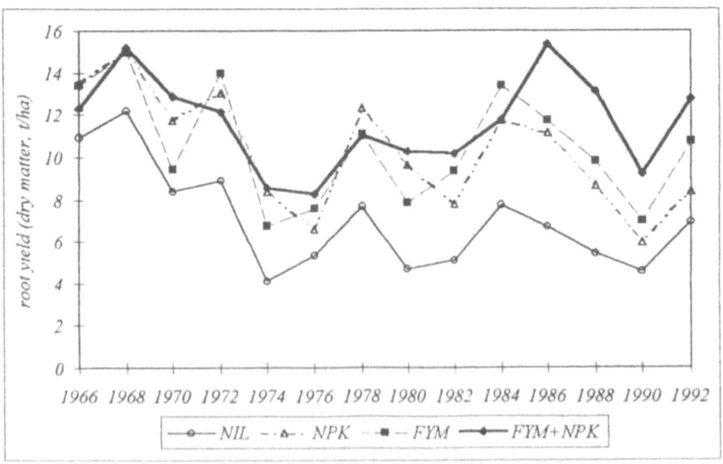

Figure 2 *Long-term effect of different fertilization on the yields of spring wheat (Praha-Ruzyně, field B, 1966-1993)*

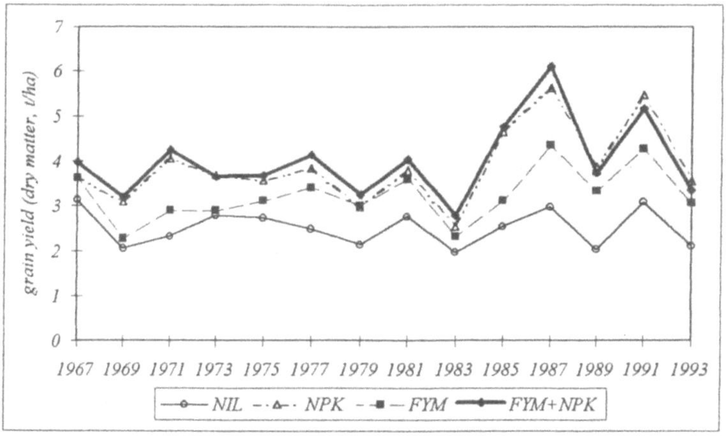

Trends in soil organic matter and carbon

Total carbon increased with the application of inorganic N (NPK), farmyard manure (FYM) and FYM + NPK. For instance at the experimental field B, analysis from 1994 showed that total carbon was 1.19 % C in the plot that had received no organic or inorganic fertilizer (NIL plot), 1.31 % C in the NPK plot, 1.47 % C in the FYM plot and 1.48 % C in the FYM + NPK plot.

Main changes over the course of the experiment

Two major changes have occurred since 1955. In 1965 the crop rotation on fields I and B was changed from 45 % cereals, 33 % root crops, 22 % fodder crops to 67 % cereals, 11 % root crops, 22 % fodder crops (field I) and to 50 % cereals, 50 % root crops (field B). In 1972 were changed compost application (fields II, III, IV), farmyard manure application (field I) and straw + nitrogen application (field B) to slurries application with or without straw.

References

Klír, J. (1994) Fertilizer nitrogen efficiency and its balance in crop rotation with legumes. *Rostlinná Výroba* ***40:*** 323-331

Klír, J., Škarda, M. & Marková, J. (1992) Effect of organic fertilization on the nutrients balances in long-term field experiment in Praha - Ruzyně. In: *Proceedings of Symposium "Long-term Field Experiments and Nutrient Dynamics"*. Bad Lauchstädt: 73-77

Klír, J., Kubát, J. & Pova, D. (1995) Stickstoffbilanzen der Dauerfeldversuche in Prag. *Mitteilgn. Dtsch. Bodenkundl. Gesellsch.* **76**: 831-834.

Škarda, M. & Marková, J. (1986) Long-term effect of straw and sugar beet thrash application on mineral fertilizing efficiency. *Rostlinná Výroba* **32:** 1103-1114.

The Static Experiment Bad Lauchstädt, Germany

Martin Körschens & Annett Müller,
Centre for Environmental Research Ltd.,
Hallesche Straße 44,
D-06246 Bad Lauchstädt,
Germany

The Location

The soil of the experimental field at Bad Lauchstädt is a loess black earth and the parent material is loess over boulder clay. The ground water level is situated at 12 m; however, water-bearing strata occur between 4 to 6 m depth, depending on the quantity and distribution of precipitation over long periods (Table 1).

Table 1 *Soil properties - Static Experiment Bad Lauchstädt*

Soil properties (Top soil)			Experimental mean value
Particle size distribution (%)			
Clay		< 2.0 µm	21.1
Fine silt	6.3 ...	2.0 µm	7.0
Silt	20.0 ...	6.3 µm	15.9
Coarse silt	63.0 ...	20.0 µm	44.8
Fine sand	200.0 ...	63.0 µm	8.6
Sand	630.0 ...	200.0 µm	2.1
Coarse sand	2000.0 ...	630.0 µm	0.5
Dry matter density (g cm^{-3})			2.56
Bulk density (g cm^{-3})			1.35
Hygroscopicity (mass % water in absolute dry soil)			4.62
Field capacity (mass %)			24.0
Sorption capacity (mval 100 g^{-1} soil)			29.4

NATO ASI Series, Vol. I 38
Evaluation of Soil Organic Matter Models
Edited by David S. Powlson, Pete Smith and Jo U. Smith

Weather Conditions

Bad Lauchstädt lies in the Central German Dry Zone with a long-term mean precipitation of only 483 mm and has an annual mean temperature of 8.7 °C (Fig. 1).

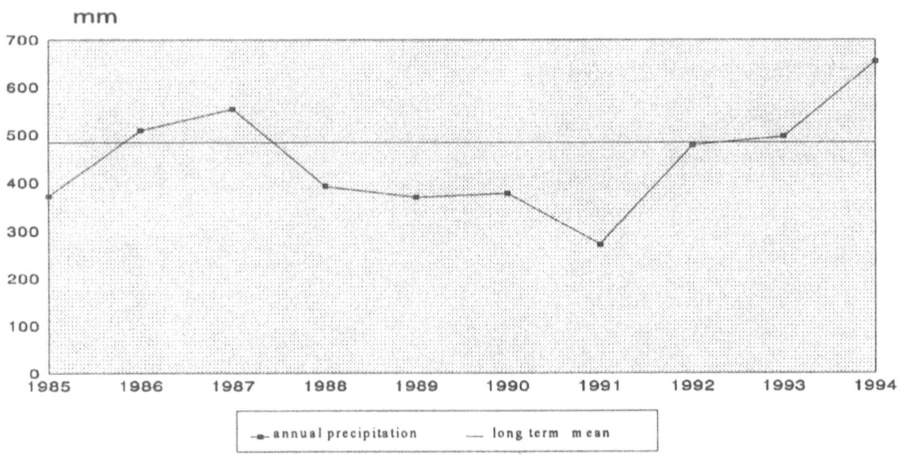

Figure 1 *Annual precipitation of the last decade (1985-1994) in comparision to the long term mean*

The Experiment

The Static Fertilisation Experiment at Bad Lauchstädt was initiated in 1902 by Schneidewind and Gröbler. They started with two manurial treatments: "without farmyard manure" (treatment a_3) and "20 t ha^{-1} farmyard manure every second year" (treatment a_2). Both treatments were subdivided into six mineral fertilisation treatments: NPK, NP, NK, N, PK and no fertilizer. In 1906 the experiment was extended to include another plot: "30 t ha^{-1} farmyard manure every second year" (treatment a_1); see Figures 2 and 3.

Originally the treatment covered four fields labelled I, II, III and IV and the crop rotation was, and continues to be sugar beet, spring barley, potatoes and winter wheat. In 1924, the four fields were each bisected and the following treatments imposed:

Field	Northern half		Southern half	
I	Strip 1	lime every 4th year	Strip 2	no lime
II	Strip 3	$(NH_4)_2SO_4$	Strip 4	$CaNO_3$
III	Strip 5	$(NH_4)_2SO_4$	Strip 6	$NaNO_3$
IV	Strip 7	no legumes	Strip 8	legumes every 4th year

The application of different fertilisers to strips 3, 4, 5 and 6 was stopped in 1977 and strips 4 and 5 have since been used to investigate effect of changes in total carbon content due to the previous fertiliser treatments. A range of N dressings, with and without farmyard manure have been applied to these plots.

Figure 2 *Experimental design and Experimental object*

Treatments and treatment levels	
A	**farmyard manure (3 levels)**
a_1	30 t ha^{-1} farmyard manure to root crops
a_2	20 t ha^{-1} farmyard manure to root crops
a_3	no farmyard manure
B	**mineral fertilization with N, P, K (6 levels)**
b_1	NPK
b_2	NP
b_3	NK
b_4	N
b_5	PK
b_6	0
C Field I liming	strip 1 with lime
	strip 2 without lime
Field IV legumes	strip 7 normal crop rotation
	strip 8 25 % legumes

Experimental object:	The 4 crops follow each other on the strips 1, 2, 3 , 6 and 7. On strip 8 alfalfa is cultivated for 2 years every 6 years
Layout method:	**systematic design with alternating land use**
number of plots:	**432**
size of each plot:	10 • 6,3 m = 63 m^2

Figure 3 *Scheme of the Static Fertilisation Experiment Bad Lauchstädt*

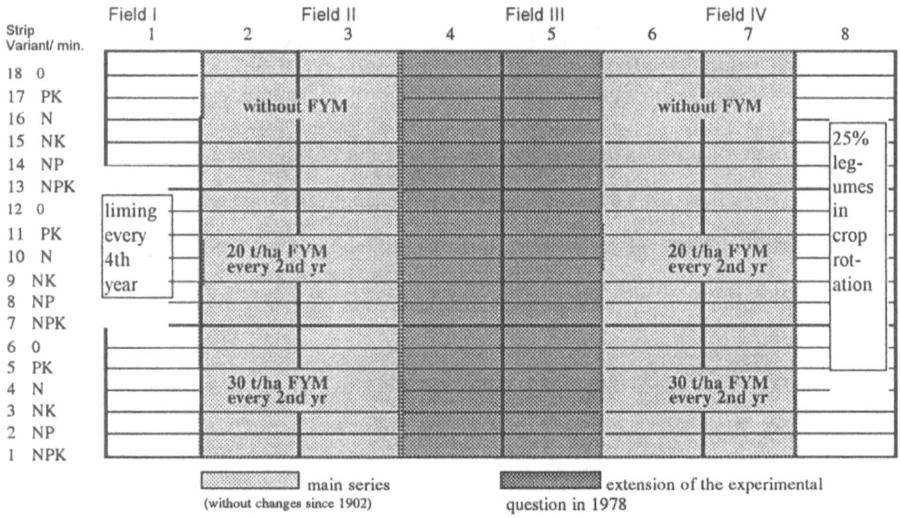

Investigations and Datasets

- The yields of all treatments and all crops since 1902 have been collected in a database.
- Daily weatherdata are available since 1956.
- The N content of main products and by-products and the uptake of nitrogen by the plants have been monitored since 1959.
- The soil carbon and nitrogen content, the P, K and Mg content and the soil pH value are determined every year on the main treatments and every fourth year on all treatments.
- Changes in physical, biological and chemical soil properties such as bulk density, different carbon and nitrogen compounds, biomass and others are monitored with respect to time.

Selected Results

Yields

The average of the yields of all crops of the rotation 1991 to 1994 are presented in Table 2. The total amount is affected by the dry weather conditions in the period from 1988 to 1992. The winter wheat yield of the nil plot is relatively high at about 5 t ha^{-1}; this is similar to the mean of the last 15 years and is caused by the N from the atmosphere. The yield differences between the variants 1-10 and 13 are small. This is because the 20 or 30 t ha^{-1} farmyard manure applied every second year supplies sufficient P and K for maximum crop growth in the absence of mineral P and K. With 30 t ha^{-1} farmyard manure, the nitrogen supply for the winter wheat also lies within the optimum range.

Yields of spring barley are also similar within treatments 1-10 and 13, although the farmyard manure variants without mineral fertilisation (treatments 5 and 6) show N deficiency, which leads to a clear decline in yield. The reason is that the largest part of the N reserve in the soil was taken up by the previous sugar beet crop. In contrast to winter wheat, yields of spring barley on the "no farmyard manure" treatment are consistently lower than in the two treatments with farmyard manure.

There are large differences in potato yield between treatments. Yield with mineral fertilisers alone is on average 38 t ha^{-1} lower than the treatments with farmyard manure plus mineral fertilisers. Large declines in yield are also found at 30 t ha^{-1} farmyard manure without mineral nitrogen. In contrast to cereals, the potato yields are very sensitive to K - and / or P - deficiency.

Sugar beet yields on the nil plot (treatment 18) are, like winter wheat, relatively high. Due to the long vegetation period and the deep rooting both crops exploit both the available N reserves and the N inputs from the atmosphere much better than potatoes and spring barley. For the same reason , there are considerable yield differences between inorganic N rates and a clear positive effect of the farmyard manure. However, sugar yield does not follow this trend as a high N supply decreases sugar content. In the long term there are no differences in the sugar yields between the variants 1, 7 and 13 , as with the winter wheat yields.

Table 2 *Yields (dt/ha) of strip 2, 3, 6 and 7 average of the last rotation*

Variant	wheat	spring barley	potatoes	sugar beets
		$30\ t\ ha^{-1}$ farmyard manure every second year		
1. NPK	69.4	55.1	419	596
2. NP	71.0	55.0	402	592
3. NK	70.9	53.9	418	578
4. N	72.3	54.2	413	574
5. PK	71.6	46.9	317	479
6. Zero	72.7	46.5	308	478
		$20\ t\ ha^{-1}$ farmyard manure every second year		
7. NPK	71.8	56.0	407	553
8. NP	73.7	55.7	389	533
9. NK	72.4	56.3	374	561
10. N	73.7	56.3	352	537
11. PK	62.9	43.1	262	391
12. Zero	64.1	43.3	230	376
		no farmyard manure		
13. NPK	71.2	51.0	380	520
14. NP	69.7	48.0	215	428
15. NK	60.2	37.3	214	342
16. N	59.9	35.1	194	321
17. PK	43.5	25.8	145	259
18. Zero	50.4	22.6	76	238

Carbon- and Nitrogen Content in the topsoil

The wide range of fertilizer treatments since 1902 has led to a great difference in the total carbon (C_t) and total nitrogen (N_t) contents of the plots, especially between the highest fertilization level and the "Nil" treatment.

Since a steady state was reached in approximately 1930, the range of the C_t-contents first increased slightly, then from 1957, remained constant at 0.65 to 0.66 %. The N_t-contents show the same tendency. The results from the years 1972 to 1992 (Table 3) allow the following statements:

- a differentiation of 0.66 % C_t and 0.059 % N_t was caused by the different fertilization treatments

– 20 t ha^{-1} FYM + NPK have had virtually the same effect on C_t and N_t as 30 t ha^{-1} FYM

– The NPK-fertilization has increased soil C and N content at all organic manuring levels amounting to:

without FYM	0.19 % C_t and 0.018 % N_t
with 20 t ha^{-1} FYM	0.11 % C_t and 0.013 % N_t
with 30 t ha^{-1} FYM	0.15 % C_t and 0.014 % N_t

– at all FYM-levels without mineral N ("PK", and "Nil") clearly lower C_t- and N_t-contents were found compared to treatments with mineral N

- 20 t ha^{-1} FYM increased the C_t-content, compared to "without FYM" by 0.39 % and the N_t- content by 0.032, meaned over all mineral levels. 30 t ha^{-1} FYM increased C_t by 0.13 and N_t by 0.012 % compared to 20 t ha^{-1}.

Table 3 *C_t- and N_t-content as well as C/N-ratio of the main series on average of the years 1972 - 1992*

30 t ha^{-1} Farmyard manure				20 t ha^{-1} Farmyard manure				no Farmyard manure			
	C_t %	N_t %	C/N		C_t %	N_t %	C/N		C_t %	N_t %	C/N
NPK	2.33	0.194	12.0	NPK	2.17	0.180	12.1	NPK	1.86	0.153	12.2
NP	2.29	0.192	11.9	NP	2.15	0.179	12.0	NP	1.81	0.148	12.2
NK	2.30	0.188	12.2	NK	2.16	0.179	12.1	NK	1.75	0.144	12.2
N	2.25	0.190	11.8	N	2.14	0.179	12.0	N	1.71	0.145	11.8
PK	2.21	0.183	12.1	PK	2.09	0.171	12.2	PK	1.65	0.136	12.1
Zero	2.18	0.180	12.1	Zero	2.06	0.167	12.3	Zero	1.67	0.135	12.4
mean	2.26	0.188	12.0	mean	2.13	0.176	12.1	mean	1.74	0.144	12.2

Nitrogen Balances

For calculation of N balances we can assume that in this experiment the soil has reached steady-state conditions; evidence for this is given elsewhere (see Körschens, 1994). On this basis we can establish a nitrogen balance by comparing fertiliser input with crop uptake, while the uptake in the zero variant is regarded as nitrogen from the atmosphere. For the period of 1983 to 1994 the N balance indicated an input of 56 kg ha^{-1} y^{-1} from the atmosphere. This large input must be included in N balance calculations. Direct measurements of N deposition on this site show comparable values (see Körschens, 1994).

The nitrogen balance, however, shows that for all variants N uptake is larger than N fertilisation, with the exclusive mineral fertiliser variant showing the highest efficiency. These results must be interpreted with regard to the locational factors and cannot be transfered to other locations.

References

Körschens, M. & Müller, A. (1993) Yields, uptake of nutrients and soil properties in the Static Fertilization Experiment Bad Lauchstädt. *Proceedings of the International Symposium "Long-term static fertilizer experiments", June 15.-18., 1993, Warszawa-Krakow, Poland*: 119-137.

Körschens, M. (ed.) (1994) Der Statische Düngungsversuch Bad Lauchstädt nach 90 Jahren.- Teubner-Verlag Stuttgart, Leipzig , 179pp.

The Park Grass Experiment, 1856-1995

P.R. Poulton
Soil Science Department
IACR-Rothamsted
Harpenden, Herts., AL5 2JQ
U.K.

Introduction

The Park Grass experiment at Rothamsted, UK, was started in 1856 and was intended to complement experiments started a few years earlier whick looked at the nutritional requirements of legumes, cereals and root crops. It was soon realised that the treatments they imposed had a dramatic effect on plant species composition and diversity which was of greater interest to the botanist than to the farmer (Lawes & Gilbert, 1880). Today, 140 years later, it is regarded as the foremost long-term ecological study in the world (Tilman *et al.*, 1994) and continues to be of particular interest to botanists and environmentalists. It is an ideal site on which to study SOM changes under long established grassland conditions.

Site, treatment and harvest details

The experiment was started on a level site which had been in grazed grassland for several centuries. The soil is a silty clay loam (Aquic Paleudalf, USDA; Chromic Luvisol, FAO) with about 23% clay in the top 20 cm (Avery & Catt 1995). The climate is cool, temperate with a long-term average rainfall of 728 mm per year. Meteorological data is available from the start of the experiment. Thirteen plots were laid out initially, seven more were added later; they ranged in size from 0.05 to 0.2 ha. Plots received either no treatment (unmanured), farmyard manure (FYM), or N, P or K, either singly or in various combinations. N was applied either as sodium nitrate or ammonium sulphate at rates up to

NATO ASI Series, Vol. I 38
Evaluation of Soil Organic Matter Models
Edited by David S. Powlson, Pete Smith and Jo U. Smith

144 kg N ha^{-1}. On some plots the treatment has changed, e.g. the original FYM treatments were stopped after 8 years because of the accumulation of undecomposed material on the surface.

A test of liming was carried out in the 1880s and a regular liming scheme was started on half-plots on some treatments in 1903. Most plots were divided again in 1965, creating four sub-plots. The present aim is to maintain three subplots at pH 7, 6 and 5 whilst the fourth receives no lime.

The treatments being considered for this modelling exercise are the unlimed areas of:

Plot 3: No treatment since 1856.

Plot 14: N_2*PKNaMg annually since 1858(1)

Plot 13: N_2PKNa Mg annually 1856-1904 plus cereal straw till 1897(2)
Farmyard manure, fishmeal since 1905(3)
(4-yr cycle; farmyard manure in yr 1, fishmeal in yr 3).

(1) N_2*, 96 kg N ha^{-1} as sodium nitrate

(2) N_2, 96 kg N ha^{-1}as ammonium sulphate. Straw, 2240 kg ha^{-1} ≡ 11 kg N ha^{-1}

(3) Farmyard manure, 35 t ha^{-1}, c. 230 kg N ha^{-1}, every 4 years. Fishmeal, dressing ≡ 63 kg N ha^{-1}, every 4 years.

Treatments are given in detail by Warren & Johnston (1964), Thurston *et al.* (1976) and in annual publications "Yields of the Field Experiments, 1994" (Rothamsted Experimental Station, 1995).

In June each year the herbage is cut and made into hay. Until 1959 samples of hay were taken for determination of yield, chemical analysis and archiving. Since 1960, part of each plot has been cut by a flail harvester, the produce weighed green and yields, unaffected by hay-making losses, recorded; this sample is also used for chemical analysis and archiving. The rest of the plot is then cut and made into hay as usual. Published yields since 1960 are not directly comparable to earlier yields, representing as they do total production rather than what is *removed* from the plot after hay making losses. Since 1875 a second cut has been taken in late autumn, weighed green and removed. Before 1875, regrowth between haymaking and late autumn was often grazed by sheep, penned on individual plots (Jenkinson *et al.*, 1994).

Inputs

There have been no measurements of carbon inputs to the site. However, some estimates can be made. The 2240 kg ha^{-1} of straw applied to plot 13 each year from 1856 to 1897 would have contained c. 900 kg C ha^{-1} yr^{-1}. The 35 t ha^{-1} fresh FYM every 4 years since 1905 has been calculated to contain, on average, 3 t C ha^{-1} per dressing and the fishmeal, to supply 63 kg N ha^{-1} every 4 years, contained 360 kg C ha^{-1} per dressing.

Other carbon inputs will be from above ground debris, dead roots and root exudates. As mentioned earlier the method of measuring the yield of the 1st cut has changed. Assuming that there has been no long-term change in yield (e.g. because of enhanced CO_2) as was concluded by Jenkinson *et al* (1994) then the difference in the reported yield pre- and post-1960, 0.27 and 0.65 t ha^{-1} DM on plot 3 and plot 14 respectively is the amount of herbage which would have remained on the plot as a result of the hay making process. Because most of the plot is still made into hay this difference is also applicable post-1960. (N.B. Based on this much simplified comparison, published yields since 1960 should be decreased by 20% on plot 3 and 12% on plot 14 to make them comparable to pre-1960 data. The method used for the second cut has not changed and yields should not be adjusted). The input from dead roots must also be taken into account. Mean values for root measurements taken in 1876 and again in 1982 show a greater root mass on the unmanured plot (plot 3) than the plot receiving NPK (plot 14), 9.7 and 4.5 t ha^{-1} DM respectively. Assuming 20% of roots die each year the following (minimum) estimate for carbon input can be made (Table 1).

Table 1. *Estimated annual carbon input to plots 3 and 14 of the Park Grass Experiment*

	Input during hay making t ha^{-1}	From dead roots t ha^{-1}	Total dm t ha^{-1}	Total C[(1)] t ha^{-1}
Plot 3	0.27	1.94	2.21	0.88
Plot 14	0.65	0.90	1.55	0.62

(1) Assuming 40% C in dm.

These values take no account of inputs from above ground debris during the rest of the year, e.g. as seed or senescing herbage or the fact that some of the debris may be lost gaseously. It also takes no account of inputs from root exudates. Thus the estimate of input during hay making (first column of Table 1) will be an overestimate, as gaseous losses are ignored but, ignoring root exudates etc will lead to an underestimate of C inputs.

Inputs of nitrogen in rainfall have been measured for many years. Present values, c. 12 kg N ha^{-1} yr^{-1}, are twice the amount measured 100 years ago. *Total N* deposition (both wet and dry) is currently c. 40 kg N ha^{-1} yr^{-1} (Goulding and Poulton, 1992) but no early measurements were made. For this modelling exercise we have assumed that total N deposition was 20 kg N ha^{-1} yr^{-1} until 1900 and 40 kg N ha^{-1} yr^{-1} since.

Soil sampling and analysis

A few plots were sampled in 1870 but the first major soil sampling was in 1876, 20 years after the experiment started. Soils were taken with an open-ended iron box 12" x 12" x 9" deep (30.5 x 30.5 x 22.9 cm). Obviously this method of sampling was fairly disruptive on grassland and only 3 holes were taken on each main plot. Since 1932 samples have been taken by straight-sided, semi-cylindrical auger, bulking a large number of cores for each plot (or subplot). All samples (except 1932) have been taken to a depth of 23 cm, either as one depth or divided into several layers. Using %N and soil weight data for the various horizons sampled in 1959, the 0-20 cm, 1932 data published by Richardson (1938) has been corrected to make it comparable to the other, 0-23 cm, samplings. Samples have been taken to a greater depth on several occasions.

Samples have been analysed for carbon by a variety of methods. These have included early "chromate" and "combustion" methods and Walkley and Black (1934). More recently, soils have been analysed for organic carbon by the Tinsley (1950) method. Where necessary, a correction factor has been used to make earlier analyses comparable to those obtained by the Tinsley method, e.g. Walkley and Black values are multiplied by a factor of 1.3. Most samples have been analysed for nitrogen by the Kjeldahl method (Bremner, 1965). Data for the 1876 samples, analysed by the soda lime method which is known to give lower values, has been corrected accordingly.

Table 2 shows organic C in the profile on the unmanured plot in 1876. The distribution within the profile is probably typical of the site when the experiment started in 1856.

Table 2. *Organic C in soil, 1876, plot 3 (unmanured) of the the Park Grass Experiment.*

Depth, cm	0-23	23-46	46-69	69-91	91-114	114-137
%C	3.38	0.76	0.37	0.29	0.23	0.21

One of the biggest problems in trying to follow changes in soil properties over an extended time-scale is in ensuring that samples taken on one occasion are comparable to those taken 10, 50 or 100 years earlier. The method of sampling may change or the depth sampled may be different, as described above. In many cases, particularly with older experiments, the original plots have been divided, e.g. to test new treatments. On Park Grass, splitting the plots to test lime means that whereas samples taken in 1876 and 1886 come from whole plots, those taken in 1913, 1932 and 1959 come from unlimed ½-plots, and those taken in 1966, 1972 and 1985 come from unlimed ¼-plots. In 1990 plot 14 was further divided and the previous N dressing withheld on one half of the plot. The original treatment is therefore continued on only ⅛ of the original plot (Figure 1).

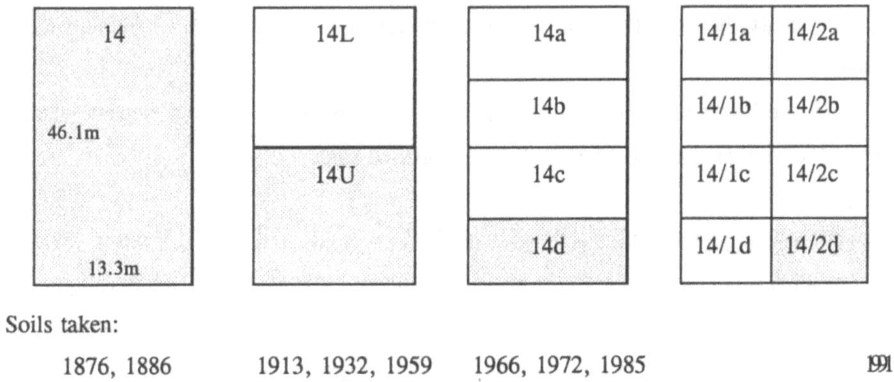

Figure 1. *This figure shows how plot 14 has been divided. The shaded area is that part of the plot on which the original treatment has continued.*

If soil in the original plot was *uniform* then measurements made on later samples will reflect any long-term change (or lack of change) on that treatment. If soil variability in the original plot was *great* then later measurements may reflect that variability rather than any other factor. There is some evidence to suggest that the plots were reasonably uniform. In 1876 the samples were taken from 3 points within each plot; these were analysed separately. On the control treatment (plot 3) total N values ranged from 0.248 - 0.260 %N. On most other plots the range was similar. Other soil measurements, e.g. pH (Table 3), Olsen P, exchangeable cations, show gradual change or no change (depending on the treatment) between one sampling and the next, rather than any marked discontinuity. Again this suggests that the original plots were fairly uniform.

Table 3[(1)]. *pH (in H_2O) on the unlimed part of selected treatments, 0-23 cm of the Park Grass Experiment.*

		Year				
Plot	Treatment	1876	1923[(2)]	1959	1984	1991
3	Control	5.4	5.7	5.2	5.0	4.8
14	N_2*PKNaMg	5.9	6.4	6.0	5.8	5.8
13[(3)]	FYM/Fishmeal	4.5	4.6	4.7	4.7	4.6

(1) Data from Johnston *et al* (1986) and P.R. Poulton (pers. comm.)

(2) The increase between 1876 and 1923 is a result of the test of lime on whole plots in the 1880s.

(3) Note, the lower pH values on plot 13 are a result of the ammonium sulphate applied before 1905 rather than the FYM or fishmeal applied since.

Organic carbon and total nitrogen data for the 0-23 cm depth are shown in Table 4. On plot 3 (unmanured) and plot 14 (N_2*PKNaMg) there has been little change in the organic matter content over 120 years. On plot 13 there was a decline between 1876 and 1932 and then an increase over the next 60 years. This plot received N_2PKNaMg until 1904 together with straw until 1897 and then farmyard manure and fishmeal on a 4-year cycle since 1905. Whether the decline and subsequent increase are directly related to the change in treatment or to soil variability is unclear.

Table 4. *Organic carbon and total nitrogen, 0-23 cm*[1] *on selected treatments of the Park Grass Experiment*

	Plot					
	3		14		13	
Year	%C	%N	%C	%N	%C	%N
1876[2]	3.38	0.257	3.12	0.253	3.51	0.265
1886[3]	3.15	0.272	-	-	-	-
1913[4]	-	0.275	-	0.269	-	-
1932[5]	3.42	0.274	2.95	0.252	2.82	0.238
1959[6]	3.29	0.268	2.88	0.266	3.06	0.250
1966[3]	3.09	0.256	-	-	-	-
1972[3]	3.09	0.247	-	-	-	-
1985[3]	3.12	0.258	-	-	-	-
1991[2]	3.07	-	2.87	-	3.33	-

(1) All data is equivalent to organic carbon by Tinsley, total nitrogen by Kjeldahl and 0-23 cm (see text)

(2) Data not previously published

(3) From Jenkinson *et al* (1992)

(4) From Thurston *et al* (1976)

(5) From Richardson (1938)

(6) From Warren & Johnston (1963). Value for %N, plot 14 corrects previously published value

References

Avery, B.W. & Catt, J.A. (1995) The soil at Rothamsted. Lawes Agricultural Trust, Rothamsted, UK.

Bremner, J.M. (1965) Total nitrogen. In *Methods of Soil Analysis* (Ed. C.A. Black) Vol 2 pp 1149-1178 *American Society of Agronomy* Madison USA.

Goulding, K.W.T. & Poulton, P.R. (1992) Unwanted nitrate. *Chemistry in Britain* December 1992, pp. 1102-1112.

Jenkinson, D.S., Potts, J.M., Perry, J.N., Barnett, V., Coleman, K. & Johnston, A.E. (1994) Trends in herbage yields over the last century on the Rothamsted Long-term Continuous Hay Experiment. *Journal of Agricultural Science, Cambridge* **122**, 365-374.

Johnston, A.E., Goulding, K.W.T. & Poulton, P.R. (1986) Soil acidification during more than 100 years under permanent grassland and woodland at Rothamsted. *Soil Use and Management* **2**, 3-10.

Lawes, J.B. & Gilbert, J.A. (1880) Agricultural, botanical and chemical results of experiments on the mixed herbage of permanent meadow, conducted for more than 20 years in succession on the same land. Part 1. The agricultural results. *Philosophical Transactions of the Royal Society* **171**, 289-415.

Richardson, H.L. (1938) The nitrogen cycle in grassland soils: with especial reference to the Rothamsted Park Grass Experiment. *Journal of Agricultural Science, Cambridge* **28**, 73-121.

Rothamsted Experimental Station (1995) Yields of the Field Experiments, 1994.

Thurston, J.M., Williams, E.D. & Johnston, A.E. (1976) Modern developments in an experiment on permanent grassland started in 1856: effects of fertilizers and lime on botanical composition and crop and soils analyses. *Annales Agronomique* **27**, 1043-1082.

Tilman, D., Dodd, M.E., Silvertown, J., Poulton, P.R., Johnston, A.E. & Crawley, M.J. (1994) The Park Grass Experiment: Insights from the Most Long-term Ecological Study. In *"Long-term Experiments in Agricultural and Ecological Sciences"* (Eds R.A. Leigh & A.E. Johnston) pp. 287-303, CABI, Wallingford, UK.

Tinsley, J. (1950) The determination of organic carbon in soils by dichromate mixtures. *Transactions of the 4th International Congress of Soil Science, Amsterdam* **1**, 161-164.

Walkley, A. & Black, I.A. (1934) An examination of the Degtjareff method for determining soil organic matter and a proposed modification of the chromic acid titration method. *Soil Science* **37**, 29-38.

Warren, R.G. & Johnston, A.E. (1964) The Park Grass Experiment. *Rothamsted Experimental Station Report for 1963*, pp. 240-262.

Acknowledgements

I thank A.E. Johnston for useful discussion. IACR receives grant-aided support from the Biotechnology and Biological Sciences Research Council of the United Kingdom.

Geescroft Wilderness, 1883-1995

P.R. Poulton
Soil Science Department
IACR-Rothamsted
Harpenden, Herts., AL5 2JQ
U.K.

Introduction

The Geescroft Wilderness at Rothamsted, UK, is one of two areas of land previously in arable cropping that were fenced off in the 1880s and allowed to run wild. The other, much smaller, area had been part of the Broadbalk Continuous Wheat experiment (Jenkinson, 1971). Lawes (1895) was interested in what plants would become established after abandoning arable cropping and what changes might occur in the soil. After 110 years, Geescroft is now a deciduous woodland and, although it is not an experiment in the conventional sense, we can use the site to follow changes in soil organic matter under undisturbed conditions.

Site details

The soil is a silty clay loam (Aquic Paleudalf, USDA; Chromic Luvisol, FAO) with about 21% clay in the top 23 cm (Avery & Catt, 1995). Unlike many other arable fields at Rothamsted, Geescroft had not received dressings of chalk and the soil contained no free $CaCO_3$. The climate is cool, temperate with a long-term average rainfall of 734 mm per year.

The field had been in arable cropping for several centuries before Lawes and Gilbert started an experiment on field beans (*Vicia faba*) which ran from 1847 till 1878. Clover was sown

NATO ASI Series, Vol. I 38
Evaluation of Soil Organic Matter Models
Edited by David S. Powlson, Pete Smith and Jo U. Smith

in 1883 and after a second cut in 1885 part of the field (1.3 ha) was fenced off and left unattended. When in arable cropping the field was regarded as one of the wettest and most unworkable on the farm. Over the next 30 years the area was colonized by many damp-loving species. Although some shrubs and trees were present the area was still fairly open (Brenchley & Adams, 1915). By 1957 the site had reverted to woodland and most grassland species had disappeared (Thurston, 1958). The site is now an oak dominated deciduous woodland with few ground-cover plants.

Soil sampling, analysis and inputs

The Wilderness covers part of the previous experiment on beans which included phosphate treatments. Where possible data has been given for soils taken from areas which had received no P fertilizer, although Jenkinson (1971) found no effect of P on the build up of organic C or N.

Soil samples were taken in 1883 whilst still in clover, using an open-ended iron box 6" x 6" x 9" deep (15.2 x 15.2 x 22.9 cm). The same method was used in 1904 and 1965 and on each occasion 3 depths (0 - 22.9, 22.9 - 45.7, 45.7 - 68.6 cm) were sampled. In 1985 the top 22.9 cm only was sampled using a straight-sided semi-cylindrical auger. Soils have been analysed for organic carbon by several methods, including: the Shaw wet combustion procedure as modified by Jenkinson (1965), a dry combustion method, and a modified Tinsley method (Kalembasa & Jenkinson, 1973). Total N was measured by Kjeldahl (Bremner, 1965). Details of sampling and analyses are given by Jenkinson (1971) and Jenkinson *et al.* (1992).

There have been no measurements of total carbon inputs to the soil. However, leaf-fall measurements in 1959, 1965 and 1966 provide an estimate of 1.57 t C ha^{-1} yr^{-1} for at least part of the input. As on Park Grass (Poulton, this volume) we have assumed that total N input from the atmosphere was 20 kg N ha^{-1} yr^{-1} till 1900 and 40 kg N ha^{-1} yr^{-1} since.

Soil pH (in H_2O) has declined markedly since 1883. Values for the 0 - 22.9 cm layer were 7.1, 6.1, 4.5 and 4.2 in 1883, 1904, 1965 and 1985 respectively. Lower depths have also become progressively more acid (Johnston *et al.* 1986).

Correction for changing bulk density

As organic matter has accumulated, the bulk density of the sampled 0 - 22.9 cm layer has decreased, i.e. the soil has expanded. For example, the weight of fine, ignited (organic matter free) soil in the top 22.9 cm in 1883 was 2.62 M kg ha^{-1}; in 1985 it was only 1.88 M kg ha^{-1}. Obviously a correction needs to be made to allow for this change. If deeper soils have been sampled, as on Geescroft, then the "equivalent" depth can be calculated, i.e. that depth which would need to be sampled to give the same weight of fine, ignited soil as in the initial sample. Once the "equivalent" depth has been calculated then the amount of carbon or nitrogen in that depth can also be calculated. This was done graphically by Jenkinson (1971), and Figure 1 shows how this was done for the 1965 samples. A similar calculation was done for the 1985 soils.

The amounts of organic C and N in the 0 - 22.9 cm layer *as sampled* and on an *equivalent depth* basis are shown in Table 1. The amounts of C and N in the top layer differ considerably. By 1985 the increase in organic C and N was 33,300 and 1,290 kg ha^{-1} respectively when calculated on an equivalent depth basis but only 26,500 and 520 kg ha^{-1} in the depth as sampled. Clearly, when modelling data from sites where soil expansion, or compaction, has occurred then corrected values should be used wherever possible.

Table 1. *Amount of organic C and N.*

Year	Depth as sampled cm	"Equiv" depth cm	In sampled depth Org. C kg ha^{-1}	Org. N kg ha^{-1}	In equiv. depth Org. C kg ha^{-1}	Org. N kg ha^{-1}
1883	0-22.9	0-22.9	28,400	3,140	28,400	3,140
1904	0-22.9	0-23.1	37,100	3,550	37,400	3,610
1965	0-22.9	0-27.8	43,400	3,640	49,300	4,240
1985	0-22.9	0-29.0	54,900	3,660	61,700	4,430

The data shows a greater rate of increase in organic matter between 1965 and 1985 than in the previous 61 years. During this time the pH has declined markedly because of atmospheric inputs (Johnston *et al.*, 1986) and there has been a transition in the humus type from mull or moder to mor, i.e. a layer of largely undecomposed organic matter at the

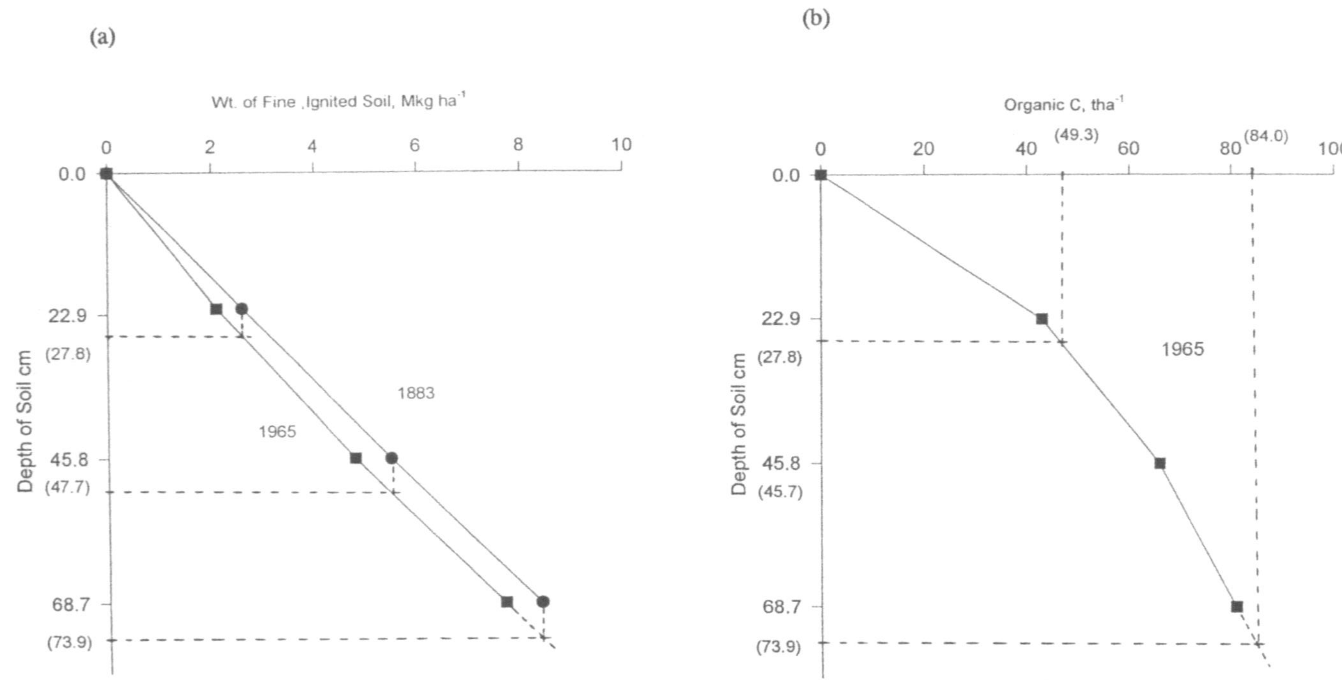

Figure 1. *(a) Weight of soil in 1883 and 1965 plotted against sampled depth; values in parentheses are the "equivalent" depth to which the 1965 samples should have been sampled to give the same weight of soil as in 1885. (b) Amount of carbon in the depth as sampled; values in parentheses are the "equivalent" depth (from Fig. 1(a)) and the amount of carbon in that "equivalent" depth.*

surface which is quite distinct from the mineral soil beneath. This layer varies in depth from c. 2-6 cm. The increase in organic matter between 1965 and 1985 is probably real but it could, in part, reflect the difficulties involved with sampling an increasingly variable site.

During its reversion to woodland there has been much accumulation of C in the standing trees. Using estimates for the dry weight of trees (Jenkinson, 1971) we can estimate that by the 1960s the trees contained c. 70,000 kg C ha^{-1}, 3 times as much as the amount that had accumulated in the soil profile to 69 cm by that time. In contrast, the amount of N in the trees, c. 700 kg ha^{-1}, was only half that accumulated in the soil.

References

Avery, B.W. & Catt, J.A. (1995) The soil at Rothamsted. *Lawes Agricultural Trust, Rothamsted, UK.*

Bremner, J.M. (1965) Total nitrogen. In *Methods of Soil Analysis* (Ed. C.A. Black) Vol 2 pp 1149-1178 *American Society of Agronomy* Madison, USA.

Brenchley, W.E. & Adam, H. (1915) Recolonisation of cultivated land allowed to revert to natural conditions. *The Journal of Ecology* **III**, No. 4, 193-210.

Jenkinson, D.S. (1965) Studies on the decomposition of plant material in soil. I. *Journal of Soil Science* **16**, 104-115.

Jenkinson, D.S. (1971) The accumulation of organic matter in soil left uncultivated. *Rothamsted Experimental Station Report for 1970*, Part 2, 113-137.

Jenkinson, D.S., Harkness, D.D., Vance, E.D., Adams, D.E. & Harrison, A.F. (1992) Calculating net primary production and annual input of organic matter to soil from the amount and radiocarbon content of soil organic matter. *Soil Biology and Biochemistry* **24**, No. 4, 295-308.

Johnston, A.E., Goulding, K.W.T. & Poulton, P.R. (1986) Soil acidification during more than 100 years under permanent grassland and woodland at Rothamsted. *Soil Use and Management* **2**, 3-10.

Kalembasa, S.J. & Jenkinson, D.S. (1973) A comparative study of titrimetric and gravimetric methods for the determination of organic carbon in soil. *Journal of the Science of Food and Agriculture* **24**, 1085-1090.

Lawes, J.B. (1895) Upon some properties of soils which have grown a cereal crop and a leguminous crop for many years in succession. *Agricultural Students Gazette, Cirencester* **7** (New Series), 65-72.

Poulton, P.R. (1995) The Park Grass Experiment, 1856-1995. *This volume.*

Thurston, J.M. (1958) Geescroft Wilderness. *Rothamsted Experimental Station Report for 1957*, p. 94.

Acknowledgements

I thank D.S. Jenkinson for valuable discussion. IACR receives grant-aided support from the Biotechnology and Biological Sciences Research Council of the United Kingdom.

[1] Long-term Residue Management Experiment: Pendleton, Oregon USA

P.E. Rasmussen[1], R.W. Smiley[2], and S.L. Albrecht[1]

[1] USDA-ARS-PWA Columbia Plateau Conservation Research Center, and [2] Oregon State Univ. Columbia Basin Agricultural Research Center, Pendleton, Oregon, USA

Introduction

The Residue Management experiment is one of six long-term studies maintained by the Columbia Basin Agricultural Research Center, Oregon State University Agricultural Experiment Station. The Center is located 15 km northeast of Pendleton, Oregon at 45° 44′ north and 118° 37′ west . It lies within the Columbia Plateau physiographic province between the Cascade and Rocky mountains. The climate is semi-arid, but partially influenced by maritime winds from the Pacific Ocean. Winters are cool and wet, the summers hot and dry. Precipitation occurs primarily during the winter. Annual precipitation is 420 mm, with 70% received between 1 September and 31 March. Winter precipitation falls mainly as rain, with limited duration of snow cover in most years. Average annual temperature is 10.2 °C, but ranges from -0.6 °C in January to 21.2 °C in July. Weather is measured daily 200 m from the site. Temperature and precipitation have been recorded since 1931, 10-cm soil temperature since 1962, wind speed and water evaporation since 1963, and solar radiation and humidity since 1982. Elevation is 455 m above sea level. Soils consist of loess from Pleistocene alluvial deposits overlying basalt flows of Miocene age. They are classified as coarse silty mixed mesic Typic Haploxerolls by the USDA classification system. Soils are well drained and depth to water table is greater than 50 m. Soil depth ranges from 1 to 2 m, depending upon landscape position. Soil texture is silt loam throughout the profile. The upper 30 cm of soil contains 18% clay, 70% silt, and 12% fine sand. The area was originally a mid-grass prairie that was first cultivated in about 1885. It was cropped to wheat either annually or alternating with summer fallow until the experiment station was established in 1929.

---

[1] Oregon State University Agricultural Experiment Station Technical Paper no. 10783.
Abbreviations: USDA-ARS-PWA = United States Department of Agriculture - Agricultural Research Service - Pacific West Area; N = Nitrogen, C = Carbon, SOM = Soil Organic Matter

NATO ASI Series, Vol. I 38
Evaluation of Soil Organic Matter Models
Edited by David S. Powlson, Pete Smith and Jo U. Smith

Experimental design

The Residue Management experiment was established in 1931, and contains 10 treatments and two replications. Crop rotation is a 2-year winter wheat/fallow system, with a 15-month fallow and a 9-month cropping season. Because of the 2-year rotation, the experiment has duplicate trials offset by one year to provide yield data each year. Plot size is 11.6 by 40.2 m.

Table 1 *Treatment history of the residue management experiment.*

Dataset Number	Trt. No	Organic Amendments	1931-1966		1967-1978		1979-Present	
			Res[1]	N[2]	Res	N	Res	N
	1	---	--	--	--	--	--	--
	2	---	fD	0	nB	45	sB	44
	3	---	sD	0	nB	90	sB	90
3	4	---	nB	34	nB	45	nB	45
4	5	---	nB	34	nB	90	nB	90
2	6	---	fB	0	fB	0	fB	0
	7	---	sB	0	sB	0	sB	0
6	8	Manure[3]	nB	0	nB	0	nB	0
5	9	Pea Vines[4]	nB	0	nB	0	nB	0
1	10	---	nB	0	nB	0	nB	0

[1] Residue treatment: fD = autumn disk, sD = spring disk, nB = no burn, fB = autumn burn, sB = spring burn.

[2] Nitrogen rate (kg ha^{-1} $crop^{-1}$), applied 1-2 weeks prior to seeding of wheat (early October).

[3] Manure = (22.40 tons ha^{-1} $crop^{-1}$ wet wt; 47.3% dry matter; 1743 kg C and 145 kg N ha^{-1} $crop^{-1}$), applied just prior to plowing (early April) in fallow year.

[4] Pea Vines=(2.24 tons ha^{-1} $crop^{-1}$, 88.5% dry matter; 830 kg C and 38 kg N ha^{-1} $crop^{-1}$), applied just prior to plowing (early April) in fallow year.

Six of the 10 treatments were selected and formatted for modeling at this workshop. The six treatments included are: (1) #10: no inorganic N fertilizer, no organic amendments; (2) #6: no inorganic N fertilizer, no organic amendments, wheat straw burned in September; (3) #4: 45 kg inorganic N ha^{-1}, no organic amendments; (4) #5: 90 kg inorganic N ha^{-1}, no organic amendments; (5) #9: no inorganic N fertilizer, 2.24 t pea vine organic residue ha^{-1} added; and (6) #8: no inorganic fertilizer, 22.40 t cattle manure organic residue ha^{-1} added. All wheat straw (burnt wheat residue in treatment #6) and organic amendments are incorporated in the soil during spring plowing. The N rate for treatments #4 and #5 from 1931 to 1966 was 34 kg N ha^{-1}, and fertilizer was applied in the spring just prior to plowing. From 1967 to present, these treatments have received 45 and 90 kg N ha^{-1}, respectively, with fertilizer applied just prior to seeding of wheat.

Winter wheat is seeded in October and harvested in July. Wheat stubble (except for the autumn burn treatment) is left undisturbed from harvest until tillage the following spring (April). Fallow tillage is conventional (moldboard plow, cultivate harrow, rodweed). Both mechanical and chemical methods are used for weed control as needed. A single moderately-tall soft white winter wheat variety was grown from 1931 to 1966. Variety selection since 1967 has been the best adapted high yielding semi-dwarf soft white winter wheat.

Experimental data

Wheat grain yield has been determined since 1931, and grain N content since 1967. Straw yield and N content, and the nutrient content of added organic amendments have been determined since 1977. Straw yield and nutrient uptake from 1931 to 1976 has been estimated by utilizing variety-trial data coupled with periodic measurements within the experiment.

Soil samples were collected in 1931, 1941, 1951, 1964, 1976, and 1986 from the 0-30 and 30-60 cm soil depths. Samples were analyzed for organic C and N. There is essentially no inorganic C or N in the upper 60 cm of soil. Soil samples from 1931 to 1964 were analyzed for total N by macro-Kjeldahl/steam distillation, and the 1976 and 1986 samples by NPN^- tube wet digestion and automated analysis. Differences in early and late N procedures were evaluated and values corrected to current procedures. The 1931 and 1941 samples were analyzed for SOM by total combustion. No method of analysis was listed for 1951, and no records exist for 1964. The 1976 and 1986 samples were analyzed for C by automated combustion CO_2 analysis. Because early SOM results were erratic, C means for 1931 to 1964 were derived based on SOM/1.72 and adjusted based on relative N values. Soil bulk density was determined in 1931 (1.17 g cm^{-3}), 1954 (1.19), and 1990-91 (1.22). Bulk density in 1990-91 was not affected by treatments. Therefore, bulk densities of 1.17, 1.18, 1.19, 1.20, 1.21, and 1.22 g cm^{-3} were assigned to the respective soil sampling dates, and C and N content calculated as t ha^{-1} for the two sample depths. Soil samples prior to 1986 are not archived.

Table 2 *Average grain and straw yields for 5 time periods between 1932 and 1986.*

Dataset Number	Treatment Identity	Period 1932-41	1942-51	1952-66	1967-76	1977-86
		Grain yield[1], (t ha^{-1} $crop^{-1}$)				
2	fB-N_0	2.41 de[2]	2.44 cd	1.92 cd	2.74 d	2.58 f
	sB-N_0	2.48 cd	2.56 c	2.08 c	2.83 d	2.67 ef
1	nB-N_0	2.35 d-f	2.48 cd	2.01 cd	2.85 d	2.78 e
	sB-N_{45}	2.23 f	2.27 e	1.92 cd	3.61 c	3.92 d
	sB-N_{90}	2.31 ef	2.38 de	1.91 d	3.96 bc	4.50 b
3	nB-N_{45}	2.66 b	2.86 b	2.58 b	3.75 c	4.10 c
4	nB-N_{90}	2.59 bc	2.83 b	2.53 b	4.17 ab	4.64 b
5	nB-PV	2.73 b	2.98 b	2.67 b	3.69 c	3.81 d
6	nB-MN	2.96 a	3.33 a	3.09 a	4.44 a	4.87 a
	Avg	2.52	2.68	2.30	3.56	3.76
		Straw yield[1], (t ha^{-1} $crop^{-1}$)				
2	fB-N_0	4.45	4.51	3.55	5.05	5.11
	sB-N_0	4.61	4.74	3.85	5.14	5.29
1	nB-N_0	4.33	4.60	3.80	5.00	5.21
	sB-N_{45}	4.26	4.20	3.56	5.96	7.04
	sB-N_{90}	4.30	4.41	3.53	6.32	7.59
3	nB-N_{45}	5.18	5.57	5.04	6.03	7.03
4	nB-N_{90}	5.06	5.51	4.93	6.37	7.46
5	nB-PV	5.39	5.91	5.19	6.21	6.95
6	nB-MN	6.40	7.22	6.71	6.78	7.94
	Avg	4.89	5.18	4.47	5.87	6.65
	$LSD_{.05}$	0.16	0.09	0.08	0.24	0.14

[1] Yield on dry weight basis.

[2] Values in each column followed by the same letter are not significantly different at P=0.05.

Table 3 *Soil organic C and N for the 0-30 cm soil zone at 6 sample dates between 1931 and 1986.*

Dataset	Trt	Year					
Number	Identity	1931	1941	1951	1964	1976	1986
		Organic C (t ha^{-1})					
2	fB-N_0	48.67	46.44	42.15	42.45	40.01	37.01
	sB-N_0	48.12	45.74	42.69	43.18	41.42	38.96
1	nB-N_0	50.21	49.23	44.93	44.00	42.47	39.65
	sB-N_{45}	49.24	48.08	43.78	42.72	41.31	38.74
	sB-N_{90}	49.41	47.08	42.80	41.81	41.52	39.95
3	nB-N_{45}	49.90	48.36	45.96	44.91	44.24	41.36
4	nB-N_{90}	48.75	48.59	46.18	44.09	43.65	41.90
5	nB-PV	49.92	50.48	50.63	47.01	46.15	44.49
6	nB-MN	48.57	50.88	53.07	49.39	50.80	50.18
	Avg	49.20	48.32	45.80	44.42	43.49	41.36
	$LSD_{.05}$	NS	NS	5.10	3.69	3.33	4.76
	CV	-	-	-	-	4.70	7.10
		Organic N (t ha^{-1})					
2	fB-N_0	3.71	3.56	3.28	3.16	3.02	2.85
	sB-N_0	3.66	3.51	3.32	3.33	3.20	3.06
1	nB-N_0	3.83	3.78	3.50	3.43	3.29	3.19
	sB-N_{45}	3.75	3.69	3.41	3.41	3.27	3.20
	sB-N_{90}	3.77	3.61	3.33	3.35	3.33	3.30
3	nB-N_{45}	3.81	3.71	3.58	3.62	3.48	3.46
4	nB-N_{90}	3.71	3.72	3.59	3.46	3.41	3.35
5	nB-PV	3.81	3.87	3.94	3.77	3.66	3.58
6	nB-MN	3.73	3.90	4.13	4.08	4.18	4.20
	Avg	3.75	3.70	3.56	3.51	3.42	3.35
	$LSD_{.05}$	NS	NS	0.40	0.29	0.32	0.26
	CV	4.80	6.50	6.80	5.10	5.70	4.70

Wheat grain yield has nearly doubled in the last 65 years (Table 2). Straw yield tended to increase slightly with the change from medium-tall to semidwarf wheat. Changes in C and N content in the 0-30 cm of soil since 1931 (Table 3) are generally related to treatment (Oveson, 1966; Rasmussen *et al.*, 1980; Rasmussen and Parton, 1994). The projected C content of the upper 30 cm of virgin soil in 1885 was 19.4 g kg^{-1}, based on differences between virgin and cultivated sites measured 1935. The C level in 1931 was 13.8 g kg^{-1}, indicating a loss of 29% in the first 50 years of cultivation. The rate of change since 1931 is highly correlated with the level of annual C input into the system (Rasmussen and Parton, 1994). Treatments that have lost the most C tend to have the least microbial biomass (Collins *et al.*, 1992) and enzyme activity (Dick *et al.*, 1988). System changes are modeled with reasonable precision by the CENTURY model (Parton and Rasmussen, 1994).

The C and N content of the 30-60 cm zone has also declined with time, but this change is not related to treatments. Soil C was 35.2, 33.6, 32.2, 30.3, 30.3, and 25.0 t ha^{-1} in 1931, 1941, 1951, 1964, 1976, and 1986, respectively. Corresponding soil N values for these sample dates were 3.02, 2.94, 2.87, 2.77, 2.82, and 2.63 t ha^{-1}. The gradual decline is probably the result of a change in oxidation equilibrium following conversion from a virgin grassland to a cultivated system that includes 50% fallow. Carbon input from root material is generally much less under cultivation than under grassland.

References

Collins, H.P., Rasmussen, P.E., & Douglas, C.L Jr. (1992) Crop rotation and residue management effects on soil carbon and microbial dynamics. *Soil Sci. Soc. Am. J.* **56**: 783-788.

Dick, R.P., Rasmussen, P.E., & Kerle, E.A. (1988) Influence of long-term management on soil enzyme activities in relation to soil chemical properties of a wheat-fallow system. *Biol. Fert. Soils* **6**: 159-164.

Oveson, M.M. (1966) Conservation of soil nitrogen in a wheat-summer fallow farming practice. *Agron. J.* **58**: 444-447.

Parton, W.J., & Rasmussen, P.E. (1994) Long-term effects of residue management in wheat/fallow: Century model simulations. *Soil Sci. Soc. Am. J.* **58**: 530-536.

Rasmussen, P.E., Allmaras, R.R., Rohde, C.R., & Roager, N.C. Jr. (1980) Crop residue influences on soil carbon and nitrogen in a wheat-fallow system. *Soil Sci. Soc. Am. J.* **44**: 596-600.

Rasmussen, P.E., & Parton, W.J. (1994) Long-term effects of residue management in wheat/fallow: Inputs, yield, and soil organic matter change. *Soil Sci. Soc. Am. J.* **58**: 523-530.

Carbon Changes During the Growth of Loblolly Pine on Formerly Cultivated Soil: The Calhoun Experimental Forest, U.S.A.

D. D. Richter and D. Markewitz

School of the Environment and Forest Resources Center

Duke University

Durham, North Carolina 27708

USA

Introduction

Although deforestation and reforestation profoundly affect the global carbon cycle (Schlesinger, 1990, 1991; Sundquist, 1993), our understanding of the forest carbon cycle from the local to the global scale is limited. This incomplete understanding is in part due to the lack of long-term forest experiments with permanent soil plots and archived soil samples. Such experiments will increase in usefulness as forest lands are managed with ever greater intensity, as forest lands continue to be cleared and converted to other uses and where cultivated soil is afforested.

The Calhoun Forest Experiment is a relatively simple, afforestation experiment with eight permanent plots and archived soil samples. It was designed to test how an acidic, formerly cultivated Ultisol meets the nutritional needs of a rapidly growing pine forest during the first four decades after planting. The permanent plots have been sampled and soils archived on six occasions since seedlings were planted in 1957. The experiment has already provided information about how pine-forest growth alters soil carbon, exchangeable acidity, cation exchange capacity, and plant-available calcium, potassium, and magnesium (Wells & Jorgensen, 1975; Binkley *et al.*, 1989; Richter *et al.*, 1994, 1995; Richter & Markewitz, 1995a, 1995b). The experiment has a bright future because foresters, agronomists, and

NATO ASI Series, Vol. I 38
Evaluation of Soil Organic Matter Models
Edited by David S. Powlson, Pete Smith and Jo U. Smith

ecologists need to understand how nutrient demands of rapidly growing, perennial vegetation are met by soil-nutrient supplies. How soil is able to sustain, gain, and lose nutrients is critical to improving the efficiency of forestry and agriculture, both economically and environmentally.

Thanks to the Calhoun experiment's initiators, especially Dr. Carol G. Wells, the experiment has provided invaluable general data about forest-soil biogeochemistry. The experimental results are also pertinent to critical uncertainties of the loblolly pine ecosystem, one of the world's most economically important forest ecosystems. The forests of the North American South, of which the loblolly pine ecosystem is most prominent, produce more industrial wood than forests in any other region of the world, about 225,000,000 m^3 per year (Jaakko Pöyry Group, 1994). Like at the Calhoun, many of these systems are often supported by infertile, acidic, formerly cultivated Ultisols (McCracken *et al.*, 1989; Richter & Markewitz, 1995b). The Calhoun forest experiment provides the experimental basis for describing sustained soil-nutrient supply and soil-nutrient depletion in a rapidly growing forest. In other words, the Calhoun experiment tests the sustainability of the forest ecosystem's biogeochemical cycles.

Methods

Study area and land use history

The research area is located at approximately 34.5° N, 82° W in the southern Piedmont of South Carolina. The research area is on the Calhoun Experimental Forest of the Sumter National Forest in Union County, about 100 km northwest of Columbia. Annual precipitation averages about 1170 mm and temperature about 16° C (1950-1987).

Before the late 1700s, a mainly deciduous forest was supported by the acidic Ultisols that dominate the southeast U.S.A. (Richter & Markewitz, 1995b). The southern Piedmont is an ancient geologic structure and a physiographic region that stretches from the states of Virginia to Alabama, and lies to the west of the Atlantic Coastal Plain and to the east of the Appalachian Mountains. By the early 1800s, the original forests had been extensively cleared, and until the mid-twentieth century, the relatively infertile Ultisols such as those at the Calhoun forest, supported crops of cotton (*Gossypium hirsutum* L.), tobacco (*Nicotiana tabacum*), corn (*Zea mays* L.), and wheat (*Triticum aestivum* L.). Before the U.S. Civil War in the 1860s, a shifting cultivation was practiced on many farms of the southern Piedmont; when crop productivity on a field faltered, cropping was moved to "fresh" soil and the fertility of the exhausted field was regenerated by a forest fallow in which the secondary growth was later recleared and burned. Following the Civil War, agricultural practices included more continuous cropping of fields and increased use of lime, fertilizers, and soil conservation practices (Trimble, 1974). In 1954, the last crop of cotton was harvested on the Calhoun research area, and after a 2-yr fallow, loblolly pine seedlings were planted in the winter of 1956-1957.

Soil profiles

The old-field soils at the Calhoun lie on two broad interfluves with < 3 % slopes. The soil materials are residual, being inherited from the granite-gneiss bedrock below. The A and E horizons are coarse-textured sandy loams and loamy sands that are also relatively deep, 34 cm on average. The nearly flat landforms and coarse surface soil texture make these soils physically stable and not apt to erode, especially under forest cover. Under the E horizons are clayey B horizons with >2-m thickness, dominated by kaolinite clay mineralogy and prominent iron-oxides. The B-horizon clay particles are arranged in aggregates and peds that range widely in size, up to about 1-m diameter. Under the B horizons lie very deep C horizons, highly weathered and acidic. The C horizons are isovolumetrically weathered (Pavich, 1986; Stolt *et al.*, 1992) and derived from the bedrock that is at a depth of > 8 m from the soil surface. Some descriptive data from these deep soil profiles are summarized in Table 1.

Ecosystem sampling and analysis

The Calhoun forest experiment has eight permanent plots, four plots of 380 m^2 area and four plots of 595 m^2 area. Mineral soils have been collected with similar procedures from these plots on six occasions, in 1962, 1968, 1972, 1977, 1982, and 1990. Composite samples are collected from at least 20 individual samples per plot with a punch tube having 2-cm diameter. Collection depths are 0 to 7.5 cm, 7.5 to 15 cm, 15 to 35 cm, and 35 to 60 cm. Thus, at each of the eight permanent plots, four composite samples are taken, one from each depth. The soil samples are archived at Duke University's School of the Environment.

Table 1. *Mean physical and chemical properties of soil profiles under the aggrading pine ecosystem at the Calhoun Forest Experiment, SC, USA. The four most surficial layers within the upper 0.6 m are those sampled in the eight permanent plots on six occasions between 1962 and 1990. Samples were collected between 1990 and 1993.*

Depth	Soil C	O ^{14}C	Clay	pH†	ECEC ‡	Acid § Saturation
m	g kg^{-1}		g g^{-1}		cmol kg^{-1}	%
0-0.075	6.99	229.6	0.100	4.21	1.24	91.1
0.075-0.15	4.45	97.2	0.140	4.21	0.83	89.2
0.15-0.35	3.23	11.2	0.180	4.21	1.01	65.3
0.35-0.60	3.45	-92.8	0.393	4.20	2.79	48.0
0.60-1.10	2.40	-301.0	0.627	4.08	4.61	74.0
1.10-2.0	1.99	-403.4	0.507	3.93	4.24	94.0
2.0-3.0	0.73	-434.0	0.331	4.01	3.44	94.5
3.0-4.0	0.38	-630.3	0.194	3.92	3.37	93.6
4.0-5.0	0.26		0.100	3.99	2.68	83.0
5.0-6.0	0.21	-448.9	0.080	4.17	2.32	67.4

† Soil pH in 0.01 M $CaCl_2$.

‡ Effective cation exchange capacity (sum of exchangeable cations plus KCl acidity).

§ Salt-exchangeable KCl acidity as a proportion of ECEC.

Soil samples are air dried and sieved through a 2-mm screen. Subsamples were pulverized and analyzed for C with a Perkin Elmer CHNS analyzer. Additional details of soil analyses are described in Richter *et al.* (1994, 1995) and Richter & Markewitz (1995a).

To estimate carbon in pine tree biomass, diameter and height of all trees in each of the eight permanent plots have been periodically measured. Tree measurements are used with allometric equations to estimate carbon contents in above- and below-ground plant parts during the course of forest development (Shelton *et al.*, 1984, Van Lear *et al.*, 1984). Carbon is taken to be 50% of biomass.

To estimate carbon contents of canopy litterfall, 40-litterfall traps were installed, five in each permanent plot. Each trap had 0.72-m^2 area and was collected monthly over a two-year period. Carbon in forest floor (the O horizon) was estimated from no less than three samples per plot, each sample having a 0.13-m^2 area. Forest floor carbon was corrected for mineral contamination and based on ash-free O horizon mass. Carbon in fine roots (< 2 mm diameter) was estimated from collections made every three weeks over an annual period; eight 6-cm diameter cores (two from each of four permanent plots) were taken per collection and subdivided between O horizon, and 0-15-cm and 15-30-cm mineral soil depths. Fine root carbon was estimated by wet-sieving soil from cores, and ashing roots to correct for mineral contamination. Lastly, dissolved organic carbon (DOC) was determined in rainfall, throughfall (rainfall that has penetrated the foliar canopy), and in drainage from the O horizon and four mineral soil depths; solution collections were made every two weeks over a two-year period. The DOC was determined by a Shimadzu infrared gas analyzer.

To examine soil carbon accumulations as affected by four land uses, 16 sites were selected in 1991 < 15 km from the Calhoun forest experiment in areas that were supporting uneven-aged, deciduous hardwood forest (n = 4), 30 to 50-year-old pine plantations on old agricultural fields (n = 5), grass-dominated hay fields (n = 5), and row crops (n = 2). All sites shared similar soils, specifically based on texture and profile characteristics; in other words, soil series were closely similar to the Appling soil series that is prominent at the Calhoun forest.

To better describe the turnover of soil C, we have examined the rate at which the forest incorporated CO_2 into soil organic C by determining how rapidly ^{14}C from aboveground thermonuclear explosions was incorporated into soil organic matter. In these analyses, soil C is oxidized to CO_2 and reduced to form a graphite target for accelerator mass spectrometry. These data are not provided in this report but give the Calhoun data another special feature for modellers.

Results

Pine forest accumulation of carbon

Following up to about 150 years of cultivation, the old-field pine forest has rapidly accumulated carbon in forest biomass and in the soil profile over a three to four decade period (Figure 1). Total accumulations of carbon in the ecosystem over the 34-yr period (1957-1990) totaled 177.5 Mg/ha, and averaged 5.22 Mg $ha^{-1} yr^{-1}$. Carbon accumulations are mainly

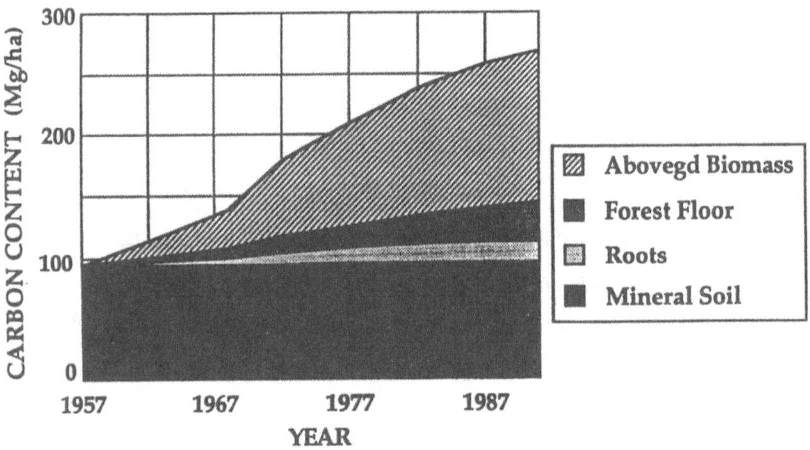

Figure 1. *Accumulation of carbon in the loblolly pine ecosystem at the Calhoun forest experiment, South Carolina, USA. Mineral soil carbon includes that in 0 to 6-m depth.*

in tree biomass and O horizons. Tree biomass carbon totaled 140.6 Mg/ha, and in the O-horizon carbon totaled 34.9 Mg/ha including large wood. In contrast, the mineral soil A horizon accumulated a total of only 1.96 Mg/ha of carbon over a 28-yr period (1962-1990). The carbon accumulation rate in forest biomass was 4.13 Mg $ha^{-1} yr^{-1}$, in the O horizon was 1.03 Mg $ha^{-1} yr^{-1}$, and in the A horizon 0.070 Mg $ha^{-1} yr^{-1}$.

Carbon inputs to O and A horizons

Changes in soil organic carbon storage depend on changes in carbon inputs to soil compartments. The O and the A soil horizons were the compartments which accumulated carbon during the growth of the experimental pine forest, and current carbon inputs were estimated for both (Table 2).

Table 2. *Comparison of current carbon inputs to O and A horizon (0-15 cm depth) with total accumulation of carbon in the two horizons under loblolly pine at Calhoun Experimental Forest, SC.*

		Horizon	
Attribute	Carbon Units	O	A (0-15 cm)
Four inputs			
Litterfall	kg ha^{-1} yr^{-1}	2450	-
Gross throughfall	kg ha^{-1} yr^{-1}	81.2	-
O horizon leachate	kg ha^{-1} yr^{-1}	-	250.7
Fine root turnover†	kg ha^{-1} yr^{-1}	-	673
Total inputs	kg ha^{-1} yr^{-1}	2531	924
Total accumulation	kg ha^{-1}	34900	1960
Residence time‡	yr	13.8	2.1

† Assumes that annual rhizo-deposition of carbon is equivalent to the mean fine root biomass in 0-15 cm soil.
‡ Calculated as if accumulations are at steady state.

In the early 1990s, the O horizon had annual inputs of carbon from litterfall and from canopy throughfall (as DOC) that averaged about 2.53 Mg ha^{-1} yr^{-1}. Currently, the O horizon contains about 34.9 Mg/ha of carbon, and over the life of the stand, we estimate that about 40% of the total carbon inputs are still present in the O horizon (calculation not elaborated here). The carbon in the O horizon has an average residence time of about 13.8 yr (Table 2).

Also in the early 1990s, the A horizon (0-15 cm) has annual inputs of carbon from fine root debris and leaching of dissolved organic carbon from the O horizon that total about 0.92 Mg ha^{-1} yr^{-1}. The average residence time of the accumulating carbon in A horizons is about 2.1 yr.

The different residence times of carbon in the O and A horizons suggest markedly different rates of decomposition in the O and A horizons of this forested soil.

Soil carbon under different land covers

In the vicinity of the Calhoun Experimental Forest, our survey of soil carbon contents under pine and hardwood stands, grass-hay fields, and row crops was able to demonstrate relatively large impacts of land use on soil carbon contents. The results strongly support the more precise results from the long-term experiment. Soils from the five pine stands near the Calhoun had 18.8 Mg/ha of carbon in the upper 30 cm of mineral soil (CV% = 14.2%), about 14.4 Mg/ha less carbon than is in the upper 30 cm of soils under hardwood forests, which averaged 33.2 Mg/ha (Table 3).

These results supported those of the long-term Calhoun forest experiment in demonstrating that pine forest ecosystems accumulated carbon in the A horizons rather modestly (Richter *et al.* 1994, 1995). Old-field pine systems are able to reaccumulate soil organic carbon very rapidly in a mor-like O horizon; but have more modest effects on carbon in the A horizons. In contrast, are soils under the fertilized grass-hay fields, which are able accumulate soil carbon to a similar extent as hardwood forests (Table 3). Since the soils and vegetation of the deciduous hardwood systems are relatively little disturbed, soil carbon loss and turnover due to forest conversion and agricultural use is estimated best by reference to these systems.

Soil carbon loss and turnover

Using soil of the deciduous hardwood ecosystems to be suggestive of soil carbon status prior to forest clearing and cultivation, cultivation appears to have reduced carbon content in the upper 30-cm of mineral soil by about 14.4 Mg/ha. In other words, cultivation has drawn down mineral soil carbon to about 56 % of that present prior to forest conversion. This provides a first-estimate of resistant soil carbon and is also useful as a measure of the mineral soil carbon that can potentially be reaccumulated in the old-field soils by converting to grass and hay management or by converting pine stands to hardwood species.

On the other hand, the accumulation of carbon in the full soil profile including the O horizon of the pine stand at the Calhoun experiment totaled 37.4 Mg/ha and averaged about 1.10 Mg ha^{-1} yr^{-1} on a 34-yr basis, a relatively large fraction of the total ecosystem accumulation during pine stand development (Figure 1). Nearly all (93%) of this carbon is in the mor-like O material which despite this carbon's is youthful age (at most 34 yr) is strongly resistant to decomposition in its present status in the O horizon.

Table 3. *Contents of organic carbon in soils derived from granite-gneiss under four land covers. Sites were located on upper interfluves with gentle slopes and with similar soils to the Appling series that underlies the nearby Calhoun forest experiment. The sampling included five individual samples per site and detailed bulk density analysis.*

Soil depth (cm)	Hardwood	Grass-Hay	Pine	Row crop
Means, Mg/ha				
0 to 7.5	15.73	14.01	7.20	6.71
7.5 to 15	8.80	6.67	4.61	4.61
15 to 30	8.70	11.31	7.04	7.04
0 to 30	33.23	31.99	18.85	18.36
CV, %				
0 to 7.5	29.1	22.1	30.5	-
7.5 to 15	40.2	26.7	20.4	-
15 to 30	6.6	42.6	24.8	-
0 to 30	24.4	20.2	14.2	-
Sites, n	5	5	4	2

Caveats about soil carbon in pine systems and coarse-textured soils

The results of the Calhoun forest experiment emphasize that during the growth of a forest, carbon accumulation by the ecosystem has mainly occurred in the aggrading biomass and O horizon and secondarily in the mineral soil. These direct observations of accumulation rate of soil carbon support estimates presented by Schlesinger (1990) based on less precise chronosequence studies about the ability of mineral soil to serve as a major rapid sink for atmospheric carbon dioxide.

Pine systems, however, such as these dominated by loblolly pine tend to build up an O horizon that is classified as a mor forest floor, a surficial organic material most notable for its slow decomposition and lack of bio-physical incorporation into the upper mineral soil. Many conifer tree species tend to accumulate a mor forest floor rather than a mull which is often more typical of deciduous hardwood forests. Mull forest floors support soil biota that physically mix O horizon material into the upper A horizon. In other words, pine systems should not be expected to rebuild carbon in A horizons rapidly (Pregitzer, 1994). A second factor important in understanding relatively low rates of carbon accumulation in the A horizon is the low clay content of these horizons. The upper A horizon is only 10% clay and thus has little ability to complex soil organic materials and protect them physically and chemically from microbial decomposition. More fine-textured surface soils have a much greater ability to stabilize organic carbon in A horizons.

References

Binkley, D., Valentine, D., Wells, C., & Valentine, U. (1989) An empirical analysis of the factors contributing to 20-yr decrease in soil pH in an old-field plantation of loblolly pine. *Biogeochemistry* **8**: 39-54.

Jaakko Pöyry Group. (1994) *Global fiber resources situation: The challenges for the 1990s.* Unpublished report, Jaakko Pöyry Consulting, Tarrytown, NY, USA.

McCracken, R.J., Daniels, R.B., & Fulcher, W.E. (1989) Undisturbed soils, landscapes, and vegetation in a North Carolina Piedmont virgin forest. *Soil Science Society of America Journal* **53**: 1146-1152.

Pavich, M.J. (1986) Processes and rates of saprolite production and erosion on a foliated granite rock of the Virginia Piedmont. In: *Rates of Chemical Weathering of Rocks and Minerals.* (ed: S.M. Colman & D.P. Dethier), Academic Press, Orlando, FL, USA: 552-590.

Pregitzer, K. (1994) *Soil carbon accumulation by old-field red pine (Pinus resinosa) in Michigan.* Unpublished report, Michigan State Univ., East Lansing, MI, USA.

Richter, D.D., & Markewitz, D. (1995a) How deep is soil? *BioScience* (in press).

Richter, D.D., & Markewitz, D. (1995b) Atmospheric deposition and soil resources in the southern pine forest. In: *Air pollution in the southeastern pine forest ecosystem.* (ed: S. Fox & R. Mikler), Springer-Verlag, Berlin: (in press).

Richter, D.D., Markewitz, D., Wells, C.G., Allen, H.L., April, R., Heine, P.R., & Urrego, B. (1994) Soil chemical change during three decades in an old-field loblolly pine (*Pinus taeda* L.) ecosystem. *Ecology* **75**: 1463-1473.

Richter, D.D., Markewitz, D., Wells, C.G., Allen, H.L., Dunscomb, J., Harrison, K., Heine, P.R., Stuanes, A., Urrego, B., & Bonani, G. (1995) Carbon cycling in a loblolly pine forest: Implications for the missing carbon sink and for the concept of soil. In: *Carbon forms and functions in forest soils.* (ed: J.M. Kelly & W.W. McFee), Soil Science Society of America, Madison, WI, USA: 233-251.
Schlesinger, W.H. (1990) Evidence from chronosequence studies for a low carbon-storage potential of soils. *Nature* (London) **348**: 232-234.
Schlesinger, W.H. (1991) *Biogeochemistry.* Academic Press, New York.
Shelton, M.G., Nelson, L.E., & Switzer, G.L. (1984) The weight, volume, and nutrient status of plantation-grown loblolly pine trees in the interior flatwoods of Mississippi. Techn. Bull. 121. Mississippi State Agric. & Forestry Exp. Sta., MS, USA.
Stolt, M.H., Baker, J.C., & Simpson, T.W. (1992) Characterization and genesis of saprolite derived from gneissic rocks of Virginia. *Soil Science Society of America Journal* **56**: 531-539.
Sundquist, E.T. (1993) The global carbon dioxide budget. *Science* **259**: 934-941.
Trimble, S.W. (1974) *Man-induced soil erosion on the southern Piedmont, 1700-1950.* Soil Water Conservation Soc., Ankeny, IA, USA.
Van Lear, D.H., Waide, J.B., & Teuke, M.J. (1984) Biomass and nutrient content of a 41-year-old loblolly pine (*Pinus taeda* L.) plantation on a poor site in South Carolina. *Forest Science* **30**: 395-404.
Wells, C.G. & Jorgensen, J. (1975) Nutrient cycling in loblolly pine plantations. In: *Forest soils and forest land management.* (ed: B. Bernier & C.H. Winget), Laval Univ. Press, Quebec, Canada: 137-158.

Acknowledgments

The authors thank Dr. C. G. Wells, retired, for the basis of this study; the National Science Foundation of the U.S.A. for financial support for the carbon research; the U.S.D.A. Forest Service, including many individuals such as Dr. W. R. Harms, for a continuity of support for tree remeasurement at the Calhoun Experimental Forest; Drs. S. Trumbore and K. Harrison for ^{14}C analyses; and Duke University for support of its research-education program in ecosystem ecology.

Soil Organic Matter Dynamics in the North American Corn Belt: The Arlington Plots

Matias B. Vanotti and Larry G. Bundy
Department of Soil Science
University of Wisconsin
Madison, Wisconsin,
USA

Site Description

The plots are located at the University of Wisconsin Arlington Agricultural Research Station in south central Wisconsin, USA, approximately 25 km north of Madison (43°18'N; 89°21'W). The climate is cool temperate and arable crops form the present ecosystem. Monthly mean temperatures are -9.1 and 21.8°C for January and July, respectively. Mean annual precipitation totals 79.1 cm. The climate of this area normally provides 47.2 cm of precipitation and 2,570 growing degree days (50°F base) from May 1 to September 30, which closely matches the normal frost-free period of 172 days (data for 1961-1990). Meteorological data from the site (1958-1991) includes daily rain and air temperature. Daily records containing solar radiation, wind speed, relative humidity, dew point and potential ET data are provided from Madison.

Prairie vegetation originally covered the area. Soil at the site is a Plano silt loam (fine-silty, mixed, mesic, Typic Argiudolls) developed from loess deposits 90 cm or more in depth over calcareous loam glacial till. These dark, well-drained soils are some of the most fertile and productive in the U.S. Corn Belt region. The soil at the site was poorly managed for approximately 25 years before establishment of the experiment in 1958; corn stalks were even burned to facilitate plowing. The site has been maintained in continuous corn with residues returned to the soil since 1958. Visual observations and penetrometer

---

M.B.VANOTTI, USDA-ARS, P.O. Box 3039, Florence, SC 29502-3039, USA
(oumvntt@clust1.clemson.edu)

NATO ASI Series, Vol. I 38
Evaluation of Soil Organic Matter Models
Edited by David S. Powlson, Pete Smith and Jo U. Smith

measurements as well as workability of the soil indicated the soil tilth improved during the continuous corn experiment.

Continuous Corn Nitrogen Fertilization Study (1958-1983)

Fertility treatments during 1958 through 1962 involved three rates of fertilizer N (0, 56, and 112 kg N ha^{-1} as ammonium nitrate) and use of differential rates of starter NPK fertilizer (6-24-24). Beginning in 1963, annual N application rates were 0, 92-140 and 184-280 kg N ha^{-1} for the control, medium and high rate treatments respectively, applied as anhydrous ammonia. Since 1963, all plots received 224 kg ha^{-1} yr^{-1} of starter fertilizer (6-24-24) providing 13 Kg N ha^{-1} yr^{-1}. These three long-term N fertilizer treatments (LTN1, LTN2 and LTN3) were continued until the 1983 growing season without lime applications. The initial average soil test values (1958 data) at the site were: pH 6.75; available P, 20 mg kg^{-1}; exchangeable K, 87 mg kg^{-1}; and soil C, 18.8 g kg^{-1}. Corn was harvested for grain each year, and the residues were returned to the soil. Yield parameters were determined all years except for the 1963-1967 period. Average grain yields in treatments receiving recommended rates were 5,700 kg ha^{-1} at the beginning of the study (1958-1962) and gradually increased with time (8,200 kg ha^{-1} for 1979-1983) due to improved corn hybrids, soil tilth, and cultural practices (Vanotti *et al.*, 1995). Corn grain yields were markedly affected by N application rate (Fig. 1). Responses to fertilizer N applied above the standard recommendation (LTN2) were usually small.

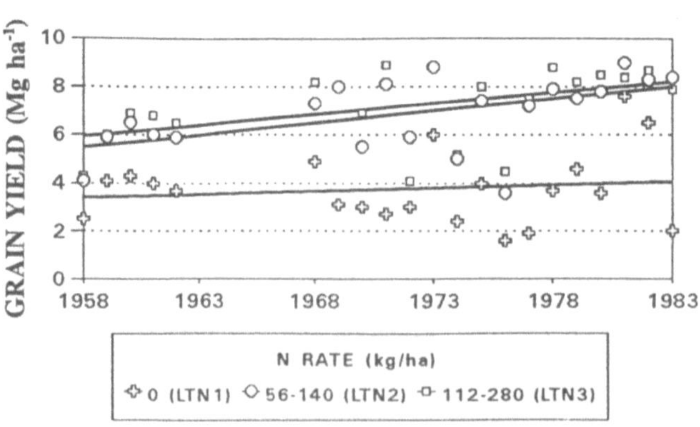

Figure 1. *Corn grain yields at three long-term N fertilizer rates. Arlington Plots. 1958-1983.*

Soil Organic Nitrogen

Total soil N was significantly affected by LTN treatment ($LSD_{0.05}$=0.14 g kg^{-1}) (Table 1). The high LTN treatment resulted in 735 and 206 kg N ha^{-1} (0-20 cm) more soil N content than LTN1 and LTN2, respectively. These values are similar to differences in total stover N returned to soil during the same period. Although net N inputs (N applied - grain N harvest) were markedly different, the final (1984) soil C/N ratios were not affected by treatment. This suggests that soil N stabilization processes in this study were regulated by C input availability.

Table 1. *Nitrogen balance for continuous corn after long-term N fertilization*

	Total Nitrogen (1958-1983), kg ha^{-1}				Soil Organic Matter(1984)	
Treatments	Applied	Grain	Stover	Net	Nitrogen, g kg^{-1}	C/N
LTN1	315	1053	676	-738	1.56	12.6
LTN2	3088	2038	1404	1050	1.74	12.7
LTN3	5825	2180	1534	3645	1.81	12.3

Soil Organic Carbon

Initial soil C was very low at 18.8 g kg^{-1} compared with an uncultivated site which had 22.6 g kg^{-1} soil C (Vanotti *et al.*, 1995). Under continuous corn, removal of corn residues can severely deplete organic matter content of Corn Belt prairie-derived soils (Larson *et al.*, 1972). After 26 years of corn monoculture with residues returned to soil, the soil C content increased in all N treatments ($LSD_{0.05}$=1.7 g kg^{-1}) (Table 2). Changes in soil C were proportional to the amount of residue returned, increasing at an annual rate of 0.045 g kg^{-1} for every 1,000 kg ha^{-1} yr^{-1} of corn stover above 5100 kg ha^{-1} yr^{-1} (Fig. 2). This is in agreement with data reported by Larson *et al.* (1972) after 11 years of continuous corn production on an Iowa Mollisol with an initial soil C content similar to that of the Arlington plots. In the Iowa experiment, soil C increased at an annual rate of 0.043 g kg^{-1} per 1,000 kg ha^{-1} yr^{-1} of corn stover returned in excess of 5,700 kg ha^{-1} yr^{-1}. It was estimated that net soil C losses will occur for quantities of corn residues returned below 5,700 kg ha yr^{-1} (Larson *et al.*, 1972).

Table 2. *Changes in soil organic carbon as affected by plant carbon inputs. 1958-1984*

Treatments	Stover Biomass, Mg ha^{-1} yr^{-1}	Total Plant C to Soil, g kg^{-1}	Soil Organic C, g kg^{-1} 1958	1984	Soil C Stabilization Efficiency, %
LTN1	5.7	23	18.8	19.6	--
LTN2	7.7	31	18.8	22.0	30
LTN3	8.0	32	18.8	22.2	29

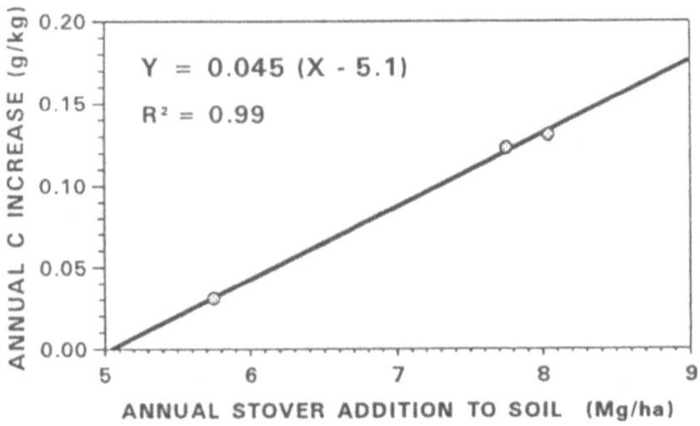

Figure 2. *Relationship between the amount of corn residue returned to soil and the annual rate of soil carbon increase.*

We calculated the soil C stabilization efficiencies for the fertilized treatments by comparing changes in soil C and plant C inputs for these treatments with changes in control (LTN1) levels over the 1958 to 1984 period. Values obtained (Table 2) are slightly higher than efficiencies of 25 to 27 % obtained by Parton & Rasmussen (1994) at the Pendleton site in plots receiving high N fertilizer treatment.

Residual N Effects of Long-term N Fertilization (1984-1991)

In 1984, the long-term N treatments were discontinued and each of the original plots was subdivided to determine the residual effects of the LTN treatments on corn response to N fertilization (Motavalli *et al.*, 1992). Urea N was applied at 0, 84, 168, and 252 kg N ha^{-1} and immediately incorporated by tillage in each of the LTN treatments from 1984 through 1991. These short-term N fertilizer treatments will be referred to as N0, N84, N168, and N252, respectively. As expected, the long history of N application increased soil acidity because of nitrification of ammonia fertilizer. The resulting pH values in 1984 were 6.1, 5.5

and 5.0 for LTN1, LTN2 and LTN3 plots, respectively. Lime was imposed as an additional split treatment in spring 1985, to study its influence on yield in response to applied and residual soil N (Bundy *et al.*, 1988). Variable lime rates were applied to raise soil pH to 6.9. Samples taken in 1990 indicated that all limed plots had a uniform pH of 6.6. Ten treatments were selected for the SOMNET dataset, which include limed and unlimed plots receiving continuous N treatment from 1958 through 1991 (Check= LTN1-N0; medium rate LTN2-N84; and high rate= LTN3-N168), and LTN plots that received no N fertilizer after 1983 (LTN2-N0 and LTN3-N0).

Effects on Corn Productivity

On plots where long-term additions were halted, the supply of mineralized N was an important factor affecting corn yields in subsequent harvests. In 1984, this response was likely a result of residual inorganic N carryover from previous fertilizer addition (86, 176 and 306 kg nitrate-N ha^{-1}, 0-90 cm, in LTN1, LTN2 and LTN3 treatments, respectively).

Table 3. *Residual effects of long-term nitrogen fertilization on corn productivity*

	Grain Yield Mg ha^{-1} yr^{-1}		Grain N Removal kg ha^{-1} yr^{-1}		Plant N uptake, kg ha^{-1} yr^{-1}		NO_3-N* kg ha^{-1} yr^{-1}
Treatments	- Lime	+ Lime	- Lime	+ Lime	- Lime	+ Lime	+ Lime
			1985-1989				
LTN1-N0	4.66	5.42	48	56	90	104	52
LTN2-N0	6.17	6.07	62	59	120	111	60
LTN3-N0	6.74	7.88	72	86	131	145	70
			1990-1991				
LTN1-N0	3.79	3.94	37	39	71	70	42
LTN2-N0	4.34	5.63	52	59	91	100	50
LTN3-N0	4.55	6.38	47	70	88	123	51

* in profile

However, differences in profile inorganic-N content in subsequent years were not large enough to account for the marked response to LTN treatments observed (Motavalli *et al.*,

1992). These responses were more likely due to changes in the soil's mineralizable N pool after long-term N fertilizer additions. Treatments with a history of high N fertilization but with a 0 N rate during 1984 through 1991 (LTN3 N0) produced average grain yields 2110 kg ha^{-1} yr^{-1} higher than plots receiving no N since 1958 (LTN1 N0) (Table 3). Total N uptake and grain N removal were also affected by the residual effects of long-term N fertilization, with an average increase for the same treatments of 40 and 26 kg N ha^{-1} yr^{-1}, respectively.

Corn productivity was also enhanced by the 1985 lime application. Responses to liming were greatest at the high LTN treatment and low 1985-1991 N rate. Addition of N fertilizer offset most of the lime effect, suggesting that yield response to liming was in part due to differences in soil N availability. Since the effect of liming on corn productivity was more pronounced in plots where corn N nutrition depended on soil N reserves, it is possible that liming provided a better environment for microbial activity and N cycling, resulting in higher N mineralization rates. Cornfield (1952) compared the mineralization rates of numerous acid soils in Great Britain having a range of pH values and found that the positive response to liming decreased as original soil pH increased. Results from field experiments on newly cultivated acid soils in Canada showed that, although liming increased mineralization, it is generally a temporary effect lasting no more than 3 years (Nyborg & Hoyt, 1978). In the Arlington plots, however, the liming effect on N cycling, as measured by corn N uptake and grain removal, was still noticeable 6 years after lime application (Table 3).

Effects on Soil Organic Matter

Split treatments imposed in 1984 (N rates) and 1985 (lime) provided information on the stability of SOM pools formed after long-term N fertilization. Total soil N was much more sensitive to changes in management practices than total soil C (Fig. 3). Soil N content was significantly reduced 6 years after discontinuing N fertilization, consistent with the residual effects on corn production observed during the same period (Table 3). In the -N treatments, soil N changes between 1984 and 1990 were similar in magnitude to those produced after 32 years of differential N fertilization. For example, 1990 soil N level in LTN3 -N plots decreased to that found in 1984 LTN2 plots, while soil N level in LTN2 -N plots approached the 1984 LTN1 level (Fig. 3b).

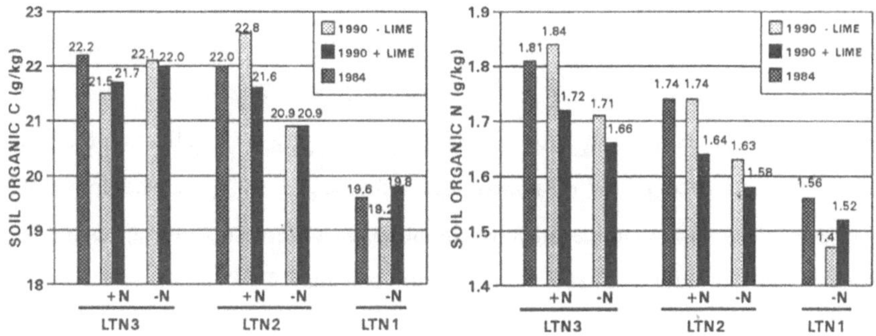

Figure 3. *Short-term N and lime treatments effect on total soil organic carbon (a) and nitrogen (b).*

In plots receiving continuous N treatments from 1958 through 1989 (LTN2 N84 and LTN3 N168), the soil N content between 1984 and 1990 was maintained in unlimed plots, but reduced by 6% where lime was added (Fig. 3b). Fu *et al.* (1987) found a two-fold reduction in N turnover from corn residues mixed and incubated with three Iowa soils that were adjusted to pH 4 relative to the same soils adjusted to pH 6. The rapid decline in soil N levels observed in the limed plots at Arlington suggest that a significant fraction of the soil N formed after a long history of NH_3-N fertilization is made up of highly labile materials, and its N turnover is limited by a fertilizer induced acid environment.

Labile Fractions of Soil Organic Matter

Fractions of SOM involved in rapid C and N turnover in soil are expected to have the greatest sensitivity to N management and cropping practices. Microbial biomass was determined by chloroform fumigation incubation method (10 d fumigated minus control, 25 °C, k_c = 0.45). Total net C and N mineralization was determined by aerobic leaching-incubation procedure at 35 °C and 80 kPa (Vanotti *et al.*, 1995). Where lime was applied, less mineralizable N was recovered in 1990, which corresponds with the higher organic N released and removed in grain during the previous 5 yr. The effect of long-term N fertilization on C and N mineralization was more pronounced than its effect on total soil C and N. The difference in soil C mineralization between the highest and lowest LTN treatments was about six times as much as the observed difference in soil C, while for N components mineralization is three times as sensitive as total N (Table 4).

Table 4. *Effect of long-term N fertilization on total and labile soil organic matter (1990 data).*

	Total Soil Organic Matter, g kg^{-1}		Net Mineralization, mg kg^{-1}	
Treatment	Carbon	Nitrogen	Carbon	Nitrogen
LTN1-N0	19.2	1.47	2234	83
LTN3-N168	21.5	1.84	3721	140
Increase (%)	12.0	25.2	66.6	68.7

The biomass-C level in 1990 was significantly affected by the liming treatment ($P < 0.01$), but it was insensitive to N treatment (limed plots = 202 mg C kg^{-1} or 0.9 % of soil C; unlimed plots = 155 mg C kg^{-1} or 0.7 % of soil C) (Fig. 4). Nitrogen content of the biomass was similar across lime and N treatment (mean = 24 mg N kg^{-1} or 70 kg N ha^{-1}, 0-20 cm).

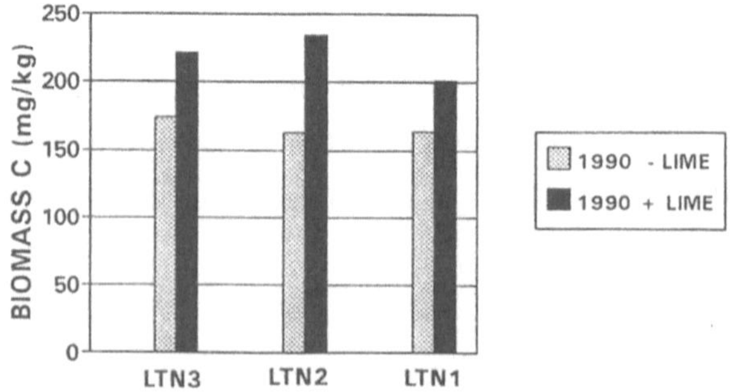

Figure 4. *Liming provided a better soil environment for microbial growth (1990 data).*

Summary

Our objective was to analyze the interactions between N fertilization and crop productivity as they affect SOM dynamics and soil N reserves in a long-term continuous corn experiment in the U.S. Corn Belt. We also contribute soil, crop production, and climatic data needed to integrate the numerous factors affecting SOM in temperate regions, and to evaluate models designed to predict land use effects on SOM and global carbon cycling. Soil management practices at the Arlington plots had a major influence on the changes in SOM content. A shift in corn stover residue management from burning to soil incorporation resulted in a net gain in SOM. Changes in soil C were linear with respect to annual C input rates, which were associated with N fertility management. Long-term NH_3 fertilizer additions significantly increased soil organic N reserves. This practice also increased soil acidity. On plots where long-term N additions were halted, the supply of mineralized N was an important factor affecting corn yields in subsequent harvests. Lime amendments resulted in higher N mineralization rates, and consequently in a more rapid decline in soil N levels. Microbial biomass was also higher in the lime amended plots. The magnitude of soil N changes after 6 years without N fertilization at optimum soil pH were equivalent to those realized after 32 years of differential N fertilization. This indicated that most of the fertilizer N was stored in relatively labile soil organic fractions. Mineralizable C and N pools measured in aerobic incubations were more sensitive indicators than total C and N of the effect of previous soil management on newly formed SOM.

References

Bundy, L.G., Andraski, T.W. & Peterson A.E. (1988) Effect of long-term nitrogen application on soil pH. In: *Proc. 1988 Wis. Fert. Aglime and Pest Management Conf.*, (ed. Kelling, K.A.), University of Wisconsin-Extension, Madison, Wisconsin, USA: 48-55.

Cornfield, A.H. (1952) The mineralization of the nitrogen of soils during incubation: Influence of pH, total nitrogen, and organic carbon contents. *J. Sci. Food Agric.* **3**:343.

Fu, M.H., Xu, X.C. & Tabatabai, M.A. (1987) Effect of pH on nitrogen mineralization in crop-residue-treated soils. *Biol. Fertil. Soils.* **5**:115-119.

Larson, W.E., Clapp, C.E., Pierre, W.H. & Morachan, Y.B. (1972) Effects of increasing amounts of organic residues on continuous corn: Organic carbon, nitrogen and phosphorus. *Agronomy Journal* **64**: 204-208.

Parton, W.J., & Rasmussen, P.E. (1994) Long-term effects of crop management in wheat-fallow: II. CENTURY model simulations. *Soil Sci. Soc. Am. Journal* **58**:530-536.

Motavalli, P.P., Bundy, L.G., Andraski, T.W. & Peterson, A.E. (1992) Residual effects of long-term nitrogen fertilization on nitrogen availability to corn. *Journal of Production Agriculture* **5**:363-368.

Nyborg, M. & Hoyt, P.B. (1978) Effects of soil acidity and liming on mineralization of soil nitrogen. *Can. J. Soil Sci.* **58**:331-338.

Vanotti, M.B., Bundy, L.G. & Peterson, A.E. (1995) Nitrogen fertilizer and legume-cereal rotation effects on soil productivity and organic matter dynamics in Wisconsin. In: *Soil organic matter in temperate agroecosystems: Long-term experiments of North America.* (ed: Paul, E.A., Paustian, K., Elliot, E.T. & Cole, C.V.), Lewis Publishers, Boca Raton, USA. (in press).

Acknowledgements

Research supported by the College of Agricultural and Life Sciences, University of Wisconsin-Madison through project 3449. The authors are grateful to Dr. A.E. Peterson for maintaining the long-term Arlington Plots and data collection prior to 1984.

Soil Evolution Under Dry Meadows in a Boreal climate: The Moscow Dry Meadow Experimental Site

Elena Yakimenko

Scientific Centre for Geoecology

Russian Academy of Sciences

Ulansky Line 13, Bd.2

Moscow 101000

Russia

Introduction

Dry meadows are known to be a constant but unstable component of the forest landscape. For the most part they appear as a result of human activity in place of woodlands or ploughed lands; more rarely as a result of secondary successions. It is known that grasslands in a boreal climate cannot persist for long without human interference because of natural reforestation. The persistence of a meadow life is determined by land management practices such as haymaking or grazing which prevent forest invasion; abandoned meadows become reforested.

To understand soil evolution under dry meadows and to estimate the influence of grasses and trees on soil properties, dry meadows of different ages and woodlands adjacent to them were studied near Moscow, Russia. The woodlands have helped reconstruct the evolution of the meadows because they have the same soil parent materials, relief and slope positions as the meadows, and the effects of these factors on differences between the meadows of different ages can be eliminated. My investigations showed that the direction and rapidity of soil transformations under grass depend on whether the meadow was either formed just after deforestation or was formed on previously ploughed land.

NATO ASI Series, Vol. I 38
Evaluation of Soil Organic Matter Models
Edited by David S. Powlson, Pete Smith and Jo U. Smith

The Moscow Dry Meadow Experimental Site

The selected dry meadows and adjacent woodlands appeared simultaneously on arable land 40, 100 and 200 years ago. After their formation these ecosystems did not undergo any treatment except irregular haymaking in the meadows.

The meadow that appeared about one hundred years ago has been the subject of long-term investigations started in 1963 which were mostly concerned with botanical changes to the grassland. The long-term existence of this dry meadow has not caused visible soil degradation, which might be expected because of the removal of harvested biomass. Moreover the meadow soils differ from the forest soils since they accumulate more organic matter and more nutrients in the root zone, due to the intensive biological accumulation under grass. However, this it typical only of the first stages in a meadow's life. The older a meadow becomes, the thinner the horizon of biological accumulation of nutrients and organic matter, because the roots are concentrated in the sod horizon.

The amount of soil organic matter does not change with time in the sod horizon and changes only slightly in the A1 horizon below. However, soil organic matter could be accumulating in the old meadows where soil water content is increasing and gaseous exchange with the atmosphere is decreasing due to the sod becoming denser. This condition could cause nutrient immobilization, but we do not see evidence for such a process in the meadows studied. Decreases in the amounts of exchangeable Calcium and Magnesium and increases in exchangeable Aluminium and Hydrogen have been observed in the meadow, but these could have resulted from nutrient removal in hay or changes in the humus composition.

Humus composition is not as constant as humus content. The ratio humic/fulvic acids declines gradually as the grassland gets older, and may be even lower than in the humus horizon of the adjacent woodland.

Dry meadow and adjacent pine forest which appeared about 100 years ago in place of rich, manured arable land, was studied in more detail. There were changes in soil morphology and in some chemical properties (humus %, exchangeable Ca and Mg; pH) resulting from the 100 years under different vegetation.

The meadow grasses have retained soil properties characteristic of arable soils for at least 100 years, but adjacent pine forest has caused distinctive soil transformations. The soil under meadow, rich in humus and nutrients, displays only gradual changes of morphology and chemical composition with depth. Forest vegetation has caused formation of the bleached eluvial A1A2 horizon (E) and a transitional A2B horizon with mixed characteristics of E, Bt and Bs horizon. Because of the forest influence, the same horizons have appeared in the meadow soil a short distance (1-2 m) from the forest boundary.

The soil properties of the forest part of the transect show greater spatial variation than those of the meadow, and changes in the soil properties through the profile are sharper. At some places the forest soil has a higher humus content in the upper horizon than the meadow soil. However, isolines of humus content show that relatively high values (more than 4%) tend to be nearer to the soil surface under the forest vegetation. Deep soil horizons are very similar in their properties along the transect.

Micromorphological comparison of meadow and forest reveal an intensive zoogenic microaggregation under woodland, whereas the humus horizon of the grassland has a weak microaggregation inherited from the arable stage.

Summary of long-term results from the Moscow dry meadow experimental site

Long-term investigations, started at this site in 1963, have shown the following:

1. The botanical composition of this dry meadow has been rich and relatively stable in time. Since 1963 *Trisetum flavescence* has dominated, and *Taraxatum officinale* and *Alchemilla vulgaris* have been widespread.

2. There have been no clear long-term trends in bioproductivity, but meteorological conditions in May of each year determine annual yields. High temperature and lack of water at this time significantly decrease the yield of the year.

3. The biomass of the dominant meadow species does not correlate with the general productivity of the grassland.

4. No evident changes in chemical or physical properties of the soil, including humus characteristics, have been observed since 1963.

To study ecosystem evolution without the haymaking regime, a small (10 x 20 m) space within this meadow was fenced off and has been investigated for 15 years. After cessation of the cutting regime productivity started to rise, but botanical diversity decreased. Gradually *Dactilus glomerata* became a dominant species. Nitrogen fixation inside the fenced plot showed a three-fold increase in rate due to accumulation of grass litter without cutting. Soil properties have been changed in response to the termination of haymaking, except that there has been a slight increase in the amount of exchangeable P.

The results from experiments conducted at the Moscow Dry Meadow Experimental Site are described in greater detail in Yakimenko (1987) and Skvoztsova & Yakimenko (1991).

References

Skvorztsova, E. & Yakimenko, E. (1991) Impact of forest and meadow vegetation on the microstructure. *Pochvovedenie (Soviet J. Soil Science)* **11**.

Yakimenko, E. (1987) Pedogenesis under dry meadows. *Pochvovedenie (Soviet J. Soil Science)*. **5**.

INDEX

The ASI Series Books Published as a Result of Activities of the Special Programme on Global Environmental Change

This book contains the proceedings of a NATO Advanced Research Workshop held within the activities of the NATO Special Programme on Global Environmental Change, which started in 1991 under the auspices of the NATO Science Committee.

The volumes published as a result of the activities of the Special Programme are:

Vol. 1: **Global Environmental Change.**
Edited by R. W. Corell and P. A. Anderson. 1991.
Vol. 2: **The Last Deglaciation: Absolute and Radiocarbon Chronologies.**
Edited by E. Bard and W. S. Broecker. 1992.
Vol. 3: **Start of a Glacial.** Edited by G. J. Kukla and E. Went. 1992.
Vol. 4: **Interactions of C, N, P and S Biogeochemical Cycles and Global Change.**
Edited by R. Wollast, F. T. Mackenzie and L. Chou. 1993.
Vol. 5: **Energy and Water Cycles in the Climate System.**
Edited by E. Raschke and D. Jacob. 1993.
Vol. 6: **Prediction of Interannual Climate Variations.**
Edited by J. Shukla. 1993.
Vol. 7: **The Tropospheric Chemistry of Ozone in the Polar Regions.**
Edited by H. Niki and K. H. Becker. 1993.
Vol. 8: **The Role of the Stratosphere in Global Change.**
Edited by M.-L. Chanin. 1993.
Vol. 9: **High Spectral Resolution Infrared Remote Sensing for Earth's Weather and Climate Studies.**
Edited by A. Chedin, M.T. Chahine and N.A. Scott. 1993.
Vol. 10: **Towards a Model of Ocean Biogeochemical Processes.**
Edited by G. T. Evans and M. J. R. Fasham. 1993.
Vol. 11: **Modelling Oceanic Climate Interactions.**
Edited by J. Willebrand and D. L. T. Anderson. 1993.
Vol. 12: **Ice in the Climate System.** Edited by W. Richard Peltier. 1993.
Vol. 13: **Atmospheric Methane: Sources, Sinks, and Role in Global Change.**
Edited by M. A. K. Khalil. 1993.
Vol. 14: **The Role of Regional Organizations in the Context of Climate Change.**
Edited by M. H. Glantz. 1993.
Vol. 15: **The Global Carbon Cycle.**
Edited by M. Heimann. 1993.
Vol. 16: **Interacting Stresses on Plants in a Changing Climate.**
Edited by M. B. Jackson and C. R. Black. 1993.
Vol. 17: **Carbon Cycling in the Glacial Ocean: Constraints on the Ocean's Role in Global Change.**
Edited by R. Zahn, T. F. Pedersen, M. A. Kaminski and L. Labeyrie. 1994.
Vol. 18: **Stratospheric Ozone Depletion/UV-B Radiation in the Biosphere.**
Edited by R. H. Biggs and M. E. B. Joyner. 1994.
Vol. 19: **Data Assimilation: Tools for Modelling the Ocean in a Global Change Perspective.**
Edited by P. O. Brasseur and J. Nihoul. 1994.

Vol. 20: Biodiversity, Temperate Ecosystems, and Global Change.
Edited by T. J. B. Boyle and C. E. B. Boyle. 1994.
Vol. 21: Low-Temperature Chemistry of the Atmosphere.
Edited by G. K. Moortgat, A. J. Barnes, G. Le Bras and J. R. Sodeau. 1994.
Vol. 22: Long-Term Climatic Variations – Data and Modelling.
Edited by J.-C. Duplessy and M.-T. Spyridakis. 1994.
Vol. 23: Soil Responses to Climate Change.
Edited by M. D. A. Rounsevell and P. J. Loveland. 1994.
Vol. 24: Remote Sensing and Global Climate Change.
Edited by R. A. Vaughan and A. P. Cracknell. 1994.
Vol. 25: The Solar Engine and Its Influence on Terrestrial Atmosphere and Climate.
Edited by E. Nesme-Ribes. 1994.
Vol. 26: Global Precipitations and Climate Change.
Edited by M. Desbois and F. Désalmand. 1994.
Vol. 27: Cenozoic Plants and Climates of the Arctic.
Edited by M. C. Boulter and H. C. Fisher. 1994.
Vol. 28: Evaluating and Monitoring the Health of Large-Scale Ecosystems.
Edited by D. J. Rapport, C. L. Gaudet and P. Calow. 1995.
Vol. 29: Global Environmental Change Science: Education and Training.
Edited by D. J. Waddington. 1995.
Vol. 30: Ice Core Studies of Global Biogeochemical Cycles.
Edited by R. J. Delmas. 1995.
Vol. 31: The Role of Water and the Hydrological Cycle in Global Change.
Edited by H. R. Oliver and S. A. Oliver. 1995.
Vol. 32: Atmospheric Ozone as a Climate Gas.
Edited by W.-C. Wang and I. S. A. Isaksen. 1995.
Vol. 33: Carbon Sequestration in the Biosphere.
Edited by M. A. Beran. 1995.
Vol. 34: Climate Sensitivity to Radiative Perturbations: Physical Mechanisms and Their Validation.
Edited by H. Le Treut. 1996
Vol. 35: Clouds, Chemistry and Climate.
Edited by P. J. Crutzen and V. Ramanathan. 1996.
Vol. 36: Diachronic Climatic Impacts on Water Resources.
Edited by A. N. Angelakis and A. S. Issar. 1996.
Vol. 37: Climate Change and World Food Security.
Edited by T. E. Downing. 1996.
Vol. 38: Evaluation of Soil Organic Matter Models.
Edited by D. S. Powlson, P. Smith and J. U. Smith. 1996.
Vol. 39: Microbiology of Atmospheric Trace Gases.
Edited by J. C. Murrell and D. P. Kelly. 1996.
Vol. 40: Forest Ecosystems, Forest Management and the Global Carbon Cycle.
Edited by M. J. Apps and D. T. Price. 1996.
Vol. 41: Climate Variations and Forcing Mechanisms of the Last 2000 Years.
Edited by P. D. Jones, R. S. Bradley and J. Jouzel. 1996

Springer-Verlag and the Environment

We at Springer-Verlag firmly believe that an international science publisher has a special obligation to the environment, and our corporate policies consistently reflect this conviction.

We also expect our business partners – paper mills, printers, packaging manufacturers, etc. – to commit themselves to using environmentally friendly materials and production processes.

The paper in this book is made from low- or no-chlorine pulp and is acid free, in conformance with international standards for paper permanency.

MIX
Papier aus verantwortungsvollen Quellen
Paper from responsible sources
FSC® C105338

If you have any concerns about our products,
you can contact us on
ProductSafety@springernature.com

In case Publisher is established outside the EU,
the EU authorized representative is:
Springer Nature Customer Service Center GmbH
Europaplatz 3, 69115 Heidelberg, Germany

Printed by Libri Plureos GmbH
in Hamburg, Germany